NUTRITIONAL COMPARISON OF PLANT-BASED MEAT ALTERNATIVES AND TRADITIONAL MEATS

NUTRITIONAL COMPARISON OF PLANT-BASED MEAT ALTERNATIVES AND TRADITIONAL MEATS

Dr. Noopur Gautam

Delve
PUBLISHING

www.delvepublishing.com

Nutritional Comparison of Plant-based Meat Alternatives and Traditional Meats

Dr. Noopur Gautam

Delve Publishing
4164 Lakeshore Road
Burlington, ON L7L 1A4
Canada
www.delvepublishing.com
Email: orders@arclereducation.com

ISBN: 978-1-77956-439-9 (Hardcover)

ABOUT THE AUTHOR

Dr Noopur Gautam is having twenty three years of experience in food technology sector and presently working as Associate Professor in Institute of Food Technology, Bundelkhand University, Jhansi UP. She has done her graduation degree from Kurukshetra University, and post graduation degree from Guru Jambeshwar University, Hisar. Her doctorate is from Bundelkhand University in Food Science and Technology. Under her supervision many students are doing their PhDs. She has also worked in Multinational Companies for five years and is a trained auditor for HACCP and ISO 9000 with wide experience in Quality Management System. She is in the panel of expert referee in Elsevier with an outstanding contribution in reviewing research papers and articles. She has authored three books and number of research papers. There is a amalgam of research and application in her approach which reflects in her work.

TABLE OF CONTENTS

List of Figures .... *xi*

List of Tables.......... *xiii*

Preface.......... *xv*

Chapter 1 Introduction to Plant-Based Meat Alternatives **1**

Introduction.......... 2

1.1. Overview of Plant-Based Meat Alternatives 2

1.2. The History of Plant-based Meats and Their Current Trends in Asia.......... 14

1.3. The Evolution of a Plant-Based Alternative to Meat.......... 17

1.4. Market Analysis of Plant-Based Meat Alternatives.......... 27

References.......... 32

Chapter 2 Nutritional Composition of Traditional Meats **39**

Introduction.......... 40

2.1. Nutritional Composition of Meat.......... 40

2.2. Nutritional Composition of Game Meat From Wild Species Harvested in Europe.......... 55

References.......... 74

Chapter 3 Nutritional Composition of Plant-Based Meat Alternatives.......... **87**

Introduction.......... 88

3.1. Nutrients in Plant-Based Meats and Meat 89

3.2. Plant-Based Meat Versus Conventional Meat: A Nutritional Stance.......... 100

3.3. Nutritional Challenges in Plant-Based Meat Products 103

3.4. Current and Potential Solutions to Improve Plant-Based Meat Nutrition.......... 113

References.......... 118

Chapter 4 Protein Quality and Digestibility **129**

Introduction 130

4.1. Role of Plant Protein on the Quality and Structure of Meat Analogs 130

4.2. Effect of Different Plant Proteins on the Quality and Structure of Meat Analog 135

4.3. Digestibility and Bioavailability of Proteins 146

4.4. Different Sources of Plant-Proteins in Meat Analogues 147

4.5. Effect of Main Development Processes on Digestibility and Bioavailability of Meat Analogues 153

References 160

Chapter 5 Fat Content and Composition **165**

Introduction 166

5.1. Dietary Fats 167

5.2. Cis/trans-Fatty Acid Content of Red Meats and the Related Effects on Meat Quality and Human Health 170

5.3. Plant-Based Fat Sources 186

5.4. The Development of Plant-Based Fat Substitutes 189

References 194

Chapter 6 Vitamins and Minerals **197**

Introduction 198

6.1. Key Vitamins and Minerals in Traditional Meats 198

6.2. Vitamin and Mineral Content in Plant-Based Alternatives 205

6.3. Bioavailability and Absorption Differences 206

References 209

Chapter 7 Safety and Nutritional Risks with Plant-Based Meat Alternatives **215**

Introduction 216

7.1. Risks Associated with Plant-Based Meat Alternatives 216

7.2. Food Safety and Nutritional Concerns 224

7.3. Plant-Based Meats, Human Health, and Climate Change 231

References 254

Chapter 8 Environmental and Ethical Considerations **257**

Introduction 258

8.1. Environmental Impacts of Traditional Meat Production 258

8.2. Sustainability of Plant-Based Meat Alternatives 262
8.3. Ethical Considerations and Animal Welfare 265
References 268

Chapter 9 Accelerate Consumer Adoption of Plant-Based Meat 275
Introduction 276
9.1. Understand the Consumer 276
9.2. Product Development 283
9.3. Choice Architecture 289
9.4. Targeted Messaging 291
References 298

Chapter 10 Practical Applications and Dietary Recommendations 307
Introduction 308
10.1. Incorporating Plant-Based Alternatives Into A Balanced Diet 308
10.2. Lower Animal to Plant Protein Intake Ratio Can Be Facilitated and Maintained with Plant-Based Meat Alternatives 313
10.3. Animal to Plant Protein Intake Ratios 317
10.4. Perspectives on PBMAs 324
10.5. Plant-Based Meat Alternatives Dietary Guidelines 330
10.6. Comparing Traditional Meat and Plant-Based Meat 333
References 337

Chapter 11 Current Research and Innovations 341
Introduction 342
11.1. Recent Studies on Nutritional Comparisons 342
11.2. Innovations in Plant-Based Meat Technology 344
11.3. Future Trends in Meat Alternatives 347
References 353

Index 361

LIST OF FIGURES

Figure 1.1. Structural characteristics of four kinds of plant-based meat analogues (PBMs): (a) SDS-PAGE of four kinds of PBMs; (b) FTIR spectra of four kinds of PBMs; (c) Free sulfhydryl content of four kinds of PBMs; (d) CD spectra of four kinds of PBMs; (e) The microstructure of four kinds of PBMs (I: HH 43; II: JY86; III: SN 52; IV: SF 5); (f) The outward appearances of four kinds of PBMs. Different lowercase letters indicate a significant difference ($p < 0.05$). HH 43: Heihe 43, JY 86: Jiyu 86, SN 52: Suinong 52, SF 5: Shengfeng 5

Figure 1.2. An overview of alternate meat products

Figure 1.3. Chinese style tofu from Buddhist cuisine is prepared as an alternative to meat

Figure 1.4. Preparing plant-based meat brands

Figure 1.5. Sandwich made of tempe, eaten at the Juice ja Cafe in Ubud in Bali, Indonesia

Figure 1.6. Vegetarian hot dog sausages from Germany

Figure 1.7. The global market value of plant-based meat alternatives and meat

Figure 1.8. Twelve most active countries in PBMA launches over the past three years (2019–2021). Note: Each bar represents the total number of PBMAs' launches between January 2019 and December 2021. Abbreviations: PBMAs, plant-based meat alternatives

Figure 1.9. Food category distribution of PBMAs launched over the past three years (2019–2021). Abbreviations: PBMAs, plant-based meat alternatives

Figure 1.10. Word cloud of the top 20 claims employed in PBMA products launched over the past three years (2019–2021)

Figure 2.1. Main wild game species harvested in Central and Mediterranean European countries

Figure 3.1. Nutrition Facts panels of 4 oz. (113 grams) of novel plant-based meat alternatives and ground beef

Figure 3.2. Methionine (A) and lysine (B) content of beef and plant proteins commonly used in plant-based meat alternatives

Figure 3.3. Phytic acid, phytate complex, and phytase enzymatic reaction representation

Figure 4.1. Types of meat analogues available on the market

Figure 4.2. Advantages, risks, technological challenges, and research gaps associated with meat analogues (PAH- polycyclic aromatic hydrocarbons; HCA- heterocyclic

aromatic amines)
Figure 4.3. Illustration of high-moisture extrusion (HME) with twin-screw extruder
Figure 4.4. Changes in protein conformation during extrusion of HMMA
Figure 5.1. Regression of trans-fatty acid content of beef loin samples (mg/g) with increasing carcass fat content in conventionally fed cattle
Figure 5.2. Regression equations of (a) mono-unsaturated fatty acid content (MUFA, w/w%) of beef loin muscle with increasing carcass fat content, and (b) conjugated linoleic acid content (CLA, w/w%) of beef loin muscle with increasing carcass fat content
Figure 5.3. Trans-fatty acid content (TFA, mg/g) of lamb loin muscle with increasing live weight
Figure 6.1. Carnosine
Figure 7.1. Alternatives to meat
Figure 7.2. Transforming plant proteins into plant-based meat alternatives: challenges and future scope
Figure 7.3. Comparison of possible greenhouse gas emissions impacts of various beef production systems and meat alternatives. All values include cradle-to-distribution LCA but exclude GHGE potential of retail, restaurant or at home use, and end-of-life stages
Figure 9.1. Foundational drivers vs. evolving drivers
Figure 9.2. Findings from an implicit purchasing study of the plant-based category
Figure 9.3. The barriers to meat reduction
Figure 9.4. Adopter categorization on the basis of innovativeness
Figure 9.5. Reducing meat consumption
Figure 9.6. Integrative model of behavior change
Figure 10.1. The beyond burger® is a newer plant-based product
Figure 10.2. Soybeans provide isolated soy protein for modern plant-based meat products, such Japanese soyabean meat
Figure 10.3. A nutritional comparison between the Impossible burger and a traditional beef burger

LIST OF TABLES

Table 1.1. Comparison of nutritional content, selected advantages, and disadvantages of conventional meat and plant-based meat alternatives

Table 2.1. Nutritional composition of meat (Pereira, et al., 2013, Sinclair, et al., 1999)

Table 2.2. Amino acid composition in fresh meat (Sinclair et al., 1999; Schweigert et al., 1956; Mahan et al., 1998)

Table 2.3. Polyunsaturated fatty acids and cholesterol in lean meat and offal (Paul et al., 1978; Williams et al., 2007; Droulez et al., 2006) (as% Total Fatty Acids)

Table 2.4. Mineral contents (mg/100 g) of meat and meat products (McCance et al., 1960; Chan et al., 1995)

Table 2.5. Mineral content of offal tissues (Paul, 1978; Williams, 2007)

Table 2.6. Vitamin content of various raw meats (McCance, 1960; Williams, 2007)

Table 2.7. Comparison of cooking losses and vitamin b1 retention in conventional and microwave cooking (McCance et al., 1960)

Table 2.8. Vitamin contents (Units/100 G Raw Tissue) of various offal tissues (Paul, 1978; Williams, 2007)

Table 2.9. Range of mean values (or mean value when only one study available) of proximate composition and energy content (per 100 g) of game meat from wild species hunted in Europe

Table 2.10. Range of mean values (or mean value when only one study is available) of fatty acids profile (g/100 g total fatty acids) of intramuscular fat from game meat from wild species hunted in Europe

Table 2.11. Range of mean values (or mean value when only one study available) of cholesterol content (mg/100 g) of game meat from wild species hunted in Europe

Table 2.12. Range of mean values (or mean value when only one study available) of minerals content (per 100 g) of game meat from wild species hunted in Europe

Table 2.13. Range of mean values (or mean value when only one study available) of vitamin content of game meat from wild species hunted in Europe

Table 3.1. Median values for nutrients per 100 g of plant-based meat versus conventional meat

Table 3.2. Comparison of the essential amino acid content between selected sources of plant proteins and animal proteins (values expressed in g/100 g of protein)

Table 3.3. Plant-based sources and effects of anti-nutritional factors

Table 3.3. Plant-based sources and effects of anti-nutritional factors

Table 3.4. Examples of high-phytate plant materials, as a percentage of dry weight

Table 3.5. Methods used to reduce phytic acid in food material

Table 3.6. Summary of clinical trials on phytase effects with statistically significant improvements in iron and zinc absorption

Table 4.1. Composition of meat analogues in scientific research

Table 5.1. Summary of saturated and unsaturated fatty acids previously reported for meat from cattle, sheep, and goats

Table 5.2. The lipid composition of beef sirloin (longissimus dorsi samples) from conventional production systems in South Africa (N = 60)

Table 5.3. The Molar (W/W%) and Gravimetric (Mg/G) Content of Cis-and Trans-Fatty Acids in Beef Sirloin (Longissimus Dorsi Samples) from Conventional Production Systems in South Africa (w/w%) (n = 60)

Table 5.4. The lipid composition of lamb loin chops (longissimus dorsi samples) from conventional production systems in South Africa (n = 80)

Table 5.5. The molar (w/w%) and gravimetric (mg/g) content of cis-and trans-fatty acids in lamb shops (longissimus dorsi samples) from conventional production systems in South Africa (w/w%) (n = 80)

Table 9.1. Diffusion of innovations theory (DOI) segments

Table 10.1. World and selected region pulse intake

Table 10.2. Protein and fiber content of selected legumes (per 100 g Cooked) from the USDA's food data central1

Table 10.3. Energy, macronutrient, and fiber content of legumes and foods made from legumes1

Table 10.4. Level of characterization of public health, environmental, and animal welfare implications in research to date on meat alternatives

PREFACE

The rising popularity of plant-based meat alternatives has sparked significant interest among consumers, researchers, and the food industry alike. As more people seek out alternatives to traditional meat for reasons ranging from health and nutrition to environmental sustainability and ethical concerns, the question of how these plant-based options stack up nutritionally against their meat counterparts has become increasingly relevant.

This book, *Nutritional Comparison of Plant-based Meat Alternatives and Traditional Meats*, aims to explore the nuances of this rapidly evolving area of food science, providing readers with a detailed understanding of the nutritional value, safety, and broader implications of both food sources.

As consumers become more health-conscious, there is a natural tendency to seek out foods that align with their personal values—whether those are related to wellness, environmental impact, or animal welfare. Plant-based meat alternatives have emerged as a compelling option for many, but there is still much to be understood about their nutritional profiles in comparison to traditional meats, particularly when considering factors such as protein quality, vitamin content, fat composition, and safety risks.

Chapter 1 serves as an entry point into the world of meat alternatives. It covers the development of these products, from early iterations to modern, highly engineered versions that mimic the texture, taste, and appearance of real meat. The chapter provides context for why these alternatives are becoming more popular, touching on factors like health trends, environmental sustainability, and ethical considerations.

Chapter 2 deals with the nutrient profiles of various meats such as beef, poultry, pork, and fish. Traditional meats are known for being rich in protein, essential amino acids, and certain vitamins and minerals like iron and vitamin B12. This chapter examines the specific contributions of these foods to human nutrition, establishing a baseline for comparison with plant-based alternatives.

Chapter 3 provides a detailed look at the nutrient content of plant-based meats, focusing on ingredients commonly used in these products, such as soy, pea protein, and other plant derivatives. The chapter explores how these ingredients are processed to replicate the properties of meat, while also analyzing how they compare nutritionally, especially in areas like protein content, vitamins, and minerals.

Chapter 4 addresses one of the most significant differences between traditional meat and plant-based alternatives—the quality and digestibility of the proteins they contain. Animal proteins are typically considered "complete" because they provide all essential amino acids in the right proportions, whereas plant-based proteins may require supplementation or combination with other plant foods to achieve a similar profile. This chapter investigates how manufacturers are attempting to close the gap in protein quality and digestibility through innovative processing techniques.

Chapter 5 contrasts the fat profiles of traditional meats and plant-based alternatives. While traditional meats can be high in saturated fats, plant-based alternatives often contain fats derived from plant oils, which tend to be lower in saturated fat and higher in unsaturated fat. The chapter discusses how these differences impact heart health, cholesterol levels, and overall dietary fat intake.

Chapter 6 compares the micronutrient content of traditional and plant-based meats. Traditional meats are excellent sources of certain vitamins and minerals, such as B_{12}, iron, and zinc. Plant-based alternatives often need to be fortified with these nutrients to match the levels found in animal products. The chapter explores the bioavailability of these nutrients and how they differ between meat and plant-based products.

Chapter 7 addresses concerns related to the safety and nutritional adequacy of plant-based meat alternatives. Topics include the use of additives, preservatives, and processing techniques that might affect the nutritional quality and potential health risks associated with long-term consumption of highly processed plant-based products.

Chapter 8 takes a broader view, discussing the environmental benefits and ethical arguments in favor of plant-based meat alternatives. It examines the impact of both traditional and plant-based meats on land use, water consumption, greenhouse gas emissions, and animal welfare, providing a holistic perspective on the broader implications of food choices.

Chapter 9 is focused on the human element, exploring how consumers perceive plant-based meats in comparison to traditional meats. The chapter discusses taste, texture, and familiarity, as well as cultural and societal influences that shape consumer choices.

Chapter 10 provides guidance on how to incorporate both traditional meats and plant-based alternatives into a balanced diet. This chapter is aimed at consumers, dietitians, and health professionals who are interested in making informed dietary decisions based on the nutritional profiles of these foods.

Chapter 11 closes the book by looking at the latest scientific advancements and innovations in the field of plant-based meats. It covers ongoing research into improving the nutritional quality of plant-based alternatives, reducing processing, and making these products more accessible and affordable to the general population.

Therefore, the book aims to provide a balanced, evidence-based comparison of plant-based meat alternatives and traditional meats, allowing readers to make informed choices based on health, nutrition, and personal values.

A nutrition professional seeking more information, or a researcher exploring the future of food, this book offers a comprehensive guide to one of the most dynamic and important topics in contemporary nutrition.

–Author

Chapter 1

Introduction to Plant-Based Meat Alternatives

Contents

Introduction 2
1.1. Overview of Plant-Based Meat Alternatives 2
1.2. The History of Plant-based Meats and Their Current Trends in Asia 14
1.3. The Evolution of a Plant-Based Alternative to Meat 17
1.4. Market Analysis of Plant-Based Meat Alternatives 27
References 32

INTRODUCTION

Plant-based meat substitutes are meals made primarily of plant-based ingredients that are intended to resemble traditional meat in terms of flavor, texture, and appearance. These substitutes are made with a range of plant proteins that supply the required protein content, such as soy, peas, wheat, and other legumes. By addressing issues with environmental sustainability, animal welfare, and human health, the goal is to provide a morally and environmentally sound meat substitute. Growing numbers of consumers are turning to plant-based meat as a healthier option or as a way to cut back on their meat consumption for ethical and environmental reasons. To produce plant-based meat substitutes that closely mimic the sensory characteristics of meat, contemporary food processing technologies are frequently employed. Food science advancements like the creation of plant-based heme-based molecules help mimic the flavor of real meat. Consumers who want to lessen their dependency on animal products now have more accessible options thanks to products like plant-based burgers, sausages, and chicken substitutes that are showing up in stores and eateries.

An important factor contributing to the growing demand for plant-based meat is environmental sustainability. Due to its resource-intensive nature, traditional livestock farming increases greenhouse gas emissions, water scarcity, and deforestation. By utilizing fewer natural resources, cutting emissions, and lessening the environmental impact of food production, plant-based meats provide an alternative. People who want to eat more sustainably are choosing plant-based alternatives to meat as their awareness of the negative effects of industrial meat production grows. The market for plant-based meat is still expanding due to technological advancements, consumer demand for healthier lifestyles, and the requirement for more sustainable food systems. Though it is still a relatively new idea in the global food industry, plant-based meat has a bright future ahead of it as businesses continue to improve their offerings and broaden the range of substitutes.

1.1. OVERVIEW OF PLANT-BASED MEAT ALTERNATIVES

The global food industry has seen a rise in the popularity of plant-based meat substitutes as a response to growing worries about animal welfare, the environment, and human health related to traditional meat consumption. With their aim to replicate the flavor, consistency, and look of meat, these substitutes have become popular worldwide and can be found in supermarkets, eateries, and fast-food franchises. The various plant-based meat substitutes are examined in this overview along with their nutritional worth, effects on the environment, and potential to influence food trends in the future.

1.1.1. Materials Used For Plant-Based Meat Alternatives

The quality characteristics of meat substitutes, such as color, taste, consistency, etc. are determined by the choice of ingredients. Meat substitutes contain between 50 and 80% water, 4 to 20% non-textured protein, 10 to 25% vegetable textured protein, 3 to 10% flavor enhancement additives, 0 to 15% fat, 0 to 5% coloring agents, and 1 to 15% binding agents. These constituents or ingredients come together to give meat substitutes the necessary textural and sensory qualities. In addition to cutting expenses, the high-water content creates the necessary juiciness, acts as a softener during the process, and facilitates emulsification. Texture, taste, and physical appearance are provided by the protein that is added for nutrition. To create fully vegetarian and vegan dishes, the textured proteins can be substituted either by blending the proteins from non-meat sources with the meat or by substituting the texturizing proteins for the meat entirely. Meat extenders lack the texture, color, and flavor of meat when cooked; however, when combined with meat, they improve the product's overall quality attributes. Conversely, meat substitutes are engineered to replicate the shape, feel, flavor, and color of entire meat when cooked and moistened without using any meat-based ingredients. Therefore, to improve the final consistency of the raw materials, very little in the way of chemicals or additives needs to be used. Concentrates of soy protein, egg white, wheat gluten, or other adhesive agents such as starches and hydrocolloids are used to increase the water-holding capacity. Consumer satisfaction is determined by the product's form, taste, and consistency.

1.1.1.1. Proteins

Plant-based proteins are becoming more and more in demand due to several factors, including price, availability, suitability for new product development, and, above all, physiological characteristics. The ability of proteins to retain water and oil as well as their ability to dissolve, emulsify, gel, and foam are crucial for the formation of the structure of meat substitutes. Depending on the type of protein, these functional characteristics vary (amino acid sequencing, chemical structure, higher or secondary level structure) environmental elements such as pH, temperature, and ionic strength, among others can alter a protein's functional properties as well as its structure. The goal of the research is to gain knowledge about the structure-function relationship of food proteins their significance and how the consistency and structure of meat are influenced by the functional qualities and quality of proteins. Because soy protein has certain desirable traits and is inexpensive, it is used in the majority of meat replacements. In addition to soybean protein, proteins derived from other oil-seed plants and proteins made by microorganisms fermenting different substrates have also been used to produce meat substitutes. These days, proteins from cereals like wheat, rice, and maize as well as defatted oil seeds, bean flours, cereal, and derivatives, and meals made from defatted soy flour, soy protein concentrates, and wheat flour are also used to make the meat substitutes. Recently, textured protein sources such as edible fungi and micro-algae have been incorporated into meat substitute formulations (Zhang et al., 2022).

1.1.1.2. Soy Protein

Soy flour, soy isolates, or soy protein concentrate are added to the recipe for meat substitutes. Because soy proteins have the desired functional qualities of absorbing fat, holding water, gelling, and emulsifying, they are frequently used as meat substitutes. Soy flour is the least processed of all the products made from soybeans.

There are many types of flour available on the market, including defatted, full fat, and toasted flour. Mashing the defatted soy flakes, which have a roughly 50% protein content, can create defatted flour. By simply fractionating the flakes of defatted soy, soybean isolates and concentrates—which have more protein than soy flour—are created.

Concentrates are extracted in liquid alcohol and produce a product with a 70% protein content. Isolates, on the other hand, are extracted in an alkaline solution, precipitate at a low pH, and produce a product with a 90% protein content. In addition to having a high protein content, soy protein isolate is light in color and has a bland flavor when compared to other proteins which is a beneficial quality for the development of meat substitutes. On the other hand, high levels of protein purity are not necessary for applications involving the replacement of meat. Nevertheless, the presence of other ingredients may be necessary or even advantageous when preparing meat substitutes (Grabowska et al., 2016)

1.1.1.3. Gluten From Wheat

Wheat gluten has the natural ability to form long, thin protein layers that are easily transformed into fibrous structures. One crucial feature of gluten is the role that protein desulfation bonding plays in the creation of a three-dimensional structure, which serves as a building block for the development of fibrous structural developments. The formation of meat-like structures in wheat gluten, as well as their retention and stabilization during product formulation, are found to be attributed to hydrogen bonds, disulfide bonds, and hydrophobic interactions. It was discovered that wheat gluten produced dough that was suitable for the development of meat substitutes (Chiang et al., 2021).

1.1.1.4. Protein From Legumes

Legumes, including lentils, chickpeas, lupines, peas, and various types of beans, are studied for their ability to form gels, stabilize foam, and have emulsifying qualities. of these, pea protein—which was extruded with high moisture—is the most promising for use as a meat substitute.

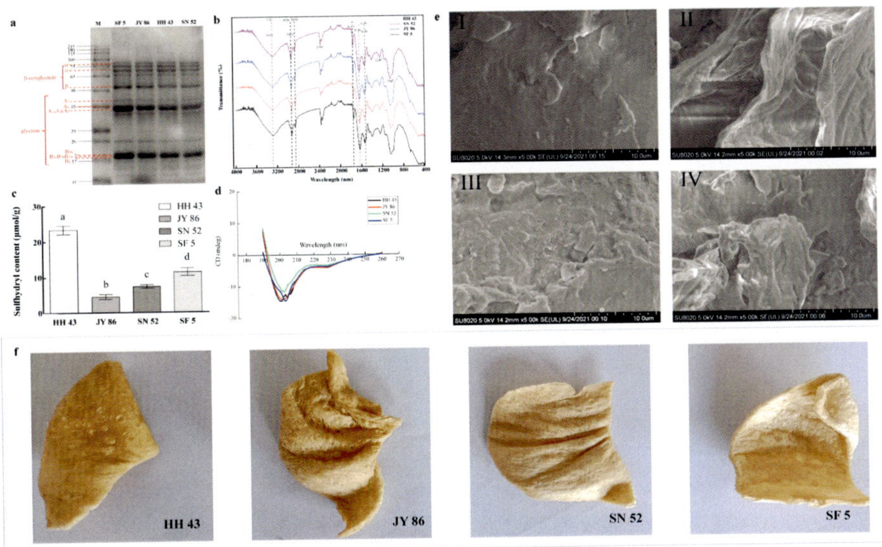

Figure 1.1. *Structural characteristics of four kinds of plant-based meat analogues (PBMs): **(a)** SDS-PAGE of four kinds of PBMs; **(b)** FTIR spectra of four kinds of PBMs; **(c)** Free sulfhydryl content of four kinds of PBMs; **(d)** CD spectra of four kinds of PBMs; **(e)** The microstructure of four kinds of PBMs (I: HH 43; II: JY86; III: SN 52; IV: SF 5); **(f)** The outward appearances of four kinds of PBMs. Different lowercase letters indicate a significant difference (p <0.05). HH 43: Heihe 43, JY 86: Jiyu 86, SN 52: Suinong 52, SF 5: Shengfeng 5.*

Source: Lyu B, Li J, Meng X, Fu H, Wang W, Ji L, Wang Y, Guo Z, Yu H. The Protein Composition Changed the Quality Characteristics of Plant-Based Meat Analogues Produced by a Single-Screw Extruder: Four Main Soybean Varieties in China as Representatives. Foods. *2022; 11(8):1112. https://doi.org/10.3390/foods11081112.*

Although pea components are softer than soy products, the strength can still be increased by altering the protein-hydrogen interaction, such as by adding salts containing chaotropic ions and by changing the temperature and size of the protein molecules, among other things. Numerous studies on chickpeas, lentils, and lupines have demonstrated their excellent emulsion and foam stabilization capabilities. On the other hand, a number of studies indicate that pretreatments of legumes may affect their gelatinization activity. It has been shown that legume proteins, especially those from peas, form fibrous meat structures. High protein isolates make meat analogs harder and chewier. It was discovered that an increase in protein content strengthened the structural strength of the meat analogs because of the network of protein-protein cross-links. Nevertheless, the temperature of the extruder also affects the meat analogs' curing strength or textural qualities. To create the fibrous meat analogs, oat fiber and pea protein isolate were mixed in different proportions and extruded at a high moisture content. The chunks that were produced had long, thin-looking fibers, as seen in Figure 1.1 and were further investigated under an electron microscope.

1.1.1.5. Proteins From Oil Seeds

Two other notable sources of protein with good emulsifying and foam formation properties are canola and rapeseed. Rapeseed proteins can start to gelatinize when used under high pressure or heat, which could result in the formation of a consistency similar to that of meat. Moreover, napin has not yet been suggested as a compound extender containing casein because it causes β-casein to aggregate in solution when prompted by ionic, hydrophobic, and salt-bridging bindings. Canola protein fractions primarily rely on globulins or albumin to form coherent gel-like structures when exposed to salt concentrations. When canola protein and k-carrageenan were combined, very strong and flexible connections were discovered, indicating that canola protein can function as a reagent for constructing structure.

1. **Oil and Fats**: The meat substitutes that are currently on the market have a lower fat content because they are typically made from defatted materials. Additionally, adding fat or oil during processing affects the formation of fibrous structures. Past studies have revealed that recipes containing more than 15% oil lubricate materials during the extrusion process, potentially cause an adverse effect on macromolecule alignment. This could make the material slippery, which would be bad for the shearing forces involved in the extrusion process. Vegetable oil and fat, on the other hand, can improve a meat substitute dish's flavor, tastiness, and softness—all of which are crucial attributes for a meat product that consumers value highly. In response, protein ingredients—such as naturally occurring fat from moderately fermented soy proteins—are added to meat alternatives. By adding them to the final product formula, these ingredients could be used in place of or in addition to the fat added to the products. These days, plant-based meat substitutes can be made with a variety of fats and oils, including canola, rapeseed, sunflower,

soy, palm, and coconut oils. Since it can improve the flavor of the meat substitutes, adding oil or fat seems essential.

2. **Binding Agents**: Compounds derived from plants or animals that function as moisture and fat binders can be used as binders in meat alternatives. These ingredients include carrageenan, wheat gluten, xanthan gum, and soy protein isolate. Depending on how much is used, many different substances can act as enhancers and adhesives. While components with low or no protein levels, like starches and flour, play the role of fillers through their ability to trap fat and water, ingredients high in protein have the primary function of binding water and forming protein networks. Research is still ongoing to determine how different concentrations of binding agents affect the nutritional values and quality characteristics of plant-based substitutes. Reports on the creation and application of binding materials in texturizing products go all the way back to the early 1980s. Protein binding for texturizing proteins is dependent on 10–20% milk and water, 1–5% gluten, and 10–20% albumin. The kind and amount of binding agents used greatly affect the final product's quality. Soy protein concentrate, xanthan gum, and casein were tested on mushroom-based sausage equivalents made with 5% fat content. Due to its adherent and elastic properties, wheat gluten has the potential to be used as a binding agent. Egg albumen is another ingredient used in meat alternatives. In addition to adding more protein, it promotes adhesion and offers advantageous physiochemical characteristics. Soy flour, concentrates, and isolates have all been utilized; however, soy isolates are the most favored and often utilized. A variety of polysaccharides, such as guar, cellulose, and pectin, are excellent choices for extenders and binders in meat substitutes (Varadan et al., 2015).

3. **Taste and Flavoring Enhancers**: Good flavor and taste play a major role in the average consumer's preference for meat substitutes. Savory scents, spices, and precursors are used with iron complexes (e.g., protein containing hemoglobin or chlorophyllin of ferrous) to replicate the flavor of the meat. As the raw materials heat up, chemical reactions occur that modify the added spices and fragrances in the premix. Furthermore, because of the nature of these compounds, complex chemical reactions at high temperatures and pressures can occur, releasing volatile constituents that significantly reduce flavor. Moreover, heat treatments like extrusion alter the flavor quality by changing the structural and textural characteristics

of the protein network and interacting with taste components like sugar, acidic compounds, and salt. It might also have an impact on other chemical reactions, such as maillard reactions, in which sugars and amino acids are used to create new flavor compounds. The most desired aroma among those produced is roasted, but there is still a chance that an off flavor could develop. As a result, improving flavor and taste is a difficult process for which careful consideration of raw material quality and manufacturing process monitoring are essential. Additionally, components of volatile aroma that are released during the cooking of meat have been studied. To mimic the scents of meat alternatives, a variety of substances are also utilized, such as thiamine, nucleotides, amino acids (cysteine, cystine, proline, lysine, serine, methionine, and threonine), sugar (reducing) (Xylose, glucose, ribose, and fructose), and nucleotides. Wu et al. (2000) reported that by adjusting the pH reactions, aromas resembling chicken and beef can be produced from soybean-dependant enzyme-hydrolyzed vegetable-based protein. During roasting, aromas resembling eggs, beans, apple sauce, and molasses are detected at reduced intensity under ideal conditions. It was observed that the hydrolysis of vegetable protein/D-xylose/L-cysteine system attributes certain sulfur-bearing substances and provides a characteristic meat-like flavor. Heterocyclic compounds containing sulfur, the primary developer of the meat's overall structure and texture formation, had an impact on the ribose/cysteine reaction. As a result, improving flavor and taste perception quality is difficult, and essential to this process are the quality of the raw ingredients and the observation of aroma formation (Lu et al., 2018).

4. **Coloring Agents**: Meat's color is thought to be the most crucial quality factor. Since a meat substitute should be the same color as the original meat, coloring additives are a necessary addition. The coloring agents are added specifically because the proteins used, such as gluten and soy protein, give the flesh a taint or yellow-brown hue that differs greatly from the brown or reddish-pink color of cooked beef. These days, thermostable coloring agents are used, like carotene, annatto or malt, cumin, turmin, and caramel colors. Similar to the process of preparing meat, nitrosylmyoglobin undergoes a color change from red to pink due to its conversion into nitrosylhemocrome. When cooking meat substitutes, a similar color change could take place. Heat-labile colorants and reducing agents are added to the finished product, depending on the desired appearance of both raw and cooked meat. At high temperatures,

colors that are unstable in the heat are deteriorated or destroyed. The recommended color extracts to resemble meat are betadine and beetroot extracts. Moreover, browning can be aided by the use of sugar (reducing), which reacts with protein groups in a manner akin to the Maillard reaction and parallels the browning of meat. Dextrose, along with other reducing sugars like mannose and arabinose, maltose, and lactose, are probably going to be added. Plant-based proteins are combined with colorants and applied as coloring solutions before the structuring process. The alternative technique involves combining colorants with the protein-containing substance post-structuring, that is, post-injection into the extrusion barrel. The task of creating a meat alternative is difficult since the color of the meat substitutes does not closely resemble the original meat, even with the availability of various colorants and embodiment techniques. This could be a result of the colorant's ideal pH range differing from that of meat substitutes. However, this can be accomplished by utilizing acidulants, such as lactic acid, citric acid, or their combination, to change the pH. Another obstacle that can be overcome is the color migration from meat substitutes. Color retention agents like maltodextrin and hydrated alginate can help with this.

1.1.2. Nutritional Value of Plant-Based Meat Alternatives

The nutritional profile of plant-based meat substitutes, which is frequently intended to be on par with or even better than that of conventional meat, is one of their main selling points. To guarantee that they contain the vital nutrients typically found in animal products, the majority of plant-based meats are fortified with vitamins and minerals like calcium, iron, and B vitamins.

1.1.2.1. Protein Content

Since consumers expect plant-based meat substitutes to provide the same levels of protein as conventional meat, protein content is crucial. Since soy and pea protein are complete proteins—that is, they contain all nine essential amino acids—they are regarded as high-quality protein sources. Another high-protein option is mycoprotein, though its amino acid composition might differ from that of soy or pea protein. Even though they are lower in protein, vegetable-based alternatives, such as beans and lentils, provide a good source of plant-based protein.

1.1.2.2. Fat and Caloric Content

Plant-based meats are a healthier choice for people trying to cut back on unhealthy fats because they typically contain less cholesterol and saturated fat than animal meats. Higher fat content may still be present in some products, especially those made to resemble beef or pork, to provide a more realistic flavor and texture. But most of the time, these fats come from plant oils like coconut or sunflower oil, which are rich in heart-healthier unsaturated fats.

Depending on the type of plant-based meat, there is a wide range in the number of calories it contains. Some products have a relatively high-calorie count because they contain fats and oils. But compared to their animal counterparts, a lot of plant-based meats have fewer calories, which makes them a good choice for people trying to control their weight.

1.1.2.3. Sodium and Additives

The sodium content of plant-based meat substitutes is one of their criticisms. A lot of products have high sodium content to prolong shelf life and improve flavor, which can be problematic for people who have high blood pressure or are trying to cut back on their salt intake. Concerns have also been raised concerning the degree of processing involved in some plant-based meats since they include preservatives and additives to enhance flavor and texture. Even though plant-based meats have many nutritional advantages, it's important for customers to carefully read labels and choose products that support their health objectives. Less processed options, like vegetable-based alternatives, might be better for people who want to stay away from additives and too much sodium.

1.1.3. Environmental Impact of Plant-Based Meat Alternatives

The fact that plant-based meat substitutes have a far smaller environmental impact than traditional meat is one of the main factors contributing to their increasing popularity. The raising of livestock is a significant cause of deforestation, water use, greenhouse gas emissions, and biodiversity loss. Consumers can lessen their impact on the environment and still enjoy the flavors and textures that they associate with meat by making the switch to plant-based meats.

1.1.3.1. Greenhouse Gas Emissions

Research has demonstrated that compared to beef, pork, or poultry, plant-based meats emit significantly fewer greenhouse gases. According to a Good Food Institute report from 2020, plant-based burgers produce 90% less emissions than conventional beef burgers. The elimination of methane emissions from cattle, a major source of greenhouse gases in the livestock industry, is primarily to blame for this decrease.

1.1.3.2. Water and Land Usage

Compared to animal agriculture, the production of plant-based meats uses a great deal less water and land. It uses far fewer resources to grow plants for protein, like peas or soy, than it does to raise animals for meat. For instance, compared to a beef burger, the production of a Beyond Burger uses 93% less land and 99% less water. This decrease in resource consumption is especially critical in areas where overgrazing is causing land degradation and water scarcity.

1.1.3.3. Biodiversity and Deforestation

The expansion of livestock farming has led to widespread deforestation, particularly in regions like the Amazon rainforest, where land is cleared for cattle grazing and soy production for animal feed. By reducing demand for animal-based products, plant-based meats can help mitigate deforestation and protect biodiversity. Additionally, plant-based protein sources like peas and beans can be grown in a way that promotes soil health and supports sustainable agricultural practices.

1.1.4. Production Technology of Plant-Based Meat Alternatives

Vegetable protein products are traditionally produced and consumed in Asian countries, making them one of the first examples of meat substitutes, i.e., seitan made from gluten, tofu, and tempeh made from soy. Unfortunately, Western consumers, who prefer vegetable-based products that mimic meat in structure, flavor, and taste, cannot find these products to replicate the sensory qualities of meat products. Modern meat substitutes have tapped into the Asian tradition of soy proteins' ability to crosslink, provided certain prerequisites are met. In fact, soy continues to be the primary raw material used in the manufacture of meat substitutes. The availability of the raw material and the techno-functional characteristics of its proteins—such as their solubility, capacity to absorb water and oil, gelling, and emulsifying abilities, and other crucial elements that define the quality of the final product—unquestionably play a role in this supremacy. However, due to concerns about GMOs, allergies, unsuitable climates for soy cultivation, and the preservation and/or value-adding of biodiversity, scientific research (and the market) is moving toward the use of raw materials other than soy. Therefore, recent research has investigated the use of proteins from various raw materials, either alone or in conjunction with soybeans, such as peas, fava beans, rapeseeds, and hemp. The most widely used raw materials are protein isolates, which have a protein content of at least 75% (typically close to 90%), regardless of the botanical source.

Wet separation methods, which use large amounts of water, alkalis, acids, or enzymes, are frequently expensive, time-consuming, ineffective, and unsustainable when producing protein isolates. Lastly, it can be difficult to standardize the technological properties of the isolates since the conditions under which proteins are isolated can have a significant impact on their functionality. Thus, protein isolates are increasingly being replaced by protein concentrates (protein content between approx. 50 and 65%), without neglecting the structural properties required in the finished product. These high-protein fractions are produced using dry separation processes. Because it doesn't use water or solvents, uses less energy, and maintains the native structure of the protein rather than causing it to aggregate, the latter kind of process is thought to be more environmentally friendly than wet methods. The various densities of the flour particles—which are higher in proteins or starch—are the basis for the air classification system. The centrifugal and gravitational forces used during the process permit the separation of the flour into a fine protein-rich fraction and a coarse starch-rich fraction. As a result, additional ingredients like lipids and fibers—which are frequently used in the creation of products based on protein isolate—are also present in the less refined protein ingredients that are produced through the air classification process. The optimization of air-classification conditions is necessary, as the lipid and fiber contents of protein concentrates can differ depending on the processing conditions and source. Although the results are promising, there are currently few examples of using high-protein fractions from legumes separated by air to produce meat analogs, indicating the need for additional research using a variety of sources.

To increase the variety of raw materials that can be utilized to produce meat substitutes while preserving the superior qualities of the final product, different colorants (e.g., red beets, red cabbage, leghemoglobin, etc.) and flavorings (e.g., spices, and herbs) have been suggested as a way to mimic the color and flavor profile of meat and to cover up the odd tastes of some legume proteins. It is also possible to achieve the juiciness, tenderness, and other sensory characteristics of meat-like products by utilizing fats or oils (such as sesame oil, coconut oil, sunflower oil, canola oil, and coconut butter). Nonetheless, the use of binding agents (e.g., hydrocolloids, fibers, oleogels, or starches) as alternatives to fat. As a lubricant, high-fat content may actually obstruct the process of protein denaturation, which is the first type of modification that proteins go through to take on the structure of meat. The process that turns the native globular structure of pulse proteins into a fibrous structure with elongated and highly ordered proteins gives rise to the meat-like structure. A recent summary of the benefits and drawbacks of various technologies for creating this structure was provided by Boukid. These technologies include extrusion, flow-induced structuring with a shear cell or a Couette cell, 3D printing, wet-spinning, and electrospinning.

Extrusion is the most widely used technology for producing meat analogs because of its high productivity, low cost, versatility, energy efficiency, and potential for scale-up. This process involves hydrating the raw materials, applying mechanical and thermal stresses during extrusion, and cooling the finished product to room temperature. Proteins experience a range of structural changes as a result of the mechanical stress, temperature, pressure, and the last cooling phase. These changes include denaturation, unfolding, crosslinking, and alignment, and the end product is a fibrous structure that resembles the properties of muscle tissues. These modifications take place in a chamber containing one (i.e., single-screw extruder) or—more commonly—two (i.e., twin-screw extruder) corotating screws that convey the material toward a die that provides the final shape to the product. The material undergoes a variety of processes (from the material inlet to the finished product outlet) in the extrusion chamber due to the unique screw profiles and temperatures. These processes include mixing, hydration, shearing, homogenization, compression, deaeration, heating, shaping, and expansion. Proteins are hydrated, unfolded, aligned, and texturized during these processes.

One can use extrusion at low moisture levels (<30%) to obtain texturized vegetable proteins (TVPs) or at a high moisture level (> 50%) to acquire meat analogs directly. Low moisture extrusion results in an abrupt drop in pressure at the extruder's end, which accelerates the evaporation of water and causes the product to expand instantly. TVPs resemble ground beef or chicken breast in texture because of their spongy meat-like structure. TVPs come in a variety of shapes (flakes, chunks, or minced), and they can maintain their structural integrity and develop the chewy, elastic texture and elasticity characteristic of meat after being hydrated and cooked through.

In the event of high-moisture extrusion, the sample is cooled to 20°C using a cooling die connected to the end of the twin-screw extruder. This stops the sample from expanding and encourages fiber alignment and stabilization, which is characteristic of the anisotropic structure that is ideal for these kinds of products. While the application of non-extrusion technologies (e.g., high-temperature-induced shearing, 3D printing) has demonstrated promising results, there are still certain obstacles to overcome before these technologies can be widely used in industry: lower costs and/or the ability to work with a variety of legume proteins.

1.2. THE HISTORY OF PLANT-BASED MEATS AND THEIR CURRENT TRENDS IN ASIA

In recent years, plant-based meat substitutes have gained a lot of attention and are still seen as novel. However, did you know that the earliest known use of plant-based meat substitutes dates back more than 2000 years? Plant-based meat has a rich and lengthy history in Asia, having been a part of many Eastern cultures since its introduction in ancient China.

A significant factor contributing to the widespread consumption of plant-based meat substitutes in China was influenced by Buddhism. Owing to Buddhism's teachings, which forbid the taking of any life, Buddhists chose to adhere to vegetarianism. As a part of their cultural practices, Buddhists integrated "originally developed" plant-based meat alternatives such as tofu, tempeh, and seitan into their vegetarian meals (Figure 1.2).

Figure 1.2. *An overview of alternatemeat products.*

Source: https://www.a-star.edu.sg/terms-and-conditions.

1.2.1. Tofu – The First Reported Plant-Based Product

Although the exact date of the creation of the first plant-based product is unknown, tofu, or coagulated soy milk, was developed in China more than 20 centuries ago, during the Han Dynasty (206 BC–220 AD). During the many ancient Chinese dynasties (Tang, Song, etc.), tofu was consumed in large quantities. Southeast Asia has been consuming tempeh, or fermented soybeans, for at least four to five centuries. It is a traditional Indonesian staple.

1.2.2. Emergence of "Second-Generation" Plant-Based Meat Products

Innovation in plant-based meat substitutes that went beyond conventional tofu or tempeh started to emerge in the 1300s. China developed mock eel meat and mock sausage, among other products. These plant-based meat substitutes are referred to as "Second-generation." These goods were widely available in many Asian nations by the 1800s, including China, Japan, and Indonesia (Figure 1.3).

Figure 1.3. *Chinese style tofu from Buddhist cuisine is prepared as an alternative to meat.*

Source: By Mr. Wabu from (optional) – Flickr, CC BY-SA 2.0, https://commons.wikimedia.org/w/index.php?curid=2534225.

Food processing techniques and technology advanced steadily in the early 20th century. The flavor and texture of plant-based meat substitutes were further improved by the addition of fried textured vegetable protein (TVP), which is a blend of soy protein and wheat gluten. In Asia, more westernized food items were also being created. One example is the Japanese deep-fried tofu burger known as "Ganmodoki." Despite coming from Buddhist temples, "Second-generation" plant-based meat substitutes were not regularly consumed because they were viewed as being too ostentatious.

1.2.3. Fast Forward to 2023: What's Trending in Asia Now?

Since their invention thousands of years ago, plant-based meat substitutes have advanced significantly. Plant-based meat substitutes are a new phenomenon in Asia thanks to ongoing innovation and technological advancements in the plant-based industry.

1.2.4. Spoilt for Choice: The Rise of Asian Plant-Based Meat Manufacturers

Asia is home to an increasing number of start-ups and manufacturers of plant-based meat. The market for plant-based foods in Asia is predicted to be valued at 17, 473 million USD and is projected to expand by roughly 10% over the next several years. Many companies offer products that are especially designed to appeal to Asian consumers in addition to the standard plant-based burger patties and nuggets. For instance, Omni Meat provides a range of plant-based foods with Asian influences, including gyoza, shrimp dumplings, and luncheon meat. However, Unlimeat has introduced vegetarian Korean-marinated barbecue meat substitutes as well as plant-based mandu (Korean meat dumplings) (Figure 1.4).

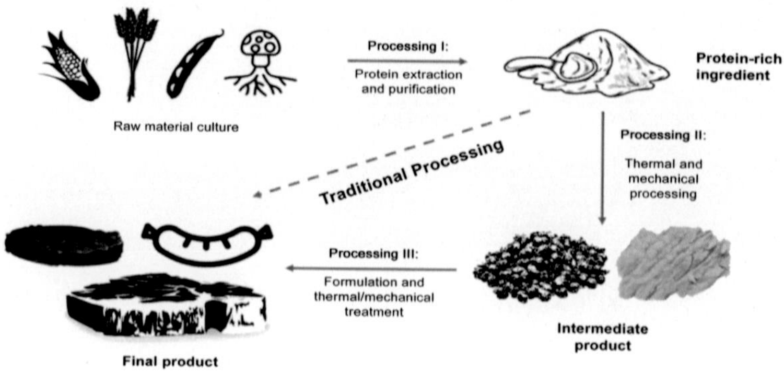

Figure 1.4. *Preparing plant-based meat brands.*

Source: Gräfenhahn M, Beyrer M. Plant-Based Meat Analogues in the Human Diet: What Are the Hazards? Foods. *2024; 13(10):1541. https://doi.org/10.3390/foods13101541.*

1.2.5. Looking Into the Future: Importance of the Asian Plant-based Market

Asian consumers should find it easier to accept and incorporate novel plant-based meat products into their diets than those in western society, even though Asians consume the most meat globally. This is because traditional plant-based meat alternatives have been an essential part of Asian diets for many centuries. Maybe we should investigate what motivates the ongoing consumption of traditional plant-based meat products; what has kept them a consistent part of the Eastern diet for so many centuries? These results may be useful in helping manufacturers of plant-based meat substitutes to create new plant-based meat products that have the desired flavor and consistency of their traditional counterparts, allowing consumers to find some familiarity with these new products.

1.3. THE EVOLUTION OF A PLANT-BASED ALTERNATIVE TO MEAT

In many European nations as well as other countries, there is a growing interest in plant-based meat substitutes due to a growing public awareness of sustainable and healthful eating practices. The new consumer category known as "flexitarians," who cut back on meat in their diet on a daily basis, is expanding quickly. This shift in consumer preferences for tasty, healthful products that can both take the place of meat in a dish and offer a comparable high protein nutritional value is necessary to meet the needs of the market. New meat substitutes were created by a group of university researchers and small and medium-sized businesses as part of the European research project "LikeMeat" (11/2010–02/2013). This collaboration simultaneously addressed consumer concerns, food safety, and technological advancement. An adapted cooking extrusion process was used to create fibrous meat-like structures from plant proteins, which were then processed into delicious food products. To ensure stable and safe food, thorough microbiological analysis and consumer research were conducted in tandem with product development iteratively.

1.3.1. Meat Alternatives

Beginning in the early 1960s, meat analogs—also known as meat substitutes, meat surrogates, and meat replacement food—were first introduced into Western markets. On the other hand, Asia has long been a consumer of goods like tempeh and tofu. Together with these traditional Asian goods, the creation of dry texturized vegetable protein (TVP), which is made by cooking and extruding typically defatted soy meal, soy protein concentrate, or wheat gluten, marked the beginning of the development of meat substitutes. These products work well in patties, stews, and sauces because of their elastic and slightly spongy texture. Another method to replicate the elastic and fibrous structures of meat is to use fibrous raw materials, like TVP that has been shredded and mycoprotein or milk protein preparations. These raw materials are combined with water, an ingredient for binding, and other ingredients to form the final product, which is then steam-cooked. These products offer significant shape and flavor differentiation over TVP and can be made with a wide variety of ingredients. They also meet the requirements for being labeled as "vegan," "organic," and "gluten-free." In the early 1990s, research on the high moisture cooking extrusion process opened up new avenues for texturizing food proteins into unique fibrous structures that resembled muscle meat. This method creates a fibrous, meat-like structure in one step that resembles meat in texture, bite, and mouthfeel and has a moisture content comparable to that of meat. Our understanding of appropriate raw materials and their properties, as well as the impact and interplay of additional functional ingredients and flavoring, has improved as a result of the LikeMeat project. A new generation of meat alternatives has entered the market thanks to improved and modified post-extrusion processes, consumer-focused product development, and innovative packaging and preservation techniques.

1.3.2. Alternative Protein Sources

As the building blocks of the human body, proteins are vital. Because animal proteins resemble human proteins in terms of amino acid composition, they are thought to be especially nutrient-dense. But the consumption of meat in industrialized nations (e.g., in Europe, which was 66.02 kg per person annually between 2010 and 2012), is significantly more than what is needed. Consuming too much meat has been connected to a higher risk of stomach cancer and cardiovascular disease. Moreover, arthropathy is encouraged by the related purines. Reducing meat consumption to two or three servings per week (300–600 g meat/week) is advised by the German Nutrition Society (DGE). For a very long time, the two most common starting materials used to make meat substitutes were soy protein and wheat gluten. In recent times, food-grade protein products have been developed using various plant-based raw materials like rice, canola, chickpeas, lupins, peas, and maize. Furthermore, protein products derived from modern sources like bacteria and fungi as well as animal-based sources like milk and eggs are readily available nowadays. Each of these proteins has distinct nutritional, technological, and functional qualities based on where they come from and how they are processed. There are many opportunities to create new and improved meat analog properties and taste sensations by utilizing new proteins, especially new combinations. Furthermore, they enable the inclusion of additional factors like being free of gluten, soy, lupine, lactose (milk), or animal-derived proteins in the meat analog category, providing customers with dietary intolerances and allergies with a wider selection of products. 10 distinct proteins were examined for their processing behavior during the cooking extrusion process and for changes in their taste profile during the hydrothermal treatments carried out in the LikeMeat project to investigate this strategy through research and development.

1.3.3. Creating a Meat-Like Structure

For proteins derived from plants and animals to form microscopic and macroscopic fibers, they must unfold, cross-link, and align. Necessary process conditions are provided by the high moisture cooking extrusion method, which has water levels as high as 70%. To process the low-viscosity mass into a protein strand with a unique fibrous structure, co-rotating twin screw extruders fitted with long screws and specially designed long cooling dies worked well. First, a long extruder barrel is continuously fed with ingredients, specifically food protein powders and water. The ingredients are thoroughly mixed by the co-rotating screws as the mass is moved toward the die section and heated gradually to temperatures between 130 and 180°C. The proteins unfold and create new covalent intermolecular bonds during the hydrothermal treatment. The proteins are aligned in the flow direction by drag and shear flows as soon as the mass enters the cooling die section. The long die section's intense cooling has a number of effects. In addition to causing the mass to cool to a core temperature below 100°C and preventing product expansion from superheated water evaporating, the temperature gradient from the strand's core to the die wall increases shear flow. Van der Waals interactions, electrostatic interactions, and non-covalent hydrogen bonds all form as a result of the cooling. The mass solidifies into a meat-like strand as the viscosity increases (Table 1.1).

Table 1.1. *Comparison of Nutritional Content, Selected Advantages, and Disadvantages of Conventional Meat and Plant-Based Meat Alternatives*

Food	Nutritional Content (%)			Advantages	Disadvantages	References
	Protein	Fat	PDCAAS[1]			
Meat				High protein content and protein digestibility. High consumer acceptance and sensory quality. Good source of iron and vitamin B_{12}	Resource-intensive production. Animal-welfare concerns. Red meat consumption linked to adverse health effects (e.g., cancer, and cardiovascular disease)	
Chicken	22.3–22.7	0.9–2.1	0.95			(Giacalone et al., 2022; Gorissen et al., 2018; Grabowska et al., 2016)
Beef	20.6–22.5	4.3–6.8	0.92			(Giacalone et al., 2022; Gorissen et al., 2018; Grabowska et al., 2016; Grahl et al., 2018.)
Pork	21.8	4.0	-			(Gorissen et al., 2018; Grandin, 2014)
Mutton	20.2–21.6	4.6–8.0	0.99			(Giacalone et al., 2022; Gorissen et al., 2018)
Meat alternatives						

Food	Nutritional Content (%)			Advantages	Disadvantages	References
	Protein	Fat	PDCAAS[1]			
Tempeh (fermented whole soybean)	61.9–56.9	8.4–23.9	0.92–0.99	Good source of protein, low in saturated fat. Good source of iron and fiber. Free of cholesterol. High digestibility. More resource-efficient production than meat. Low allergenicity, fermentation breaks down allergenic proteins	Lack sulfur-containing amino acids, including methionine and cysteine. Lack of vitamin B_{12}, except if vitamin B_{12}-producing bacteria present during fermentation. Low consumer acceptance. Sensory quality is different from meat	(Grossmann and Weiss, 2021; Hunt, 2003; Kumar et al., 2022)
Tofu (made from soymilk)	11.3	7.84	0.56–0.70	Rich in B vitamins, and low in sodium. Net protein utilization (NPU) is estimated to be around 65%, making it comparable to chicken meat in terms of assimilation and digestion. More resource efficient production than meat	Lack sulfur-containing amino acids. Lower digestibility than meat. Loss of nutritional and nutraceutical contents during processing. Presence of anti-nutritive factors. Lack flavors causing a low consumer acceptability	Nezlek and Forestell, 2022; Nishinari et al., 2014; Oussalah et al., 2020; Pereira and Vicente, 2013.

Food	Nutritional Content (%)			Advantages	Disadvantages	References
	Protein	Fat	PDCAAS [1]			
Seitan (made from wheat gluten)	34.3	0.78	0.23	Consumption of 100 g provides 61.2–74.5% of recommended daily protein	Low in lysine	Oussalah et al., 2020; Pereira and Vicente, 2013; Pointke and Pawelzik, 2022; Rotz et al., 2010.
				Its fibrous structure and high protein content make it an excellent meat substitute	Low digestibility	
				Its sensory properties can be easily modulated by spices and flavors during manufacturing due to its neutral taste and aroma	Sensory quality is closer to meat than tempeh and tofu but still not a perfect real meat analogy	
Quorn (mycoprotein, made from *Fusarium venenatum*)	9.4–11.5	2.6	0.91	High protein digestibility, low in saturated fat. Low antinutrient content. More resource-efficient production than meat. High fiber content. Texture more like meat compared to plant proteins	Lower levels of iron and vitamin B_{12} than real meat. May cause allergies and/or gastrointestinal symptoms. Possible presence of mycotoxins after inoculating *F. venenatum* into rice culture	Schutyser et al., 2015; Sha and Xiong, 2020; Tso and Forde, 2021; Tukker et al., 2011; Verain et al., 2016.

Food	Nutritional Content (%)			Advantages	Disadvantages	References
	Protein	Fat	PDCAAS [1]			
Texturized plant protein [2] Soybean isolates Wheat gluten Pea protein concentrates	87.0 80.0 50.0–85.0	<0.1 <1	~1.0 0.26 0.73	High protein content, low in saturated fat, free from cholesterol. Fibrous structure and texture like meat. Possible to blend protein sources to achieve a more complete amino acid profile	Deficient in micronutrients that are common in meat (e.g., vitamin B_{12} and iron). Considered as ultra-processed foods associated with adverse health effects. Usually not clean label as additives are added to modulate the sensory properties (e.g., texture, color, and flavor). These additives may not diffuse in the product homogeneously, leading the worse sensory quality than meat	Wittek et al., 2021; Zahari et al., 2022; Zhang et al., 2021

[1] PDCAAS = Protein Digestibility Corrected Amino Acid Score;

[2] Most common protein sources, which are used in plant-based meat analogs.

Source: Elhalis H, See XY, Osen R, Chin XH, Chow Y. Significance of Fermentation in Plant-Based Meat Analogs: A Critical Review of Nutrition, and Safety-Related Aspects. Foods. 2023; 12(17):3222. https://doi.org/10.3390/foods12173222.

Such cooking extruders process meat analogs using a wide range of machine and process parameters. In addition, the final product is greatly influenced by the water content, ingredient variety, and matrix composition. Nevertheless, there is a dearth of literature on the precise effects of each parameter and how they interact to affect fiber formation and the quality of the finished product, with the exception of soy, wheat, and pea protein. It has been demonstrated that using proteins with sufficient cross-linking ability and including high protein levels in the recipe are beneficial. Thus, the goal of the LikeMeat project was to broaden this understanding to include intricate recipes and the impact of different ingredients on the formation of fiber. The strategy of combining protein ingredients with additional ingredients like starches and fibers worked very well to prevent overpowering, distracting flavors from single ingredients and to produce products with a juicy mouth feel and bite reminiscent of meat. Additionally, it made it possible to modularly modify recipes to suit particular requirements like being gluten-free, vegetarian, or vegan. To make vegetarian dishes, the formed strand is used as a raw material for different post-extrusion processing steps. It is frequently possible to use equipment that is normally used for processing meat. Because LikeMeat is precooked, preparing it is simple and, depending on the product, comparable to preparing the matching meat product.

1.3.4. Flavor and Taste

In addition to texture, a desirable flavor is a necessary condition for consumer acceptance and the introduction of new meat substitutes into the market. These goods must have flavors that live up to consumer expectations and be free of off-putting flavors. However, the lack of flavor and aroma in many meat substitutes created in the past may have contributed to their poor commercial success. Foods can be flavored in two ways: (i) by adding spices or flavorings; or (ii) by using odorless, tasteless ingredients that become sensorially active compounds through heat processing.

The Maillard reaction is one naturally occurring process that is well known. Almost all foods undergo this reaction when they are heated, which bioconverts sugars and amino acids into a variety of flavorings, tastants, and colorants. Thus, the goal of the LikeMeat project was to create meat substitutes with superior flavor and aroma by utilizing and researching every potential flavoring application.

Due to the careful selection and mixing of the protein raw materials, the pure LikeMeat products were unaffected by an off-flavor. The meat analogs had almost no taste, a very slight scent, and a texture that was similar to that of meat. Certain strands even had a faint chicken meat scent, depending on the ingredient combinations. Many products with extremely nice aromas and tastes were produced by adding salt, spices, and/or aroma ingredients to the mixture before extrusion, as well as by flavoring it after extrusion. In any other case, the fibrous texture remained unchanged. A variety of meat substitutes with wildly disparate meat tastes (e.g., chicken, beef) were obtained, demonstrating the potential for LikeMeat to be used to create a variety of products. Another method of flavoring food is to add flavor precursors, like sugars and amino acids, but there are certain restrictions on the kinds and quantities of precursors that can be added.

1.3.5. Microbial Control and Food Safety

Meat analogs, with their high protein and moisture content and nearly neutral pH, are prone to spoiling. Microbial metabolites and enzymes are secondary to microbiological activity as the primary factor. Scientific research does not generally contain data describing the microflora of high-moisture meat analogs made from plant proteins. Therefore, to identify and characterize pertinent spoiling and potentially hazardous microorganisms, a thorough microbial analysis of the ingredients and the newly developed products was required. Microorganisms cannot grow in powdered plant-based proteins due to their low water activity. Yet, bacteria that form endospores (e.g., Bacillus species and species of Clostridium) might possibly make it through the extrusion process. Relevant concentrations of microorganisms may be present in the pre-mixtures and recipes, contingent upon the microbial load of the chosen raw materials. This includes the potential for hazardous and pathogenic microbiological species to exist. Both vegetative microbial cells and bacterial endospores are highly inactivated during a typical cooking extrusion process, depending on the process temperature and pressure conditions (130–180°C, 20–50 bar for seconds to minutes) as well as the water content. When it comes to product hygiene, every post-process step matters, and it's important to reduce the possibility of recontamination from handling and packaging the meat analog. Consequently, it is important to take into account various post-packaging procedures like pasteurization, freezing, and sterilization of finished goods.

The LikeMeat project characterized the species composition and quantitative level of microflora in a wide range of potential protein ingredients and extruded products. To identify indicator and pathogenic/toxinogenic microorganisms, a set of established, ISO-based standard procedures was employed. A variety of molecular techniques, such as 16S rRNA gene sequencing and PCR-based methods, were used to identify isolates to the species and strain levels. Moreover, inoculation experiments were carried out to show how bacterial endospores are rendered inactive during extrusion cooking. The findings demonstrated that the raw ingredients' microbiological quality was acceptable. In specifics, all extruded products had a generally low total microbial load (<100 CFU/g). A similar sporicidal and bactericidal effect was discovered for both pilot plant-scale and lab-scale extrusion processes, based on culture-based microbiological methods. There were very few Gram-positive bacteria (<<100 CFU/g) and no Gram-negative bacteria in the microbiota of the extruded meat analog. The refined products' microbial analysis results showed that post-process heat treatment was necessary to guarantee a few weeks of shelf life in chilled storage. As an alternative, the intermediate product could be frozen. Even after several weeks of storage at 6°C, pasteurized or sterilized intermediate and final meat analog products did not exhibit any microbiological activity. Intermediate LikeMeat products without post-process heat treatment and without preservatives have a shelf life that is similar to that of fresh meat products.

1.3.6. Consumer Motivation

In recent years, there has been a rise in the number of vegetarians and consumers in Europe cutting back on their meat consumption. Depending on the kind of customer, this is mostly due to ethical, health, and environmental concerns. Additional factors include the growing interest of consumers in "new" and varied foods. Therefore, one strategy to attain a significant decrease in meat consumption is to create meat substitutes that directly rival meat products. Introducing protein products that address the need for greater variety and/or new products is another strategy.

Analogs of meat are still relatively rare on the market. In the Netherlands, for instance, the market share of meat substitutes is only approximately 1% of the total market for meat and meat products. This is most likely because the current meat analogs do not satisfy consumer preferences regarding sensory quality. Specifically, the bite, flavor, and juiciness are rated lower than those of meat. Meat substitutes must outperform real meat to gain a greater market share. A consumer-focused approach to product development is especially well-suited to the search for new product strategies. The LikeMeat project's consumer research focused on meat eaters who already avoid meat for their main meal at least once a week. Four distinct segments were identified within this group based on their attitudes and behaviors regarding protein products. The four primary categories of consumers were categorized as convenience-oriented, hedonistic, animal-friendly, and health-conscious. These consumer groups served as the basis for the development of conceptual products, recipes, and practical safety considerations related to packaging, storage, and handling. All of this input was used to create (almost) final products. 88 (mostly Dutch) consumers tested a selection of these products in a restaurant setting at the Restaurant of the Future Wageningen UR, at their homes, and at a central location. The goal at the latter testing site was to replicate real-world conditions as closely as possible. The goal of this real-world test in the Restaurant of the Future was to identify the areas that needed improvement to maximize the positive customer experience. The consumers' repeated use of the products was one way to add realism to the testing. The testing process included a free selection of products, and the customer received product details regarding the meat substitute. Thus, the product testing included several elements of actual consumer behavior.

The 88 participants in the out-of-home and in-home real life tests tested a variety of food preparations made with the meat substitute, including burgers, schnitzels (Figures 1.5 and 1.6), Italian pasta sauces, Mexican dishes with minced meat analogs, curry-flavored baked strips, and unflavored LikeMeat material that was presented as a "chicken fillet." Based on the participants' sensory evaluation, choice behavior, experiences at home, and remarks, the Mexican dish was deemed the most successful product. Additionally, customers were most willing to purchase this LikeMeat dish from a store and were most likely to eat it again. As a percentage of all meals consumed, the amount consumed for the Mexican dish was the highest, making it likely one of the best predictors of future success. Participants also thought the Mexican dish was creative. Since the Mexican dish appeared to be appealing to a wide range of consumer demographics, this product has potential from a business and marketing standpoint. Furthermore, the fact that the schnitzel received such high marks from the taste test participants in the central location suggests that it may have the makings of a successful product. When details were given, some of the products were somewhat more favored. It's likely that the information made customers more accepting of products that deviated from "normal."

Figure 1.5. *Sandwich made of tempe, eaten at the Juice ja Cafe in Ubud in Bali, Indonesia.*

Source: By christian r from Singapore – Tempe Burger!, CC BY-SA 2.0, https://commons.wikimedia.org/w/index.php?curid=9465294.

The results validate our belief that LikeMeat products don't have to be exact replicas of meat; they can vary to cater to the distinct demands and preferences of various consumer segments. Expectations must always be carefully controlled to prevent perception of contrast effects. Based on the results of the consumer study, consumer workshops, and limited global market analysis conducted as part of the LikeMeat project, it was determined that meat substitutes that are successful in the market have acceptable looks, good sensory qualities (particularly taste and texture), are convenient, and possess one or more qualities that consumers value but which they may not be able to confirm (e.g., environmental friendliness or healthiness).

Figure 1.6. *Vegetarian hot dog sausages from Germany.*

Source: By Sargoth – Own work, CC BY-SA 3.0, https://commons.wikimedia.org/w/index.php?curid=4900952.

1.4. MARKET ANALYSIS OF PLANT-BASED MEAT ALTERNATIVES

Since the most recent market trends for plant-based meat substitutes have not been thoroughly examined, we carried out market research to determine the current trajectory of these goods. The information obtained from this analysis could be useful to retailers and other industries in designing new products (in terms of ingredients, claims, labeling, etc.). to enhance the design of their marketing plans.

We used Mintel's Global New Product Database (GNPD), an online database for new products introduced in particular nations, to examine market trends for PBMAs. Previous studies have made use of the same database. For example, many writers used Mintel's GNPD to look into ingredient profiles, food labeling practices, front-of-package information, and the introduction of new alternative meat products into the international market.

Our analysis's goal was to extract and investigate various data on the most recent PBMA market trends using the Mintel database. We looked for new meat substitute launches over the previous three years (January 2019 to December 2021) to get a sense of recent years.

The PBMA market has experienced a notable surge in product launches over the last three years (2019–2021), with 4965 products being released globally. The steady increase in PBMAs at the start of 2020 (the year the COVID-19 pandemic started) and the minor decline at the end of 2021 were noticed. As was the case with SARS-CoV in 2003 and the African Swine Flu in 2019, this shift may be explained by the typical short-term reductions in meat intake that occur during zoonotic outbreaks (Attwood and Hajat, 2020). Thus, during the early stages of the coronavirus outbreak, this reduction in meat consumption may have caused consumers to hunt for new substitutes. The overall direction of PBMA launches is positively growing, despite a slight negative trend during the third and fourth quarters of 2021. This new dietary pattern may present an opportunity to promote these products. More specifically, the introduction of new PBMA products (n = 1822; 36.7%) and varieties (n = 1910; 38.5%) is the main focus of this encouraging market trend. New packaging, re-launches, and new formulations are among the remaining launches (n = 1232; 24.8%) (Figure 1.7).

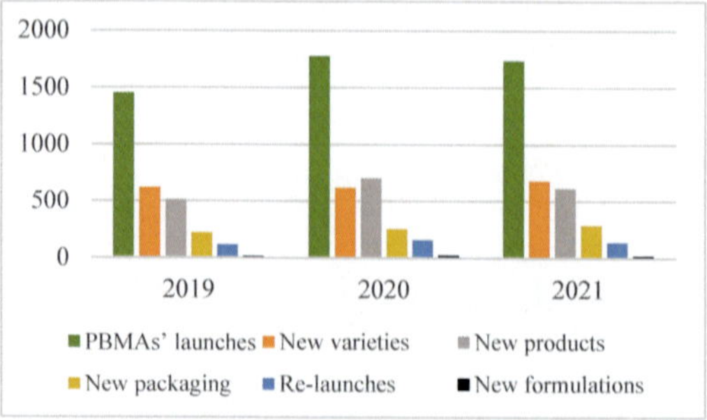

Figure 1.7. *The global market value of plant-based meat alternatives and meat.*

Source: Andreani G, Sogari G, Marti A, Froldi F, Dagevos H, Martini D. Plant-Based Meat Alternatives: Technological, Nutritional, Environmental, Market, and Social Challenges and Opportunities. Nutrients. 2023 Jan 15;15(2):452. doi: 10.3390/nu15020452. PMID: 36678323; PMCID: PMC9861156.

The global market revenue of plant-based meat substitutes is predicted to be worth USD 33.99 billion in 2027 (Global: Meat Substitutes Market Revenue 2016–2027/Statista, 2022), while the meat sector is expected to be valued at USD 1354 billion by 2027 (Global Meat Industry Value Projection, 2021–2027|Statista, 2022). These facts are noteworthy although the market for PBMA products is growing. As a result, PBMAs' market share is currently and is predicted to stay much smaller than that of the meat market.

With 417 new launches (8.4%), France led the way in the 2019–2021 period for PBMA product launches, followed by the UK (n = 393; 7.9%) and Germany (n = 391; 7.9%). Figure 1.8 shows the top 12 most active markets in this industry. This graph highlights the increased activity in the past few years in the introduction of plant-based meat substitutes in Brazil, Australia, and Europe, indicating a growing interest in meat alternatives in these regions.

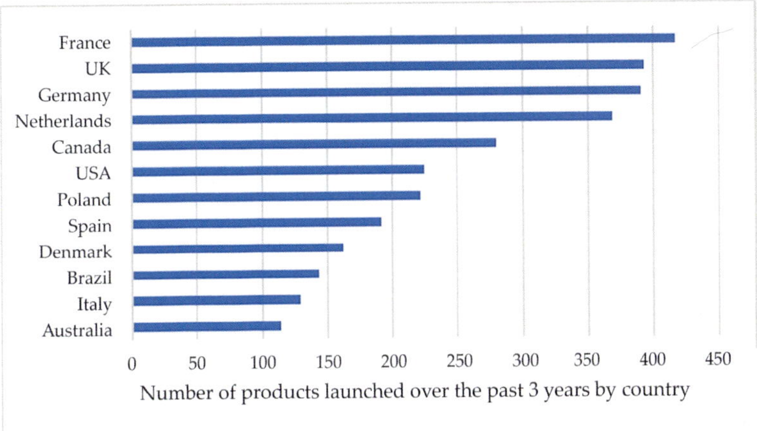

Figure 1.8. *Twelve most active countries in PBMA launches over the past three years (2019–2021). Note: Each bar represents the total number of PBMAs' launches between January 2019 and December 2021. Abbreviations: PBMAs, plant-based meat alternatives.*

Source: Andreani G, Sogari G, Marti A, Froldi F, Dagevos H, Martini D. Plant-Based Meat Alternatives: Technological, Nutritional, Environmental, Market, and Social Challenges and Opportunities. Nutrients. 2023 Jan 15;15(2):452. doi: 10.3390/nu15020452. PMID: 36678323; PMCID: PMC9861156.

General plant-based proteins (n = 1469; 29.6%) – that is, foods that don't aim to replicate an already-existing meat product – were the most represented food categories in the global market meatballs, nuggets, burgers, or sausages) but can still be regarded as plant foods that are high in protein as alternatives to meat. G. patty/burger substitutes (n = 1331; 26.8%)—such as "fried gluten with peanuts" and "teriyaki tofu." The combined share of all other food categories, such as sausage, mince, and nugget substitutes, is less than 9% of the total launches (Figure 1.9).

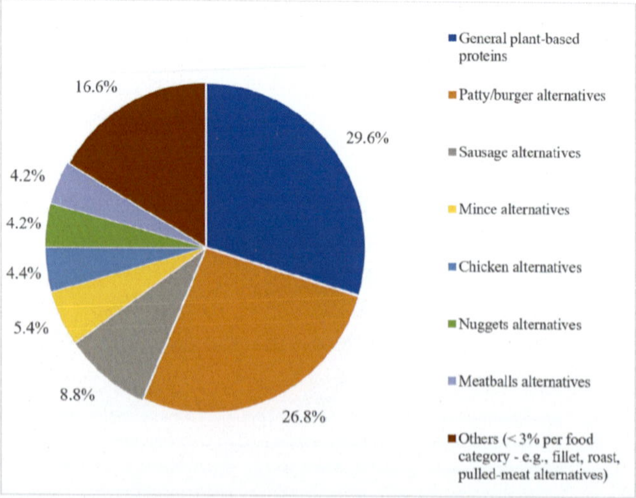

Figure 1.9. *Food category distribution of PBMAs launched over the past three years (2019–2021). Abbreviations: PBMAs, plant-based meat alternatives.*

Source: Andreani G, Sogari G, Marti A, Froldi F, Dagevos H, Martini D. Plant-Based Meat Alternatives: Technological, Nutritional, Environmental, Market, and Social Challenges and Opportunities. Nutrients. 2023 Jan 15;15(2):452. doi: 10.3390/nu15020452. PMID: 36678323; PMCID: PMC9861156.

According to a recent Smart Protein project study using Nielsen Retail Scanning Data, the UK and Germany lead the PBMA industry in terms of highest sales value in EUR and growth rate, i.e., cold cuts, burger patties, and sausages. However, there are variations in the categories between the nations. For instance, in the UK, plant-based sausages are the most popular category, while in Germany, plant-based chilled meat (such as burger patties, nuggets, minced meat, etc.) is the top category followed by sausages made without meat and plant-based cold cuts and meat spreads.

In terms of ingredients, we determined which foods are most frequently used as the first ingredient by using data from Mintel. We took into consideration the second element when it was reported that water was the first one in the list (n = 1605; 32.3%). By employing this method, we were able to identify 1914 products (38.6%) that contained soy-based ingredients—e. g., proteins, flour, or soybean curd—as the primary component. Following soy, the most common first ingredient was wheat (n = 520; 10.5%) and other pulses (n = 702; 14.1%), such as kidney beans, black beans, peas, chickpeas, and lentils. Mushrooms (n = 134; 2.7%) and jackfruit (n = 86; 1.7%) were the next most common ingredients. A total of 120 distinct claims were found to be made in the information on the packaging. of the 4965 products, 2099 (42%) reported being "plant based," and 2849 (57%) included the claim "Vegan/No Animal Ingredients." Furthermore, in keeping with Cutroneo et al. the "High/Added Protein" statement was the most frequently occurring nutrition claim (n = 1616; 33%). Figure 1.10 depicts a visual representation of the claims.

Figure 1.10. *Word cloud of the top 20 claims employed in PBMA products launched over the past three years (2019–2021).*

Source: Andreani G, Sogari G, Marti A, Froldi F, Dagevos H, Martini D. Plant-Based Meat Alternatives: Technological, Nutritional, Environmental, Market, and Social Challenges and Opportunities. Nutrients. 2023 Jan 15;15(2):452. doi: 10.3390/nu15020452. PMID: 36678323; PMCID: PMC9861156.

Ultimately, Mintel's Global New Product Database has proven to be a useful resource for gaining a worldwide perspective of PBMA industry patterns. The information taken from the database and examined revealed that there can be significant variations in the food category, ingredients, and/or claims made for plant-based meat substitutes. Nevertheless, despite these various variations, a positive global trend to support the shift to a plant-based diet is highlighted by the growing trend in product launches, particularly in Western countries. Nonetheless, there are still significant market share gaps between the meat and PBMA industries, and projections for meat revenue do not indicate any downward trends. These figures highlight the fact that the market for meat alternatives is expanding but has little impact on the meat market itself. Because they frequently complement meat rather than serve as a replacement for it, PBMAs are therefore still only marginal alternatives to animal-based products. Therefore, research, and testing strategies that could influence meat eaters' choices toward plant-based diets and support the expanding market for plant-based meat alternative supplements (PBMAs) are essential to support a dietary shift toward meat reduction (Andreani et al., 2023).

REFERENCES

1. Aaslyng, M. D., & Meinert, L. (2017). Meat flavor in pork and beef—From animal to meal. *Meat Science*, *132*, 112–117.
2. Afshin, A., Sur, P. J., Fay, K. A., Cornaby, L., Ferrara, G., Salama, J. S., Mullany, E. C., Abate, K. H., Abbafati, C., Abebe, Z., et al. (2019). Health effects of dietary risks in 195 countries, 1990–2017: A systematic analysis for the Global Burden of Disease Study 2017. *The Lancet*, *393*, 1958–1972.
3. Andreani, G., Sogari, G., Marti, A., Froldi, F., Dagevos, H., & Martini, D. (2023). Plant-based meat alternatives: Technological, nutritional, environmental, market, and social challenges and opportunities. *Nutrients, 15*(2), 452.
4. Asgar, M., Fazilah, A., Huda, N., Bhat, R., & Karim, A. (2010). Non-meat protein alternatives as meat extenders and meat analogs. *Comprehensive Reviews in Food Science and Food Safety*, 9(5), 513–529.
5. Babu, P. D., Bhakyaraj, R., & Vidhyalakshmi, R. (2009). A low cost nutritious food "tempeh"—A review. *World Journal of Dairy & Food Sciences, 4*(1), 22–27.
6. Bhat, Z. F., & Bhat, H. (2011). Tissue-engineered meat: Future meat. *Journal of Stored Products and Postharvest Research*, 2(1), 1–10.
7. Bhat, Z. F., & Fayaz, H. (2011). Prospectus of cultured meat—Advancing meat alternatives. *Journal of Food Science and Technology, 48*(2), 125–140.
8. Bohrer, B. M. (2019). An investigation of the formulation and nutritional composition of modern meat analogue products. *Food Science and Human Wellness, 8*(4), 320–329.
9. Boukid, F. (2021). Plant-based meat analogues: From niche to mainstream. *European Food Research and Technology, 247*, 297–308.
10. Bryant, C. J. (2022). Plant-based animal product alternatives are healthier and more environmentally sustainable than animal products. *Future Foods, 6*, 100174.
11. Bryant, C. J., Szejda, K., Deshpande, V., Parekh, N., & Tse, B. (2019). A survey of consumer perceptions of plant-based and clean meat in the USA, India, and China. *Frontiers in Sustainable Food Systems, 3*, 11.
12. Bryngelsson, S., Moshtaghian, H., Bianchi, M., & Hallström, E. (2022). Nutritional assessment of plant-based meat analogues on the Swedish market. *International Journal of Food Sciences and Nutrition, 73*, 889–901.

13. Buller, H., Blokhuis, H., Jensen, P., & Keeling, L. (2018). Towards farm animal welfare and sustainability. *Animals, 8*(6), 81.
14. Burghardt, G. M. (2009). Ethics and animal consciousness: How rubber the ethical ruler? *Journal of Social Issues, 65*(3), 499–521.
15. Cantwell, M., & Elliott, C. (2017). Nitrates, nitrites and nitrosamines from processed meat intake and colorectal cancer risk. *Journal of Clinical Nutrition & Dietetics, 3*(4), 27.
16. Chang, K., & Hou, H. J. (2003). Science and technology of tofu making. In Y. H. Hui, S. Ghazala, D. M. Graham, K. D. Murrell, & W. K. Nip (Eds.), *Handbook of Fruits and Vegetable Processing* (pp. 443–478). Marcel Dekker.
17. Chen, C. C., Wang, Y. H., & Wu, K. Y. (2013). Consumption of bovine spongiform encephalopathy (BSE) contaminated beef and the risk of variant Creutzfeldt-Jakob disease. *Risk Analysis, 33*(11), 1958–1968.
18. Chen, F. L., Wei, Y. M., Zhang, B., & Ojokoh, A. O. (2010). System parameters and product properties response of soybean protein extruded at wide moisture range. *Journal of Food Engineering, 96*(2), 208–213.
19. Cole, E., Goeler-Slough, N., Cox, A., & Nolden, A. (2022). Examination of the nutritional composition of alternative beef burgers available in the United States. *International Journal of Food Sciences and Nutrition, 73*, 425–432.
20. Corrin, T., & Papadopoulos, A. (2017). Understanding the attitudes and perceptions of vegetarian and plant-based diets to shape future health promotion programs. *Appetite, 109*, 40–47.
21. Crimarco, A., Landry, M. J., Carter, M. M., & Gardner, C. D. (2022). Assessing the effects of alternative plant-based meats vs. animal meats on biomarkers of inflammation: A secondary analysis of the SWAP-MEAT randomized crossover trial. *Journal of Nutritional Science, 11*, e82.
22. Cuny, C., Layer, F., Hansen, S., Werner, G., & Witte, W. (2019). Nasal colonization of humans with occupational exposure to raw meat and raw meat products with methicillin-susceptible and methicillin-resistant *Staphylococcus aureus*. *Toxins, 11*(4), 190.
23. Cutroneo, S., Angelino, D., Tedeschi, T., Pellegrini, N., Martini, D., SINU Young Working Group, Dall'Asta, M., Russo, M. D., Nucci, D., Moccia, S., et al. (2022). Nutritional quality of meat analogues: Results from the Food Labelling of Italian Products (FLIP) Project. *Frontiers in Nutrition, 9*, 852831.
24. Dagevos, H. (2016). Exploring flexitarianism: Meat reduction in a meat-centred food culture. In T. Raphaely & D. Marinova (Eds.), *Impact of*

Meat Consumption on Health and Environmental Sustainability (pp. 233–243). IGI Global.

25. De Angelis, D., Kaleda, A., Pasqualone, A., Vaikma, H., Tamm, M., Tammik, M. L., Squeo, G., & Summo, C. (2020). Physicochemical and sensorial evaluation of meat analogues produced from dry-fractionated pea and oat proteins. *Foods, 9*(12), 1754.
26. de Boer, J., Schösler, H., & Aiking, H. (2014). "Meatless days" or "less but better"? Exploring strategies to adapt Western meat consumption to health and sustainability challenges. *Appetite, 76*, 120–128.
27. De Marchi, M., Costa, A., Pozza, M., Goi, A., & Manuelian, C. L. (2021). Detailed characterization of plant-based burgers. *Scientific Reports, 11*, 2049.
28. de Vries, M., & de Boer, I. J. (2010). Comparing environmental impacts for livestock products: A review of life cycle assessments. *Livestock Science, 128*(1–3), 1–11.
29. Dekkers, B. L., Boom, R. M., & van der Goot, A. J. (2018). Structuring processes for meat analogues. *Trends in Food Science and Technology, 81*, 25–36.
30. Dekkers, B. L., Emin, M. A., Boom, R. M., & van der Goot, A. J. (2018). The phase properties of soy protein and wheat gluten in a blend for fibrous structure formation. *Food Hydrocolloids, 79*, 273–281.
31. Dekkers, B. L., Hamoen, R., Boom, R. M., & van der Goot, A. J. (2018). Understanding fiber formation in a concentrated soy protein isolate-pectin blend. *Journal of Food Engineering, 222*, 84–92.
32. Detzel, A., Krüger, M., Busch, M., Blanco-Gutiérrez, I., Varela, C., Manners, R., Bez, J., & Zannini, E. (2022). Life cycle assessment of animal-based foods and plant-based protein-rich alternatives: An environmental perspective. *Journal of the Science of Food and Agriculture, 102*(12), 5098–5110.
33. Djekic, I. (2015). Environmental impact of the meat industry: Current status and future perspectives. *Procedia Food Science, 5*, 61–64.
34. Elzerman, J. E., Hoek, A. C., Van Boekel, M. A., & Luning, P. A. (2011). Consumer acceptance and appropriateness of meat substitutes in a meal context. *Food Quality and Preference, 22*(3), 233–240.
35. Elzerman, J. E., Hoek, A. C., van Boekel, M. J., & Luning, P. A. (2015). Appropriateness, acceptance and sensory preferences based on visual information: A web-based survey on meat substitutes in a meal context. *Food Quality and Preference, 42*, 56–65.

36. Elzerman, J. E., Van Boekel, M. A., & Luning, P. A. (2013). Exploring meat substitutes: Consumer experiences and contextual factors. *British Food Journal, 115*(5), 700–710.

37. Engel, E., Ratel, J., Bouhlel, J., Planche, C., & Meurillon, M. (2015). Novel approaches to improving the chemical safety of the meat chain towards toxicants. *Meat Science, 109*, 75–85.

38. Englefield, B., Blackman, S. A., Starling, M., & McGreevy, P. D. (2019). A review of Australian animal welfare legislation, regulation, codes of practice, and policy, and their influence on stakeholders caring for wildlife and the animals for whom they care. *Animals, 9*(6), 335.

39. Fraser, R. Z., Shitut, M., Agrawal, P., Mendes, O., & Klapholz, S. (2018). Safety evaluation of soy leghemoglobin protein preparation derived from *Pichia pastoris*, intended for use as a flavor catalyst in plant-based meat. *International Journal of Toxicology, 37*(3), 241–262.

40. Froldi, F., Lamastra, L., Trevisan, M., Mambretti, D., & Moschini, M. (2022). Environmental impacts of cow's milk in Northern Italy: Effects of farming performance. *Journal of Cleaner Production, 363*, 132600.

41. Fu, D., Zhang, D., Xu, G., Li, K., Wang, Q., Zhang, Z., & Qu, L. (2015). Effects of different rearing systems on meat production traits and meat fiber microstructure of Beijing-you chicken. *Animal Science Journal, 86*(7), 729–735.

42. Gerbens-Leenes, P. W., Moll, H., & Uiterkamp, A. S. (2003). Design and development of a measuring method for environmental sustainability in food production systems. *Ecological Economics, 46*(2), 231–248.

43. Giacalone, D., Clausen, M. P., & Jaeger, S. R. (2022). Understanding barriers to consumption of plant-based foods and beverages: Insights from sensory and consumer science. *Current Opinion in Food Science, 48*, 100919.

44. Gorissen, S. H. M., Crombag, J. J. R., Senden, J. M. G., Waterval, W. A. H., Bierau, J., Verdijk, L. B., & van Loon, L. J. C. (2018). Protein content and amino acid composition of commercially available plant-based protein isolates. *Amino Acids, 50*, 1685–1695.

45. Grabowska, K. J., Zhu, S., Dekkers, B. L., de Ruijter, N. C., Gieteling, J., & van der Goot, A. J. (2016). Shear-induced structuring as a tool to make anisotropic materials using soy protein concentrate. *Journal of Food Engineering, 188*, 77–86.

46. Grahl, S., Palanisamy, M., Strack, M., Meier-Dinkel, L., Toepfl, S., & Mörlein, D. (2018). Towards more sustainable meat alternatives: How technical parameters affect the sensory properties of extrusion products

derived from soy and algae. *Journal of Cleaner Production, 198*, 962–971.

47. Grandin, T. (2014). Animal welfare and society concerns: Finding the missing link. *Meat Science, 98*(3), 461–469.
48. Grossmann, L., & Weiss, J. (2021). Alternative protein sources as technofunctional food ingredients. *Annual Review of Food Science and Technology, 12*, 93–117.
49. Hunt, J. R. (2003). Bioavailability of iron, zinc, and other trace minerals from vegetarian diets. *American Journal of Clinical Nutrition, 78*, 633S–639S.
50. Kumar, M., Tomar, M., Punia, S., Dhakane-Lad, J., Dhumal, S., Changan, S., Senapathy, M., Berwal, M. K., Sampathrajan, V., & Sayed, A. A. S., et al. (2022). Plant-based proteins and their multifaceted industrial applications. *LWT, 154*, 112620.
51. Nezlek, J. B., & Forestell, C. A. (2022). Meat substitutes: Current status, potential benefits, and remaining challenges. *Current Opinion in Food Science, 47*, 100890.
52. Nishinari, K., Fang, Y., Guo, S., & Phillips, G. O. (2014). Soy proteins: A review on composition, aggregation, and emulsification. *Food Hydrocolloids, 39*, 301–318.
53. Oussalah, A., Levy, J., Berthezène, C., Alpers, D. H., & Guéant, J. L. (2020). Health outcomes associated with vegetarian diets: An umbrella review of systematic reviews and meta-analyses. *Clinical Nutrition, 39*, 3283–3307.
54. Pereira, P. M. d. C. C., & Vicente, A. F. d. R. B. (2013). Meat nutritional composition and nutritive role in the human diet. *Meat Science, 93*, 586–592.
55. Pointke, M., & Pawelzik, E. (2022). Plant-based alternative products: Are they healthy alternatives? Micro- and macronutrients and nutritional scoring. *Nutrients, 14*, 601.
56. Rotz, C. A., Montes, F., & Chianese, D. S. (2010). The carbon footprint of dairy production systems through partial life cycle assessment. *Journal of Dairy Science, 93*, 1266–1282.
57. Schutyser, M. A. I., Pelgrom, P. J. M., van der Goot, A. J., & Boom, R. M. (2015). Dry fractionation for sustainable production of functional legume protein concentrates. *Trends in Food Science and Technology, 45*, 327–335.
58. Sha, L., & Xiong, Y. L. (2020). Plant protein-based alternatives of reconstructed meat: Science, technology, and challenges. *Trends in Food Science and Technology, 102*, 51–61.

59. Tso, R., & Forde, C. G. (2021). Unintended consequences: Nutritional impact and potential pitfalls of switching from animal- to plant-based foods. *Nutrients, 13*, 2527.
60. Tukker, A., Goldbohm, R. A., de Koning, A., Verheijden, M., Kleijn, R., Wolf, O., Pérez-Domínguez, I., & Rueda-Cantuche, J. M. (2011). Environmental impacts of changes to healthier diets in Europe. *Ecological Economics, 70*, 1776–1788.
61. Verain, M., Dagevos, H., & Antonides, G. (2016). Flexitarianism: A range of sustainable food styles. In L. A. Reisch & J. Thøgersen (Eds.), *Handbook of Research on Sustainable Consumption* (pp. 209–223). Edward Elgar Publishing.
62. Wittek, P., Ellwanger, F., Karbstein, H. P., & Emin, M. A. (2021). Morphology development and flow characteristics during high moisture extrusion of a plant-based meat analogue. *Foods, 10*, 1753.
63. Zahari, I., Östbring, K., Purhagen, J. K., & Rayner, M. (2022). Plant-based meat analogues from alternative protein: A systematic literature review. *Foods, 11*, 2870.
64. Zhang, T., Dou, W., Zhang, X., Zhao, Y., Zhang, Y., Jiang, L., & Sui, X. (2021). The development history and recent updates on soy protein-based meat alternatives. *Trends in Food Science and Technology, 109*, 702–710.

Chapter 2

Nutritional Composition of Traditional Meats

Contents

Introduction 40

2.1. Nutritional Composition of Meat 40

2.2. Nutritional Composition of Game Meat From Wild Species Harvested in Europe 55

References 74

INTRODUCTION

Customary meats, which have been consumed for centuries in many different cultures, are an important part of the human diet because they provide nutrients that are necessary for development, growth, and general health. These meats are prized for their rich nutritional profiles. They are frequently obtained from domesticated animals like cattle, sheep, goats, and poultry as well as wild game like venison, rabbits, and wild fowl. In addition to being a substantial source of high-quality protein, traditional meats offer a variety of vitamins and minerals that are necessary for internal body processes.

Traditional meats have a lot of protein, which is important for immune system function, muscle repair, and enzyme synthesis. This is one of their main benefits. These meats' proteins are complete proteins since they include each of the nine necessary amino acids. Additionally, traditional meats are high in B vitamins like niacin, riboflavin, and B_{12} as well as micronutrients like iron, zinc, and B_{12}. Heme-form iron is present in meats and is readily absorbed by the body, aiding in the blood's ability to carry oxygen. B vitamins are necessary for energy metabolism and neurological function, and zinc helps the body heal wounds and maintain a healthy immune system.

Conventional meats are a great source of vital nutrients, but the type, cut, and method of preparation all affect how much fat is in the meat. For example, compared to lean meats like poultry and wild game, red meats like beef and lamb typically contain higher levels of saturated fats. An additional nutritional benefit of grass-fed or pasture-raised meats is the presence of omega-3 fatty acids, which lower inflammation and promote heart health.

Conventional meats offer important micronutrients, high-quality protein, and, depending on the source, healthy fats. They thus make a valuable addition to a diet that is balanced. It's essential to comprehend their nutritional makeup to make wise dietary decisions.

2.1. NUTRITIONAL COMPOSITION OF MEAT

Human health is largely dependent on the consumption of fresh, wholesome, and healthful food products. A balanced diet has become incredibly popular all over the world as people become more conscious of the need to maintain their health. A balanced diet guarantees that all the vital nutrients—which the body needs to carry out daily tasks—are consumed. In this situation, knowing the nutritional makeup of food items has become crucial to ensuring a balanced meal, which in turn protects people's health. Nutritional composition provides energy values and a comprehensive framework of information about essential nutritional components of food items. The components that supply the nourishment required for growth and life maintenance are known as nutrients, and they include both macro- and micronutrients. The macronutrients—proteins, fats, and carbohydrates—are those that the body needs in significant quantities. Micronutrients, which include vitamins, minerals, and fiber, are substances that the body needs in minute amounts. A variety of food items, such as meat, cereal grains, milk, fruits, and vegetables, provide all of these. Meat has a special place among them since it provides the majority of the protein needed by humans. There are various meats available, such as beef, mutton, lamb, chicken, and fish. With only minor variations in composition, every variety of meat has importance of its own.

One of the most important, nutrient-dense, and energy-packed natural food products that humans use to meet their daily needs is meat. It is thought to be highly significant for keeping up a nutritious and well-balanced diet, which is necessary for achieving the best possible human growth and development. Even though a small number of epidemiological studies have suggested a potential link between its consumption and increased risks of metabolic disorders, cancers, and cardiovascular diseases, its significance for the evolution of the human species—particularly about brain and intellectual development—cannot be overlooked.

Meat is defined by European legislation as the edible parts of domestic animals, such as caprine, bovine, ovine, and porcine, as well as poultry, farmed, and wild animals. It is a great source of high-quality proteins, zinc, iron, selenium, potassium, magnesium, sodium, B-complex vitamins, and folic acid, among other fats, including omega-3 polyunsaturated fatty acids. Its composition changes depending on the breed, kind of feed consumed, weather, and cut of meat, all of which have a significant impact on the animal's sensory and nutritional qualities.

In terms of nutrition, meat is regarded as a rich source of essential amino acids and, to a lesser extent, minerals. In addition to that, vitamins, and essential fatty acids are also components of it. Vitamin B_1, vitamin A, and nicotinic acid are all abundant in organ meats like liver. Research is still being conducted to gain a better understanding of the potential variations in the nutritional value of various meat cuts, breeds, and animal species. Previous research has clearly shown that meat with lower levels of connective tissue is more likely to have lower levels of absorption and digestion. Additionally, meat with higher connective tissue is thought to contain fewer essential amino acid contents, making it less nutritious than meat with lower connective tissue, which has higher digestibility and nutritional value. Table 2.1, which follows, provides information on the nutritional makeup of various types of meat products.

2.1.1. Water

One of the most vital components of all food materials is water. Generally speaking, food products can be divided into three categories based on how much moisture they contain: perishables (which have a moisture content of more than 70%), non-perishables (which have a moisture content of between 50 and 60%), and stable food materials (which have less than 15% moisture). Any food item's likelihood of having a longer shelf life decreases with its water content because microorganisms have a higher chance of growing on it and shortening its life.

Meat is classified as a perishable food item because it contains more than 70% moisture. In addition to shortening the shelf life, it significantly affects the color, texture, and flavor of meat's muscle tissues. Because the animal's adipose (abdominal) tissues have a lower moisture content than other tissues, a fatter animal will have a lower water content in its corpse and vice versa. The moisture content of the younger, leaner animals was about 72%.

A larger percentage of the water found in meat tissues is found in free form in the muscle fibers, while a lesser amount is found in the connective tissues. A tiny amount of water is left in the muscle fiber during processing procedures like curing and heat treatment, which is then followed by storage. This water is referred to as "bound water" The majority of the water "lost" during these conditions is known as "free water," but the three-dimensional structure of muscle fiber strengthened with pressure and temperature helps the water to remain in the muscles during processing conditions. Disruptions to meat's muscle fibers may change the meat's capacity to retain water, which helps extend the shelf life of meat products. The processes that can affect the final water contents of meaty products include chopping, grinding, salting, freezing, thawing, breaking down connective tissues by chemical or enzymatic means, applying heat, and using chemicals or organic additives to change the meat's acidity (pH).

2.1.2. Carbohydrates

The liver is the primary source of carbohydrates in an animal's body, holding approximately 0.5 of the total amount of carbohydrates found in the body. They are primarily stored in the muscles and liver as "glycogen," though they are also somewhat stored in glands and other organs. Blood contains significant amounts of it in the form of glucose. The color, texture, tenderness, and ability of the meat to hold water are all indirectly influenced by the glycogen. The intricate process of converting glycogen stored in the body to glucose and then back again to lactic acid is controlled by hormones and enzymes. The lactic acid content of muscles rises in the early stages of aging, lowering pH. The water-holding capacity, color, texture, and tenderness of muscles are all significantly influenced by pH. The muscle's normal pH is thought to be approximately 5.6. Before being killed, if an animal experiences extreme stress or exercise and is unable to replenish its glycogen stores, a small quantity of glycogen will remain and be converted to lactic acid, which will raise the pH (i.e., Consequently, meat muscles become firm, dark, and dry (DFD). This kind of meat is the end result of fatigue, which leads to the breakdown of glycogen prior to killing. This affects the other cattle that are referred to as "Dark Cutters" as well, though it happens less frequently in beef (2%). The greater water-holding capacity is the primary cause of the dark-colored meat with a high pH. As a result, the muscles take in more water, which causes them to absorb light instead of reflecting it off the surface of the meat, giving the meat a darker appearance. Because of this DFD defect, which negatively affects the sensory and nutritional qualities of the product and is highly disliked by retailers and customers, stress, and rough handling of animals should be avoided shortly before slaughter.

A very quick postmortem lowers the pH of the muscles (i.e., pale, soft, and exudative condition (PSE), which is fairly common in pork meat, is used to identify 5.0. A muscle portion affected by PSE can be identified by its soft texture, pale yellow color, and low water-holding capacity. Because PSE meat has softer muscles, it has a lower water-holding capacity, which increases its reflectance of incident light and gives the meat a pale yellow color.

The carbohydrate content of the meat is related to all of the DFD and PSE requirements listed above, and this has a big impact on the meat's nutritional value.

2.1.3. Proteins and Their Amino Acids

Meat is a food that is high in protein and has a high biological value for the general population. Proteins are complex nitrogenous molecules that are found naturally. They have a very high molecular weight and are primarily made up of nitrogen, but they also contain carbon, hydrogen, and oxygen. Only a small percentage of the proteins' structures contain sulfur and phosphorus. These elements all worked chemically to create various kinds of unique proteins with unique characteristics. These differ in corresponding tissues of different species as well as from one tissue to another within the same living organism. Because of their size and composition, proteins are more complex than fats and carbohydrates. The proportion of protein in meat varies greatly amongst meat varieties. Meat protein generally has an average value of 22%, although it can vary greatly, ranging from as high as 34.5% in chicken breast to as low as 12.3% in duck meat. Meat has a high score of 0.92 on the protein digestibility-corrected amino acid scores (PDCAAS), which show the protein digestibility, compared to other protein sources like lentils, pinto beans, peas, and chickpeas, which score 0.57–0.71. The availability of amino acids in protein is the primary factor affecting its quality.

The building blocks of proteins are amino acids. The presence or lack of several amino acids can significantly affect the nutritional value of meat. There are 192 known, but only 20 are utilized to make protein preparations. Out of the twenty amino acids, eight are deemed essential because the human body is unable to produce them and therefore needs to obtain them through diet. The human body can manufacture the remaining 12 non-essential amino acids provided specific dietary sources are consumed. Failure to do so may lead to protein malnutrition. All of the essential and non-essential amino acids found in meat are listed in Table 2.2.

Table 2.1. *Nutritional Composition of Meat (Pereira, et al., 2013, Sinclair, et al., 1999)*

Meat Cut	Protein (g)	Sat. fat (g)	Fat (g)	Energy (kcal)	Vit. B_{12} (mcg)	Na (mg)	Zn (mg)	P (mg)	Fe (mg)
Chicken breast, raw	24.2	0.2	8.5	178	0.39	71	0.9	199	1.2
Beef, steak cuts, raw	21	1.9	4.5	123	1.9	59	1.7	167	1.3
Chicken, raw	22.8	0.6	1.9	113	0.70	78	1.4	202	0.7
Beef, calf, loin, raw	20	3.4	7.3	146	1.1	22	3	193	0.10
Beef, loin, raw	20.9	1.5	3.2	115	2	59	3.7	142	1.6
Pork, chop, raw	18.1	10.8	31.7	353	1	60	1.8	190	1.4
Pork, loin, raw	21.9	1.7	4.9	134	1.1	55	1.9	220	0.7

Pork, leg, raw	20.8	2.8	7.8	155	1.2	84	2.6	164	0.8
Turkey, skinless, raw	19.9	1.8	7.1	136	1.9	42	1.5	209	2.1
Duck meat, skinless, raw	19.4	1.8	6.6	130	2.8	90	1.8	201	2.5
Turkey, breast, skinless, raw	23.6	0.5	1.6	106	1	62	0.5	208	0.6
Chicken breast, skinless, raw	23.8	0.4	1.28	109	0.40	59	0.7	218	0.4
Mutton, chop or meat, raw	20	2.4	4.8	122	2	63	3.6	221	1.9

Source: Ahmad, R. S., Imran, A., and Hussain, M. B. (2018). Nutritional Composition of Meat. InTech. doi: 10.5772/intechopen.77045.

Table 2.2. *Amino acid composition in fresh meat (Sinclair et al., 1999; Schweigert et al., 1956; Mahan et al., 1998)*

Essential Amino Acids				
Amino acids	**Category**	**Beef**	**Lamb**	**Pork**
Lysine	Essential	8.2	7.5	7.9
Leucine	Essential	8.5	7.2	7.6
Isoleucine	Essential	5.0	4.7	4.8
Cystine	Essential	1.5	1.5	1.2
Threonine	Essential	4.2	4.8	5.2
Methionine	Essential	2.2	2.4	2.6
Tryptophan	Essential	1.3	1.2	1.5

Phenylalanine	Essential	4.1	3.8	4.3
Arginine	Essential	6.4	6.8	6.6
Histidine	Essential	2.8	2.9	3.1
Valine	Essential	5.6	5.1	5.2
Non-Essential Amino Acids				
Amino acid	**Category**	**Beef**	**Lamb**	**Pork**
Proline	Non-essential	5.2	4.7	4.4
Glutamic acid	Non-essential	14.3	14.5	14.6
Aspartic acid	Non-essential	8.9	8.6	8.8
Glycine	Non-essential	7.2	6.8	6.0
Tyrosine	Non-essential	3.3	3.3	3.1
Serine	Non-essential	3.9	3.8	4.1
Alanine	Non-essential	6.3	6.2	6.4

Source: Ahmad, R. S., Imran, A., and Hussain, M. B. (2018). Nutritional Composition of Meat. InTech. doi: 10.5772/intechopen.77045.

Compared to lamb and pork, the contents of valine, lysine, and leucine in beef meat seem to be higher. Research has shown that the breed, age, and muscle location of the animal account for the majority of the variation in the essential amino acid proportion. According to earlier studies, the age of an animal's meat affects its levels of valine, isoleucine, phenylalanine, arginine, and methionine. The contents of essential amino acids also vary depending on the area of the carcass. The use of processing methods like heat and ionization radiations may also have an impact on their composition, but only in extreme, prolonged situations. These amino acids are occasionally unavailable for use by humans. In one study, the researchers observed that at 160°C, only 50% of the lysine was available, whereas at 70°C, 90% of it was. On occasion, the availability of essential amino acids has been impacted by the interactions between the proteins and other components. The meat's smoking and salting processes have also contributed to this. In the case of canned meat, storage has had an impact on amino acids in addition to the processing conditions.

2.1.4. Fat and Fatty Acids

One of the three main macronutrients, along with proteins and carbohydrates, is fat. Triglycerides, which are esters of three fatty acid chains and the alcohol glycerol, are the term for fat contents. Fatty tissues, or fat cells packed with lipids, are found in meat and vary in fat content. The fat in meat serves many purposes, including energy storage, skin, and organ protection (particularly around the heart and kidney), and insulation against temperature loss. Animal carcasses can have anywhere from 8 to 20% fat (the latter found only in pork). There are notable differences in the fatty acid and fat composition of fatty tissue between poultry and other meat products like ham, sausages, and offal in different parts of the body. Due to the higher concentration of unsaturated fat in external animal parts, external body fat has a softer texture than the internal fat that envelops the organs. The primary source of fat in poultry meat is the skin. Turkey and chicken have fat contents in the main retail cuts that vary from 1 to 15%, while meat cuts with skin have a higher percentage. Meat fat content and the makeup of fatty acids can both be significantly impacted by cooking. Research findings indicated that broiling, grilling, and pan-frying meat cuts without additional fat resulted in significant fat losses.

In the structure of fats, meat comprises unsaturated fatty acids; these include oleic (C-18:1), linoleic (C-18:2), linolenic (C-18:3), and arachidonic (C-20:4) acids, which are considered essential. These components are vital components of mitochondria, cell membranes, and other metabolically active sites. Linoleic acid (C-18:2) is highly prevalent in vegetable oils like soya and corn oils, boasting a concentration 20 times higher than in meat, while linolenic acid (C-18:3) is commonly found in the leafy parts of plants. Eicosapentaenoic acid (C-20:5) and docosahexaenoic acid (C-22:6) have low concentrations in meat tissues, but they are abundant in fish and fish oils. Table 2.3 displays the concentrations of polyunsaturated fatty acids and cholesterol in the muscle and offal tissues of various meat species.

Table 2.3. *Polyunsaturated Fatty Acids and cholesterol in Lean Meat and Offal (Paul et al., 1978; Williams et al., 2007; Droulez et al., 2006) (as% Total Fatty Acids)*

Meat Source	**Cholesterol (mg/100 g)**	**C-18:2**	**C-18:3**	**C-20:3**	**C-20:4**	**C-22:5**	**C-22:6**
Mutton	81	2.4	2.4	Nil	Nil	Trace	Nil
Beef	62	2.1	1.4	Trace	1.1	Trace	Nil
Pork	71	7.5	1.0	Nil	Trace	Trace	1.1
Brain	2200	0.5	Nil	1.6	4.1	3.5	0.4
Pig's Kidney	415	11.6	0.4	0.5	6.72	Trace	Nil
Sheep's Kidney	399	8.2	4.1	0.6	7.2	Trace	Nil
Ox's kidney	401	4.9	0.6	Trace	2.7	Nil	Nil
Sheep's Liver	429	5.1	3.9	0.7	5.2	3.1	2.3
Pig's Liver	262	14.8	0.4	1.2	14.4	2.4	3.9
Ox's Liver	271	7.5	2.4	4.5	6.5	5.4	1.3

Source: Ahmad, R. S., Imran, A., and Hussain, M. B. (2018). Nutritional Composition of Meat. InTech. doi: 10.5772/intechopen.77045.

It is clear that pig lean meat has a higher concentration of linoleic acid than meat from oxen or sheep. The fatty acid profiles of the kidney and liver also show these differences in fatty acid composition concentrations between species. All of the aforementioned animal species' liver tissue is thought to be a rich source of polyunsaturated fatty acids. However, the concentration of C-22 polyunsaturated fatty acids is notably higher in the brain. It has been calculated that the brain and other offal tissues have higher cholesterol concentrations than muscle tissues.

The number of polyunsaturated fatty acids indicates that omega-3 fatty acids merit special attention because they protect against cardiovascular diseases and other conditions that affect human health generally. The primary source of omega-3 fatty acids is seafood. However, eating meat can account for up to 20% of your daily intake of long-chain omega-3 polyunsaturated fatty acids. Meat's polyunsaturated omega-3 content varies depending on the feeding source; a diet high in grass and forages has a higher level of this fatty acid. Additionally, it's been proposed that animal fat's polyunsaturated fatty acids are essential for brain development, especially in fetuses. Animal liver can break down ingested linoleic and linolenic acids to produce polyunsaturated fatty acids. Furthermore, prostaglandins, which are crucial for controlling blood pressure, are produced when linoleic acid chains lengthen. Prostaglandins are primarily found in tissues and organs, where they are produced by the cell from necessary fatty acids. They act on endothelium, uterine, and platelet cells and are produced by all nucleated cells. They are referred to as autocrine and paracrine lipid mediators.

It is necessary to increase the potential for unsaturation in the fats and fatty tissues of ruminant animals to prevent any potential negative health effects from eating their meat. In general, rumen bacteria's reduction or condensation of vegetable fats renders them useless for feeding sheep and cattle. However, there would be resistance to reduction if formaldehyde was applied initially, which would raise the possibility of unsaturation in the ruminant fat reserves. Numerous studies have focused on various methods of improving the fatty acid composition of meat because of the meat's significant role in the human diet, rising rate of consumption over time, and significant impact on human health. The composition of meat fatty acids can be altered by feeding an animal, especially single-stomach poultry and pigs, whose long-chain polyunsaturated, linoleic, and alpha-linolenic fatty acid contents react quickly to increased dietary applications. A notable distinction was observed in the fatty acid composition of animals fed grain versus pasture, with the pasture-fed animal groups exhibiting a higher concentration of polyunsaturated fatty acids. The fatty acid composition of meat may be influenced by an animal's digestive traits.

Microbial enzymes promote the hydrolysis of unsaturated fatty acids, which raises the concentration of stearic acid and causes it to enter and be absorbed in the small intestine. Rumen bacteria's process of biohydrogenation results in the formation of trans-fatty acids in beef. Conjugated linoleic acid (CLA), the most prevalent and well-known component of meat from ruminant animals, has been shown to protect against diabetes, obesity, and cardiovascular disease.

2.1.5. Minerals

The nutrients found in food materials that do not include carbon are called minerals, and they are necessary for the healthy growth, development, and maintenance of the human body. They are separated into two groups: i. e. minerals, both macro and micro, according to what the human body needs. The minerals that the body needs in greater quantities are known as macro-minerals. Magnesium, calcium, phosphorus, sodium, potassium, chloride, and sulfur are examples of these, whereas iron, zinc, iodine, copper, cobalt, manganese, selenium, and fluoride are considered micro-minerals because they are needed in smaller amounts. The micro- and macromineral content of meat and meat products is shown in Table 2.4.

It's clear that potassium dominates other minerals in terms of quantitative content. e. Phosphorus, sodium, and magnesium come next. Additionally, meat is an excellent source of zinc, selenium, and iron. The various roles that these minerals play in the growth, development, and maintenance of the human body are as follows.

- **Potassium:** Potassium aids in the body's metabolism, nerve impulse transmission, growth, development of muscles, and preservation of the acid-base balance.
- Phosphorus
- An essential mineral element, phosphorus helps to form phospholipids with calcium and is involved in the development of teeth and bones.
- **Sodium:** Maintains the osmotic pressure of bodily fluids, controls the body's water content, and helps transport carbon dioxide.
- **Magnesium:** Magnesium helps to keep bones strong, regulates blood pressure, prevents tooth decay, and repairs and enhances human growth.
- **Zinc:** Zinc is an essential component of many enzymes that the immune system of the body needs for cell division, growth, and wound healing.
- **Selenium:** Prevent cancer, and the harmful effects of heavy metals, and assist the body following immunization.

- **Iron:** One of the essential minerals found in meat is iron, which is crucial to human health and a shortage of it can impede many bodily processes, most notably the growth and development of children. Since iron is excreted and over 90% of it is used internally by the body, its metabolism differs significantly from that of other mineral contents. The skin, intestines, and urinary tract are required sources of disruption or loss of iron and red blood cells, as are female menstrual bleeding episodes. The diet could be the main means of overcoming its deficiency. Heme and non-heme iron are two types of iron that are found in various food items. The former is only found in animal foods and has a high degree of bioavailability, making it simple for the intestinal lumen to absorb. It originates from hemoglobin and myoglobin.

2.1.5.1. Organ Meat as a Mineral Source

It is clear that, in comparison to the minerals found in muscle tissues, the offal organs have higher concentrations of minerals like iron, zinc, and copper. It has been noted that children who follow a strictly vegetarian diet may experience cognitive impairment due to zinc deficiency, which is why eating meat has been recommended. The mineral composition of offal organs is shown in Table 2.5.

Table 2.4. *Mineral contents (mg/100 g) of meat and meat products (McCance et al., 1960; Chan et al., 1995)*

Meat Source	K	Cu	Fe	P	Zn	Mg	Na	Ca
Chopped Mutton, (raw)	244	0.15	0.99	174	4.2	18.8	74	12.5
Chopped Mutton, (grilled)	303	0.25	2.5	205	4.2	22.7	101	17.9
Beef, Steak (raw)	335	0.1	2.4	275	4.2	24.4	68	5.5
Beef, Steak (grilled)	369	0.22	3.8	302	5.8	25.1	66	901
Bacon (raw)	267	0.2	1.0	95	2.4	12.2	976	13.6
Bacon, (fried)	516	0.2	2.7	228	3.7	25.8	2792	11.6
Pork (raw)	399	0.1	1.5	224	2.5	26.2	44	4.2
Chopped Pork, (grilled)	259	0.1	2.5	179	3.6	14.8	60	8.2

Source: Ahmad, R. S., Imran, A., and Hussain, M. B. (2018). Nutritional Composition of Meat. InTech. doi: 10.5772/intechopen.77045.

Table 2.5. *Mineral content of offal tissues (Paul, 1978; Williams, 2007)*

Meat source	Fe	P	Na	Ca	Cu	Mg	Zn	K
Ox (Kidney)	5.6	231	182	9	0.5	16	1.8	232
Ox (Liver)	7.1	362	80	6.1	2.4	19.2	4.1	321
Sheep (Kidney)	7.5	242	221	10.2	0.5	17.1	2.5	272
Sheep (Liver)	9.5	371	75	7.1	8.8	19.1	4.0	291
Pig (Kidney)	5.1	272	191	8.1	0.7	19.1	2.7	291
Pig (Liver)	21.2	372	88	6.2	2.8	21.3	7.0	319
Brain	1.5	341	142	12.2	0.4	15.1	1.3	269

Source: Ahmad, R. S., Imran, A., and Hussain, M. B. (2018). Nutritional Composition of Meat. InTech. doi: 10.5772/intechopen.77045.

2.1.6. Vitamins

A class of organic compounds known as vitamins have multiple roles in the human body. Even though the human body needs these components in very small amounts, they are crucial for healthy growth, development, and maintenance. Children especially need them in their early years of life. They engage in a range of metabolic activities that include chemical and biological reactions. One characteristic that sets them apart is that, in most cases, mammalian cells are unable to prepare them; therefore, they must be obtained through diet. Based on their solubility in fat and water, they are typically divided into two groups. e. vitamins that are soluble in fat and water. The B-complex vitamins (thiamin, riboflavin, nicotinic acid, pyridoxine, choline, biotin, folic acid, cyanocobalamin, inositol, vitamins B_6 and B_{12}) and vitamin C are examples of water-soluble vitamins. Vitamins A, D, and K, which are fat-soluble vitamins found in meat, also contribute to the food value of meat.

Five of the B-complex vitamins—thiamine, riboflavin, nicotinic acid, vitamin B_6, and vitamin B_{12}—are found in meat. In addition, it has biotin and pantothenic acid, but not much folacin. Table 2.6 shows the different raw meats' vitamin contents.

2.1.6.1. Water Soluble Vitamins

- **Thiamin:** In concert with other B-complex vitamins, it facilitates a variety of chemical reactions necessary for the development and upkeep of the human body. They are a part of the metabolic processes that produce the energy needed to carry out different bodily functions. A thiamine deficiency may result in depression, irritability, weariness, constipation, and appetite loss. Meat in general, especially fish, which has higher levels of thiamine than other meat sources (apart from pork), is a good source of the mineral.

- **Riboflavin:** Energy from the main food components, such as proteins, fats, and carbohydrates, must be released. It supports maintaining healthy skin and eyesight. It also facilitates iron absorption and utilization. Furthermore, it is necessary for the conversion of tryptophan to niacin. Beef, lamb, and poultry meat are regarded as some of the best sources of riboflavin.
- **Niacin:** Niacin participates in several intracellular enzyme systems, including those that are involved in energy production, along with other B-vitamins. Its sources include fish, poultry, and meat. The condition known as "pellagra," which is characterized by rough or raw skin, is brought on by its deficiency. Diarrhea, vomiting, and memory loss are other issues.
- **Vitamin B_6:** About 100 enzymes in the human body catalyze vital chemical reactions, and vitamin B6 is essential to their function. It is crucial for the synthesis of heme iron I and aids in the production of neurotransmitters. e.g., part of the hemoglobin molecule. Furthermore, it facilitates the conversion of tryptophan into niacin. Fish, poultry, and meat are significant meat-based sources of vitamin B_6.
- **Vitamin B_{12}:** The synthesis of deoxyribonucleic acid (DNA), a gene-containing component of the cell nucleus and essential for the healthy growth and development of the human body, depends on this vitamin. Since vitamin B_{12} can only be found in foods derived from animals, vegans—vegetarians who abstain from eating animal products—may have needed to add extra vitamin B_{12} to their diet. Vitamin B_{12} injections can be used to treat people with pernicious anemia, which is the inability to absorb vitamin B_{12} from food if they do not consume vitamin B_{12}. Pork, lamb, beef, and liver are excellent providers of this vitamin. Cheese, oysters, fish, and egg yolks are a few more sources.

2.1.6.2. Loss of B Complex Vitamins During Meat Processing

Both conventional heating and microwave heating cause the loss of some of the vitamins found in meat, particularly vitamin B_1. The table below illustrates how conventional cooking preserves vitamins B_1 and B_2 in various types of meat. The primary method used to detect vitamin B_1 loss was leaching. About 15–40% are lost during boiling, 40–50% are lost during frying, 30–60% are lost during roasting, and 50–70% are lost during canning. The problems with B_1 are also present in other vitamins of the B complex family, such as B_6, B_{12}, and pantothenic acid. In contrast, vitamin A can be retained at temperatures as high as 80°C. Table 2.7 shows how B complex vitamins are lost or retained during conventional and microwave cooking.

Table 2.6. *Vitamin Content of Various Raw Meats (McCance, 1960; Williams, 2007)*

Vitamin Units/100 G Raw Meat	Beef	Bacon	Mutton	Veal	Pork
A (Inter. Unit.)	Trace	Trace	Trace	Trace	Trace
D (Inter. Unit.)	Trace	Trace	Trace	Trace	Trace
B_1 (mg)	0.06	0.39	0.14	0.11	1.2
B_2 (mg)	0.21	0.16	0.24	0.26	0.21
Nicotinic acid (mg)	5.1	1.6	4.99	7.1	5.2
Pantothenic acid (mg)	0.5	0.4	0.6	0.5	0.5
Biotin (μg)	2	8	4	6	5
Folic acid (μg)	9	Nil	2	6	2
B_6 (mg)	0.2	0.3	0.3	0.4	0.4
B_{12} (μg)	2	Nil	2	Nil	2
C (mg)	Nil	Nil	Nil	Nil	Nil

Source: Ahmad, R. S., Imran, A., and Hussain, M. B. (2018). Nutritional Composition of Meat. InTech. doi: 10.5772/intechopen.77045.

Table 2.7. *Comparison of Cooking Losses and Vitamin B_1 Retention in Conventional and Microwave Cooking (McCance et al., 1960)*

Meat Samples	Cooking Method Involved	Cooking Losses Water And Fat (% Initial Weight	Vitamin B_1 Retention In Meat And Dripping (%Initial)	Internal Temperature (°C)
Beef	Conventional	19–20	82–87	62.5
Beef	Microwave	28–38	70–80	70.5
Beef loaves	Conventional	24.2	76.5	85.5

Beef loaves	Microwave	27.3	79	84.5
Pork	Conventional	34.1	80.3	85
Pork	Microwave	36.7	90.8	86
Ham loaves	Conventional	18.4	91.4	85
Ham loaves	Microwave	27.8	87.2	84

Source: Ahmad, R. S., Imran, A., and Hussain, M. B. (2018). Nutritional Composition of Meat. InTech. doi: 10.5772/intechopen.77045.

2.1.6.3. Fat-Soluble Vitamins

Fat-soluble vitamin A is essential for maintaining normal vision and eyesight as well as the health of tissues. The majority of vitamin A is found in green and yellow vegetables, where it takes the form of carotene, a precursor that the body transforms into vitamin A. Vitamin A is frequently added fortification to milk and margarine. It is suggested that the best food source of vitamin A is liver. It is also a good source of vitamin K and vitamin D, two other fat-soluble vitamins. Table 2.8 displays the water and fat-soluble vitamin contents of the different offal organs.

Table 2.8. *Vitamin Contents (Units/100 G Raw Tissue) of Various Offal Tissues (Paul, 1978; Williams, 2007)*

Meat Source	B_1 (mg)	B_2 (mg)	B_3 (mg)	B_6 (µg)	B_9 (µg)	B_{12} (µg)	Vit. C (mg)	Vit. D (µg)	Vit. A (I.U.)
Brain	0.06	0.02	2.99	0.10	6.0	8.9	23.0	Trace	Trace
Sheep's kidney	0.5	1.9	8.4	0.32	31.0	54.9	6.9	Nil	99
Ox's kidney	0.38	2.2	6.1	0.33	77.2	31.2	10.1	Nil	150
Pig's kidney	0.33	2.0	7.4	0.24	42.1	14.2	14.3	Nil	110
Sheep's liver	0.28	3.4	14.1	0.43	220	83	9.9	0.49	20,000
Ox's liver	0.22	3.2	13.5	0.84	330	109.7	23.0	1.14	17,000
Pig's liver	0.32	3.1	14.7	0.69	110	24.8	13.2	1.14	10,000
Sheep's lung	0.13	0.5	4.8	Nil	Nil	4.8	31.2	Nil	Nil
Ox's lung	0.10	0.4	4.1	Nil	Nil	3.2	38.7	Nil	Nil
Pig's lung	0.10	0.3	3.3	Nil	Nil	Nil	13.1	Nil	Nil

Source: Ahmad, R. S., Imran, A., and Hussain, M. B. (2018). Nutritional Composition of Meat. InTech. doi: 10.5772/intechopen.77045.

Research has shown that a healthy diet rich in fats, proteins, and carbohydrates along with minerals, vitamins, and other useful substances can help prevent both major and minor diseases caused by nutrient deficiencies. To achieve the necessary health benefits, this food item needs to be consumed in a balanced diet in a significant amount. Amino acids and proteins aid in the development and maintenance of human muscle. Because meat contains saturated fats, which can raise cholesterol levels and cause coronary heart disease if consumed in excess, there is reason for concern regarding meat consumption due to its fat and fatty acid profile composition. Therefore, eating meat in a balanced amount needs to follow the advice of nutritionists and medical professionals. Furthermore, meat is thought to contain significant amounts of several minerals and vitamins that are good for human health overall, including zinc, iron, selenium, sodium, copper, magnesium, calcium, potassium, phosphorus, and vitamin A. Additionally, meat also contains a good amount of B complex vitamins (Ahmad et al., 2018).

2.2. NUTRITIONAL COMPOSITION OF GAME MEAT FROM WILD SPECIES HARVESTED IN EUROPE

The demand for game meat has grown in Europe over the past few decades, primarily due to consumer perceptions of the meat as "natural and sustainable." The majority of the food fed to free-range animals is natural (i.e., pasture), and there are no hormones, antibiotics, or other products in their meat. Meat from wild game species satisfies ethical standards for animal welfare, and compared to farmed species, the rearing of animals in the wild has little to no impact on ecosystems. As a result, consumers view this meat differently than they do traditional meat that comes from domestic species. It is considered a seasonal product that is available in fresh form during the colder months and can also be consumed in other seasons as frozen or cured meat, including pates and conserves like cecina and salchichón. Consumers in Europe view game meat as a novel food and are willing to pay a premium for it. If game meat improves in quality and becomes more widely available, they may even increase their consumption.

Nevertheless, only 2–4% of Europeans regularly eat game meat, making it an extremely low consumption level. This can be explained by several factors, including its high cost, seasonal availability, the fact that consumers who are not involved in hunting do not typically use game meat in recipes, and even safety concerns. According to a recent study, those who are unfamiliar with hunting may consume less game meat because those who hunt and their families consume it at higher rates than those who are not. Therefore, to increase the consumption of game meat and products derived from it, sellers, and consumers should be informed about the benefits of hunting for rural economies and wildlife conservation as well as the nutritional value of game meat and how it may fit into a balanced diet. Through the initiative "European wild meat, nature at its purest," the European Union (EU) is encouraging the consumption of game meat in certain nations. About half of the French producers of game meat advertise under the "French game meat brand," showcasing the meat's quality and culinary possibilities in a potent marketing campaign aimed primarily at chefs and consumers. The National Association of Game Meat Producers (ASICCAZA) in Spain is running a campaign to raise consumer awareness of game meat.

Wild or farmed animals can be used to produce meat from game species. Game farming is becoming more popular in Europe, and countries like New Zealand, Australia, China, and Canada view the meat from farmed game as a significant meat subsector. Nonetheless, hunting wild game, such as ungulates, lagomorphs, and birds, is still a popular pastime in Europe. This is especially true in Spain, where an estimated 20,000 tons of game are killed annually, valued at 45 million euros.

Figure 2.1. *Main wild game species harvested in Central and Mediterranean European countries.*

Source: Soriano, A., and Sánchez-García, C. (2021). Nutritional Composition of Game Meat from Wild Species Harvested in Europe. IntechOpen. doi: 10.5772/intechopen.97763.

According to Regulation (EC) No 853/2004, meat from wild birds, ungulates, and lagomorphs, as well as other land mammals hunted for human consumption—including creatures that live in enclosed spaces with freedom comparable to that of wild game—is classified as wild game meat. Wild ungulates such as red deer (*Cervus elaphus*), fallow deer (*Dama dama*), roe deer (*Capreolus capreolus*), and wild boar (*Sus scrofa*) are among the most hunted species in Europe. Additionally, birds like ring-necked pheasants (*Phasianus colchicus*), partridges (*Alectoris* spp. as well as Perdix species.), ducks, and swans, as well as mammals like hares (*Lepus* spp.) and wild rabbits (*Oryctolagus cuniculus*). According to European regulation (EC) No 853/2004, meat from most hunted species may be sold lawfully in the European market as long as it has successfully completed an official post-mortem inspection at game-handling facilities as cutting plants (Figure 2.1).

2.2.1. Factors Affecting Chemical Composition and Characteristics of Wild Game Meat

Game meat is a heterogeneous product because it is derived from a variety of species, including birds and mammals, as well as from different species within the same taxonomic category. These variations can be seen in physicochemical and sensory characteristics as well as microbiology. Moreover, several ante- and post-mortem factors can influence game meat from species that are hunted in the wild, making it challenging to standardize and regulate quality attributes.

2.2.1.1. Ante-Mortem Factors

Chemical meat composition is influenced by many antemortem factors, such as species, sex, age, body condition, muscle type, hormone levels, site, climate, harvesting period, and hunting style.

Since hunting is practiced throughout Europe, a species can be harvested in a variety of environmental settings, including varying climates and habitats, which can affect the types of food available and their availability. For most game species, however, hunting seasons are extended, usually outside of the breeding season, from late summer to early spring (the coldest time of year). Hunting targets adults, juveniles (the latter in small game species), males, and females. Depending on the location, the targets may have eaten natural or supplemented food, as prolonged droughts occur in many parts of Europe, particularly the Mediterranean basin.

The quality of meat may vary depending on the type of hunting. For instance, compared to stalking, where animals are frequently shot while they are still, driven hunting with dogs results in higher levels of stress. Meat quality may also be impacted by other variables, such as whether you hunt with a rifle or a bow, at night or during the day, and how stressed you are. High levels of stress before the animal being shot or captured result in a higher consumption of muscle glycogen before the animal dies, which lowers the amount of lactic acid generated from the depleted glycogen reservoir after death and alters the pH value of the meat. The meat can attain the required color, tenderness, water retention capacity, and other technological characteristics for transformation into meat products with an appropriate pH decrease and a final value ranging from 5.5 to 5.7.

2.2.1.2. Post-Mortem Factors

Since not all European nations have standard operating procedures, the methods used to obtain meat from animals after they die vary depending on the location, as do the features of the cutting plant. Actually, there are a lot of differences between nations when it comes to how the meat is dressed, how long it takes to refrigerate after death, what kind of vehicles are used to deliver the meat to the cutting plant or slaughterhouse, how the meat is matured to achieve the desired texture and flavor, and how the meat is packed and stored. The physicochemical, microbiological, and sensory qualities of game meat are all impacted by these variables.

Small game is typically kept unskinned and uneviscerated after shooting, then hung in a cool, dry location until being transported to a cutting plant or game dealer. Only a small percentage of small game is sold; the majority is usually retained by the hunters for their own use.

The evisceration of large game species is done by the hunter or the hunting area staff. The length of time between death and evisceration varies, and the hunting environment receives little consideration, both of which have the potential to significantly reduce the quality of the meat. There are also significant variations in the time between hunting and the refrigerated truck delivery of the decapitated carcass (which has been cleared by a veterinary inspection) to the cutting plant. Last but not least, cutting plant managers frequently fail to consider the ideal circumstances for carcass aging for a particular species, which presently occurs over a broad range of temperatures and times, using various carcass ripening techniques (primarily skinned vs. Without a skin.

Standardized processes must be put in place in the cutting plant and in the field to guarantee the traceability, safety, and quality of game meat.

2.2.2. Nutritional Composition of Wild Game Meat

The following provides information on the macronutrients, micronutrients, cholesterol level, and fatty acid profile of meat from the most popular hunted species in Central and Mediterranean Europe. The body of research from regions where hunting is highly prevalent has been examined. Italy, Spain, Portugal, Greece, Poland, Hungary, Romania, Croatia, Germany, Slovenia, Czech Republic, Slovak Republic, and the Republic of Lithuania are the nations where those regions are situated. The studies that have been cited only deal with wild game (never farmed), which includes both adult and juvenile, male, and female, animals that are harvested from late summer to spring.

2.2.2.1. Proximate Composition

The proximate composition and energy content of meat from large wild game species, such as wild boar (Sus scrofa), fallow deer (Dama dama), and wild red deer (Cervus elaphus), are displayed in Table 2.9. Small game, such as brown hare (Lepus europaeus) and wild rabbit (Oryctolagus cuniculus), have been the subject of more studies in Europe. The protein and fat contents were used to calculate the energy content in accordance with Regulation (EU) No 1169/2011. The most studied parts of the carcass, the loin and legs, are shown in Table 2.9, along with the range of mean values or mean value when only one study is available.

Table 2.9. *Range of Mean Values (or Mean Value When Only One Study Available) of Proximate Composition and Energy Content (per 100 g) of Game Meat From Wild Species Hunted in Europe*

		Moisture (g)	**Protein (g)**	**Fat (g)**	**Ash (g)**	**Energy (kcal)**
Red Deer	Loin	75.22–77.11	21.41–22.20	0.10–0.96	1.10–1.34	90–98
	Legs	77.90	20.50–21.50	1.00–1.72	0.98	91–101
Fallow Deer	Loin	74.30–75.40	21.80–22.80	0.30–0.89	1.07–1.10	91–96
	Legs	74.70–75.50	21.90–22.60	0.41–0.81	1.01	94–95
Wild Boar	Loin	70.50–74.72	21.24–25.87	0.69–2.80	1.03–1.26	101–117
	Legs	71.28–76.74	19.71–23.73	1.30–2.80	0.88–1.05	95–111
Hare	Loin	72.48–75.15	21.53–25.30	1.23–2.73	0.98–1.27	100–112
	Legs	74.35–75.43	20.28–21.58	1.79–3.23	0.97–1.27	102–110
Wild Rabbit	Loin	75.35	22.05	1.55	1.00	102
	Legs	75.08–75.55	20.28–21.43	2.26–3.48	1.02–1.09	106–113

Source: Soriano, A., and Sánchez-García, C. (2021). Nutritional Composition of Game Meat from Wild Species Harvested in Europe. IntechOpen. doi: 10.5772/intechopen.97763.

2.2.2.1.1. Large Wild Game

Research has focused primarily on the loin of red deer and fallow deer carcasses, examining the entire longissimus thoracis et lumborum muscle (also known as longissimus dorsi) or just its two sections, the longissimus lumborum and longissimus thoracis muscles. The legs, specifically the semimembranosus, semitendinosus, and triceps brachii muscles, have also been studied. Since the two parts are generally similar in composition, deer meat, both red and fallow, has a high protein content (20.5–22.8%), indicating that it is a relatively rich source of proteins and amino acids, and a low fat content (0.1–1.7%). The portion with the lowest fat content (less than 1%) is the loin. It is commonly known that the fat content of game meat varies greatly because of many factors, such as age and sex, season, feeding schedule, physiological, and sexual status, and hunting season.

In the wild, seasonal changes, an individual's nutritional and sexual status, and the types of vegetation present all have an impact on fat content. Given that deer species—particularly those found in the wild—are known to experience seasonal variations in body condition, theories regarding seasonal variations in meat composition are permissible. Winter has a higher fat content than autumn and summer because body condition is typically at its peak at the beginning of winter and at its lowest at the beginning of spring. However, the species, the severity of the winter, and the population density all affect this variation. Furthermore, in a number of deer species, adult males have been observed to lose body weight during the rutting season, whereas adult females' ability to regain body condition during the summer may be restricted due to the expense of lactation. For red deer, the meat energetic value (per 100 g) ranges from 90 to 101 kcal, and for fallow deer, it is 91.5 to 96 kcal. Due to their higher fat content, these values are lower than those of meats like beef, pork, lamb, or poultry, whose energetic values range from 114 to 231 kcal/100 g of muscle tissue. The amount of ash in all muscles and for both species of deer is relatively similar, with mean values ranging from 1 to 1.3%.

Contradictory findings have been found when comparing the fat content of meat from wild and farmed deer: wild fallow deer have been found to have a higher fat content than farmed ones, while farmed red deer have been found to have similar or slightly higher values than wild ones.

The diet of the wild boar varies depending on the resources available because it is a generalist omnivore that can eat a wide variety of foods. The meat's chemical composition is ultimately influenced by its wide geographic distribution, which includes arid zones (semi deserts), wetlands, high mountain environments, forests, and farmland ecosystems. High protein (21.2–25.9%) and low fat (0.7–2.8%) are found in the proximate composition of the loin (longissimus thoracis et lumborum and longissimus thoracis) and legs (semimembranosus and teres major) from European wild boars. Numerous factors have been studied in relation to wild boar meat, with the following being the most important: (i) the food that is available on the hunting grounds, which ultimately determines the meat composition, particularly the high levels of fat and protein. Thus, meat from wild boars that are harvested from fenced hunting grounds, from areas where cereals are supplemented, and from cultivated forests has been found to have a higher fat content than meat from hunting grounds where only natural food is available; (ii) age, as meat from adults has a higher fat and protein content and a lower water content when compared to juveniles; (iii) carcass weight, as there is a positive correlation between the protein content and the carcass weight; (iv) hunting period, as it has been shown that the highest amount of intramuscular fat is reached in wild boar harvested in winter as opposed to spring and summer, likely because there is more food available from cultivated crops during this time. Interestingly, research indicates that sex has no discernible impact on the proximate composition of meat. However, younger females were found to have higher protein contents than males, and male meat was found to have higher protein contents overall. The ash content, with mean values ranging from 0.9–1.3%, is the component that is less affected and more consistent among the various components that were studied.

The primary distinction between the macronutrient composition of wild boar and pork is the former's higher protein content and lower fat content. Therefore, the average intramuscular fat percentage and protein content in pork loin are 4.6 and 21.4%, respectively. According to one study, wild boar meat had a mean fat content of 5.3%, which is comparable to pork fat.

2.2.2.1.2. Small Game

Being herbivorous, wild rabbits and hares eat a wide range of plants, such as fruits, grains, and herbs. The kind and quantity of these plants differ from place to place and throughout the season, which can lead to significant differences in the meat's composition. The nutritional makeup of wild hare and rabbit in Europe is largely unknown; in fact, only one study has been found that looks into the nutritional value of wild rabbit meat collected in Spain. The proximate composition has been assessed in various regions, including the legs (semimembranosus and triceps brachii) and the loin (longissimus thoracis et lumborum). Legs show an overall high protein content (20.3–25.3%) and low fat content (1.2–3.5%) for both rabbits and hares, with slightly higher fat and lower protein content than loin. This results in an energetic value of 100–113 kcal/100 g. Further research is necessary to confirm that the meat composition of the two lagomorphs harvested in Europe is similar.

It is evident from the variability for each of the aforementioned parameters that the availability of food at the location of animal harvesting greatly affects the composition of meat. However, other research has indicated that many factors affect the amount of protein and fat, and therefore the energetic value. of these, the most significant are: (i) sex, since meat from females, has a higher fat content and a lower protein content; (ii) the hunting season, as winter has a higher fat and protein content than spring; and (iii) muscle type, as legs have a higher fat content than loin.

Domestic rabbit meat has a higher fat content (2.6–3.0%) and a lower protein value (21.5–22.2%), according to several studies comparing it to hare meat. Hare meat has a higher protein content when fed the same diet as domestic rabbits and farmed brown hares. When the proximate composition of meat from wild and farmed brown hares was compared, the farmed hares' fat and mineral content was found to be similar or slightly higher.

Despite having a balanced macronutrient content, meat from wild rabbits and hares is not as popular as other domestic meats due to its dark red color, distinct strong flavor, compact texture, and dryness.

Further research is needed on game birds because no comprehensive studies on their chemical composition have been found. To summarize, the primary attributes of game meat derived from wild species are its high protein content and low fat and calorie content. These qualities can support both a healthy diet and an effective marketing strategy.

2.2.2.2. Fatty Acids Composition

It is generally acknowledged that fatty acid composition and quantity have a significant impact on human health. Furthermore, for human nutrition, the FA composition matters more than fat content. The total fat content, the distribution of particular FA, the ratio of polyunsaturated to saturated FA, and the n-6/n-3 FA ratio are the four interrelated factors that have significant health implications. It has been demonstrated that each of these variables affects how coronary diseases develop.

The health authorities advise increasing the intake of PUFA and MUFA (monounsaturated fatty acids) and decreasing the intake of fat, SFA, trans-FA, and the n-6/n-3 ratio. While MUFA and PUFA lower the concentration of the "bad" LDL cholesterol and increase the concentration of the "good" HDL (high-density lipoproteins) cholesterol, which lowers the risk of heart disease and atherosclerosis, SFA are thought to be the primary cause of cardiovascular diseases because they raise blood pressure and the concentration of "bad" LDL (low-density lipoproteins) serum cholesterol concentrations. In addition to lowering thrombic tendencies and decreasing the incidence of ventricular arrhythmias, the n-3 PUFA lowers serum triacylglycerols. The recommendation is <4.0, but the n-6/n-3 ratio in human diets today is frequently over 10, indicating a lack of healthy n-3 FA. However, to help lower the risk of coronary diseases in people, the British Department of Health recommends a PUFA/SFA ratio of at least 4.0.

As previously stated, game meat is lean and contains little intramuscular lipids, which are mainly made up of phospholipids (polar lipids) and triglycerides (neutral lipids), with the polar lipids being less saturated. Because a higher fat content causes changes in FA, with a higher content of neutral lipids that tend to be more saturated, the intramuscular fat composition and fat amount are related.

Table 2.10 provides an overview of the range of mean values for the total content of SFA, MFA, PUFA, n-6, and n-3 FA found in the literature. It also displays the PUFA/SFA and n-6/n-3 ratios for the intramuscular fat of various muscles in the legs (semimembranosus, semitendinosus, psoas major, and triceps brachii) and loin (longissimus et lumborum, longissimus lumborum, and longissimus lumborum).

Table 2.10. *Range of Mean Values (or Mean Value When Only One Study is Available) of Fatty Acids Profile (g/100 g total fatty acids) of Intramuscular Fat From Game Meat From Wild Species Hunted in Europe*

SFA: Saturated Fatty Acids; MUFA: Monounsaturated Fatty Acids; PUFA: Polyunsaturated Fatty Acids

	Red Deer		**Fallow Deer**	**Wild Boar**		**Hare**	**Wild Rabbit**
	Loin	**Legs**	**Loin**	**Loin**	**Legs**	**Legs**	**Loin**
SFA	30.4–38.2	35.6–43.3	55.4–63.1	31.6–44.7	28.5–40.1	33.7–35.1	28.8
MUFA	15.3–22.7	22.1–37.2	25.7–29.4	30.2–46.8	27.8–50.4	17.5–21.8	13.6
PUFA	37.6–50.1	25.5–37.4	11.2–14.9	17.3–30.5	14.7–25.5	41.5–45.5	39.7
PUFA/SFA		0.44–1.09	0.18–0.27	0.38–0.84	0.52–0.62	1.20–1.89	1.38
n-6	30.2–38.2	18.7–29.3		15.7–28.0	17.8–24.0	39.3–41.1	28.8
n-3	7.2–11.7	4.6–8.1		0.9–5.7	1.4–6.2	2.0–4.4	9.2
n-6/n-3	3.5–4.3	2.6–4.8		3.1–22.3	3.0–17.6	14.2–25.1	3.8

Source: Soriano, A., and Sánchez-García, C. (2021). Nutritional Composition of Game Meat from Wild Species Harvested in Europe. IntechOpen. doi: 10.5772/intechopen.97763.

2.2.2.2.1. Large Wild Game

The total amount of FA and its ratios found in wild game meat harvested in Europe have a fairly wide range of mean values. SFA and PUFA are generally the most significant components of wild red deer meat; fallow deer, on the other hand, have higher levels of SFA and MUFA. There aren't many studies, though, because there is only one available for fallow deer harvested in Poland. Taking into account all research done on wild red deer, the n-6/n-3 ratio has demonstrated values that are either below or nearly at the suggested maximum value of 4.0, and the PUFA/SFA values are either marginally higher from 0.63–1.09 or slightly higher than 0.40, from 0.44–0.49, which is the healthier range. This ratio was lower in fallow deer because of their higher SFA and lower PUFA contents. Most studies indicate that wild boar exhibit higher levels of MUFA than ruminants. They also exhibit a higher n-6/n-3 ratio, ranging from 6.9 to 17.0, which is two to four times higher than the maximum nutritional recommended ratio of 4.0, making them less favorable. The wild boar loin and legs exhibit lower values, ranging from 3.0 to 4.7, according to a single Polish study. Most research on wild boars captured in Europe reveals PUFA/SFA ratios higher than the minimal ratio of 0.40 suggested to lower the risk of coronary heart disease in humans.

Based on research on domestic animals, the most abundant FA in large game are oleic (C18:1), stearic (C18:0), and palmitic (C16:0). Generally speaking, game species have high concentrations of the beneficial long chained ($C > 20$) unsaturated fatty acids. It is noteworthy to draw attention to the elevated levels of arachidonic acid (C20:4n-6), which can be attributed to the phospholipidic origin of this fatty acid and the lower meat content of wild animals compared to domesticated animals.

Despite PUFA's biohydrogenation in the rumen, the FA composition of wild red and fallow deer is primarily influenced by the food composition. The higher α-linolenic acid (ALA, C18:3n-3) content in pasture results in a more favorable n-6/n-3 ratio when compared to feed. On the other hand, linoleic acid (C18:2n-6) is found in feed, and both are primarily found in the phospholipid fraction of deer intramuscular fat. It follows that feed given to farmed deer raises the content of total fat and SFA while lowering the content of PUFA, particularly n-3 PUFA. Because ALA is present in pasture, meat from wild deer is typically a reasonably good source of n-3 PUFA. The very low intramuscular fat content and the fact that both species are grazers may be the cause of the high levels of ALA found in the meat of red deer and fallow deer.

The FA profile may also be influenced by other variables besides diet, such as age, sex, and harvesting region. For instance, the PUFA/SFA ratio is higher in the meat of stags because their hindquarters have a higher PUFA concentration than those of red deer or fallow deer. With respect to the age effect, younger animals exhibit a higher content of polyunsaturated fatty acids (PUFAs) than older animals. This is likely because older animals have higher levels of intramuscular fat and their muscle lipids contain different proportions of triacylglycerol and phospholipid fractions. Older animal meat has a higher n-6/n-3 ratio (> 4.0), indicating that it is less nutritious.

Wild deer meat has a high PUFA content when compared to other red meats from domestic animals; it is especially rich in C18:2 n-6, C18:3 n-3, and C20:4 n-6. Thus, compared to cattle breeds, venison has a noticeably higher content of C18:3n-3 (ALA). Red deer meat, on the one hand, has a better fat composition than grain-fed beef; it has a higher content of long-chain n-3 PUFA, a lower amount of SFA (two to three times lower), and better n-6/n-3 and PUFA/SAF ratios. Cattle fed grain that has been supplemented with C18:2n-6 experiences a decrease in n-3 PUFA concentrations and an increase in n-6 PUFA concentrations. Conversely, deer meat contains conjugated linoleic acid, or CLA (C18:2 cis-9 trans-11 isomer), which is an element that is a cancer inhibitor and offers significant protection based on several models of carcinogenesis, particularly in diets high in dairy and beef, the two known food sources. Because of its potent antioxidant qualities, CLA may also slow the advancement of atherosclerosis; however, the amount of CLA in deer meat is lower than that in meat from domestic ruminants.

In wild boar, double bonds of FA are not hydrogenated during digestion, in contrast to ruminants. Like other monogastric animals, wild boars have different FA profiles depending on whether they are fed natural resources exclusively or supplements, with the latter being a fairly common practice carried out by managers during hard winters. This means that the FA composition of wild boar meat is strongly influenced by diet. In their natural habitat, wild boars consume a wide range of native plants, including grains, seeds, roots, fruits, insects, earthworms, slugs, small mammals, and carrion, though the majority of their diet is made up of plant matter. Compared to wild boars that were not supplemented, which had a higher content of n-3, higher amounts of SFA were found in the former group. Furthermore, the meat of wild boars inhabiting forest-dominated habitats has a higher PUFA n-3 content and a lower n-6/n-3 ratio than meat from wild boars inhabiting farmland habitats, as the habitat also affects the intramuscular FA. It has been established that FAs C18:1 and C18:2n-6 are more impacted by a wild boar diet, whereas SFAs C16:0 and C18:0 do not exhibit appreciable changes. The biosynthesis of intestinal bacterial flora in wild boar meat may account for the higher amounts of CLA in the meat compared to pork. As would be expected given the variety of food sources, the hunting area also affects the FA profile.

It has been proposed that the hunting season has a significant impact on the FA profile in addition to diet. When comparing autumn (October–November) to winter (December–January), research has revealed lower MUFA proportions, higher PUFA proportions, more favorable PUFA/SFA ratio values, and a less favorable n-6/n-3 ratio. Additionally, compared to summertime, meat from wild boar harvested in the winter has been found to have higher PUFA and lower SFA proportions, as well as PUFA/SFA and n-6/n-3 PUFA ratios.

Contradictory findings about the impact of age have been observed, which may be related to the variations in age groups employed in the various studies. There were only marginal differences found between the sexes, indicating that men had a higher overall proportion of SFA than women. An additional factor was discovered to be the carcass weight; consequently, SFA rose with weight and the PUFA proportion rose with a decrease in carcass weight.

When examining the intramuscular FA profile of pigs and wild boars, it is generally observed that the former have lower n-6/n-3 ratios and higher levels of PUFA.

2.2.2.2.2. Small Game

When compared to large game species, the FA profile of hares and rabbits differs slightly from that of wild deer and wild boars because of the higher concentration of C20:4 n-6, which raises the proportion of PUFA and lowers the proportion of MUFA. The most common FA in hares and wild rabbits are C18:2 n-6, 16:0, 18:1, 20:4 n-6, and 18:0; however, the n-6/n-3 ratio is higher in hares than in rabbits. Nevertheless, with only three studies carried out in Europe, the fatty acid profile of hares and rabbits has not been thoroughly investigated.

Vegetation's seasonal and geographical variations have an impact on the FA composition. Accordingly, an ideal FA profile has been found, especially in the early spring when newly sprouting vegetation is abundant. This results in a higher concentration of PUFA, especially n-3, and a lower concentration of SFA and MUFA, as evidenced by studies conducted on wild rabbits that are hunted in Greece. Furthermore, sex has an impact on differences in some FA.

In comparison to livestock meats, hares, and wild rabbits have been found to contain higher amounts of C18:2, C18:3n-3, and C20:4n-4. Higher amounts of PUFA, especially n-3, and lower amounts of MUFA and SFA have been observed in reared rabbits and farmed hares fed the same diet.

2.2.2.2.3. Fatty Acid Profile and Organoleptic Characteristics

In addition to its effects on human health, the FA profile also has an impact on the sensory qualities of deer meat, such as texture, flavor, and odor. Individual

FAs have significant but varied effects on the firmness or softness of the fat and meat because of their varying melting points. FA, particularly unsaturated FA, produces volatile lipid oxidation products during heat treatment or during the production of cured meat, which has an impact on the flavor of meat. PUFA can be found in natural grazing, which is also thought to play a significant role in the development of "wild," "gamey," and "grassy" flavors in meat. On the other hand, a high PUFA content could be detrimental to meat and its products' oxidation stability and other technological aspects.

2.2.2.3. Cholesterol

Due to its role in the structure of cell membranes, production of adrenal and sexual hormones, and its structural makeup, cholesterol is an essential component of bodily function. The liver produces around two thirds of the cholesterol that is synthesized; the remaining portion comes from diet, and the amount of cholesterol is self-regulated. Elevated serum cholesterol levels raise the risk of cardiovascular disease; however, there has been debate regarding the effectiveness of reducing cholesterol-rich food consumption as a preventative measure.

The research that is currently available on wild game meat reveals that only the muscles of wild red deer and boar, specifically the triceps brachii, semimembranosus, psoas major, and longissimus thoracis et lumborum, have cholesterol. The large range of mean values observed in those studies may be explained by variations in the techniques employed to determine cholesterol, while also taking into account the influence of age, diet, and the type of muscle under investigation. The impact of age remains unclear as meat from younger red deer was found to have lower cholesterol values, while another study found the opposite. Similar to the findings regarding total fat content, muscle also affects cholesterol levels, with the loin having a lower level than the legs.

Table 2.11. *Range of Mean Values (or Mean Value When Only One Study Available) of Cholesterol Content (mg/100 g) of Game Meat From Wild Species Hunted in Europe*

Red Deer		Wild Boar	
Loin	**Legs**	**Loin**	**Legs**
45.3–52.8	55.2–94.6	34.4	20.9–56.9

Source: Soriano, A., and Sánchez-García, C. (2021). Nutritional Composition of Game Meat from Wild Species Harvested in Europe. IntechOpen. doi: 10.5772/intechopen.97763.

Using the same method and in the same lab, the amount of cholesterol found in red deer meat was compared to that of beef and found to be at the same level in the meat of the deer as well as the loin and legs of the beef. The values found in wild boar meat (20.9–56.9 mg/100 g) are lower or comparable to those found in pork, which are roughly 45.3–80.0 mg/100 g, depending on the muscle. In general, game meat has cholesterol levels that are comparable to those of lamb, chicken, pork, or beef. When comparing the amount of cholesterol in wild and farmed deer, different results have been observed. Specifically, wild red deer have been found to have lower values, while wild fallow deer have shown higher values than farmed deer.

2.2.2.4. Minerals

The human body uses macro- and microminerals for a variety of purposes. They serve as the building blocks for teeth, skin, hair, and bones. They are also essential for maintaining acid-base equilibrium, controlling the metabolism of water and electrolytes, and carrying out metabolic processes. Furthermore, the minerals found in meat are easier for the body to absorb than those found in plants, making them better for human health.

Macro and microelement concentrations in the loin (longissimus thoracis et lumborum, longissimus thoracis and longissimus lumborum) and legs (semimembranosus) of wild red deer, fallow deer, and wild boar are displayed in Table 2.12. Despite species-specific variations, K, P, and Na are the macrominerals that are most prevalent in the meat of all species. Conversely, Fe, and Zn are the most common micro-minerals, with Cu coming in second. The individual values for each species exhibit broad variations because the research was done in various nations and locales, which led to varying mineral availability. Because soil microelements affect the vegetation that wild game eats, the mineral content of game meat—particularly the micro-minerals—is closely related to their availability in the environment. A species's muscle mineral concentration may also be influenced by other variables, such as varying levels of physical activity, the makeup of the muscle fibers, the time of year, sex, age, and other environmental factors. As of right now, it is understood that age has little effect on the mineral profile; the only variations found in red deer and wild boar pertain to differences in Na and Fe, respectively, with older animals having higher Fe concentrations. While higher amounts of Na and Zn were found in stags of red deer and higher amounts of Cu in female wild boar, neither sex significantly affects wild board. To completely comprehend the elements influencing the mineral profile, more investigation is required.

Table 2.12. *Range of Mean Values (or Mean Value When Only One Study Available) of Minerals Content (Per 100 G) of Game Meat From Wild Species Hunted in Europe*

	Red Deer		Fallow Deer		Wild Boar	
	Loin	Legs	Loin	Legs	Loin	Legs
Macro-Minerals						
Ca (mg)	6.0–25.1	9.1	21.0	9.0	14.8–20.8	10.1–11.8
P (mg)	166.1–289	187.9	226.9	216.8	202.3	214.3–223.2
Mg (mg)	17.5–38	18.6	21.4	21.3	20.1–55.7	19.6–26.9
K (mg)	217–326.9	282.9	302.0	303.3	299.7–1123.9	324.3–328.7
Na (mg)	45.7–121	83.5	63.5	76.2	69.4–157.2	60.9–68.1
Micro-Minerals						
Cu (mg)	0.17–0.93	0.17	0.18	0.21	0.15–0.58	0.13–0.18
Zn (mg)	1.36–7.05	4.86	2.90	3.41	2.3–11.7	4.1–4.5
Fe (mg)	4.17–5.36	4.02	1.66	3.52	1.9–8.2	2.9–3.3
Mn (μg)	17–27	21.8	49.1	48.2	36.8–19.1	32.2–64.0
Se (μg)	3.7				13	

Source: Soriano, A., and Sánchez-García, C. (2021). Nutritional Composition of Game Meat from Wild Species Harvested in Europe. IntechOpen. doi: 10.5772/intechopen.97763.

The micro-mineral content of game meat is generally higher than that of beef and pork, and these variations may be caused by environmental and genetic factors. The P content of wild ruminant muscles is higher than that of corresponding bovine muscles, but the K and Na contents are lower. Compared to pork, wild boar meat has been found to have a higher content of Ca, P, Cu, Fe, and Zn, but a lower content of Na and Mg, with inconsistent results regarding K. Wild boar have been found to have a two-fold higher content of Fe.

2.2.2.5. Vitamins

Vitamins are necessary for human nutrition because they participate in many physiological processes, regulate the metabolism of macronutrients, and act in enzymatic reactions. Certain vitamins also have antioxidant qualities.

The vitamin content of game meat has not been the subject of many studies; instead, the research that has been done focuses on vitamin E (tocopherols and tocotrienols), which has antioxidant qualities that are essential for maintaining the quality of meat by delaying lipid oxidation and color loss. The concentrations of B-group vitamins and some fat-soluble vitamins, such as retinol and vitamin E homologs (α- and γ-tocopherols), which are present in red deer and wild boar meat at high levels, are displayed in Table 2.13.

Table 2.13. *Range of Mean Values (or Mean Value When Only One Study Available) of Vitamin Content of Game Meat From Wild Species Hunted in Europe*

	Red Deer		**Wild Boar**	
	Loin	**Legs**	**Loin**	**Legs**
B-group (per 100 g)				
B_1 (thiamine, mg)	0.18			
B_2 (riboflavin, mg)	0.40			
B_3 (niacin, mg)	5.53			
B_6 (pyridoxine, mg)	0.14			
B_9 (folic acid, µg)	3.17			
B_{12} (cyanocobalamine, µg)	5.87			
Retinol (µg /g)			0.01–1.11	
Tocopherols (µg /g)				
Α		5.24–6.46	0.52–2.6	15.5–19.2
Γ		1.42–2.06	0.03–0.30	1.14–1.75

Source: Soriano, A., and Sánchez-García, C. (2021). Nutritional Composition of Game Meat from Wild Species Harvested in Europe. IntechOpen. doi: 10.5772/intechopen.97763.

Only one study on B-group vitamins from red deer loin (longissimus thoracis) has been done. Deer loin has higher contents of thiamine (2 times higher than from poultry and lamb, and 3 times higher than from beef), riboflavin (about 3 times higher), and especially vitamin B_{12} (between 3 and 5 times higher than from all the meats mentioned). In a red deer study, higher values of α-tocopherols were found in hinds when tocopherols and tocotrienols were measured in the psoas major muscle.

Concentrations of vitamin E homologs and retinol in the longissimus and psoas major muscles of wild boars have been investigated. Studies conducted in Italy and Germany have found significant differences in the levels of retinol, α- and γ-tocopherol, which can be strongly influenced by the animals' location. Due to the variations in crops, pasture, and acorns between habitats, a study on wild boar harvested in farmland and forest areas revealed higher levels of α-tocopherol in farmland areas and higher levels of γ-tocopherol in forest areas. Pigs are routinely supplemented with vitamin E because it decreases oxidation and lengthens the shelf life of meat. When pigs were given supplements up to 700 mg/kg of feed, the amount of α-tocopherol in their psoas major muscle increased to 15.1–16.3 μg/g of meat. These levels are comparable to the average values observed in wild boars.

2.2.3. Nutrition and Health Claims in Wild Red Deer Meat

Soriano et al. recently conducted a study. 2020 has provided information on the nutritional value and health claims associated with wild red deer meat following Commission Regulation (EU) No. 1047/2012, as amended, and Regulation (EC) No. 1924/2006. These statements are permissible to use on the packaging, labeling, and market advertising of this meat. Using both stags and hinds that were hunted in the fall and winter, the nutritional quality of loin (longissimus thoracis) from 71 wild Iberian red deer (*Cervus elaphus hispanicus*) was examined in that study.

Concerning the nutrition claims, it can be assumed that all samples had a protein content of at least 73%, a fat content of less than 2 g/100 g, and a Na content of less than 0.12 g/100 g. Therefore, the claims of "high protein," "low fat," and "low sodium/salt" are true. This meat can be referred to as a "source of P, Fe, Cu, riboflavin, and niacin" because it contains significant amounts of B_2 (riboflavin), B_3 (niacin), and B_{12} (cyanocobalamin), among other minerals and vitamins. At least 15% of the reference daily intake for adults of these micronutrients is provided by 100 grams of deer loin. Deer loin may be classified as "high in Zn and vitamin B_{12}" because it has at least twice the amount of these nutrients as the typical source. Permitted health claims listed in the Annex of Commission Regulation (EU) No 432/2012 could be used in conjunction with the previously mentioned nutrition claims for wild deer meat regarding the minerals and vitamins present in significant amounts. Similar to this, claims regarding the development of children may also be made (Regulation (EC) No 1924/2006; Commission Regulation (EU) No 432/2012). These claims are predicated on the bioactivity and presence of specific components, like protein and iron, which must be present in significant amounts because deer loin complies as a source of these nutrients.

2.2.4. European Wild Game Meat Production and Its Challenges

The production of wild game meat offers a healthier and more profitable alternative to other red meats, but it is still a niche market because of supply constraints and low demand.

Although it makes up a very small portion of the total meat supply in Europe, hunting is a sustainable source of meat. Hunting is permitted and, with appropriate management, can support both the rural economy and the preservation of wildlife. The majority of hunting techniques satisfy regulations about animal welfare; however, to reduce moral dilemmas, sustainable hunting practices ought to be encouraged.

Because game meat is produced differently than meat from farmed animals, certain steps need to be taken up the food chain. While game species, pre-mortem conditions, hunting techniques, and post-mortem handling may vary, there are some universal standards for meat safety and quality. Ensuring that wild game meat is safe for consumption through official inspections is crucial to the growth of the wild game meat industry. Regulations (EC) Nos., a set of general regulations on food hygiene, constitute the entirety of European legislation as of yet. 852/2004, 853/2004, 854/2004, 882/2004, and 2075/2005), as modified by state members' legislation. This implies that all parties involved in the production process—hunters, cutting plants, and owners of hunting grounds—must adhere to the best practices advocated in Europe. These best practices are essential because everyone involved in this issue needs to be trained, including hunters, staff members with training, and official veterinarians. They also serve as a guide for inspectors when game carcasses or food companies that trade in wild game meat are inspected. However, there are still certain issues and problems that need to be addressed.

Nonetheless, lead (Pb) ammunition is still frequently used in hunting techniques, and it is well known that Pb fragments can be discovered after striking an animal, both at the impact site and in adjacent areas. Removing the impact site is a standard procedure for carcass dressing. Trading companies also have metal detectors, which stop game meat from being distributed if there are bullet remnants in it. Nevertheless, Pb residues found in some surface waters and soils are unintentionally consumed by wild animals, leading to the presence of Pb in their viscera. As a result, some people in the population—mainly hunters and their families—are at risk of being exposed to lead. In this regard, EFSA's research on the dietary exposure of Europeans to lead from meat and meat products has shown that wild boar meat has especially high levels of lead exposure. Larger food consumptions, such as grains and their products, milk, and dairy products, non-alcoholic drinks, vegetables, and vegetable products, have the biggest effects on dietary exposure to lead; however, this is not the case with meat from wild game. While chronic toxicity from repeated intakes has a toxic effect on the body, primarily on the central nervous system (especially on the developing brain of young children and fetuses), no negative effects from a single high intake of Pb have been reported. However, the International Agency for Research on Cancer (IARC) has determined that inorganic Pb compounds are likely carcinogenic to humans.

When comparing wild game meat from Central and Mediterranean European countries to livestock meat, the general trends are as follows: (i) lower fat content (<3 g/100 g for large and <4 g/100 g for small wild game species), which translates to a lower energy content; (ii) higher or similar protein content; (iii) a positive fatty-acids profile, showing a higher proportion of PUFA, especially n-3, and consequently more favorable PUFA/SFA ratio; and in the specific case of wild boar, a relatively adequate content of conjugated linoleic acid (CLA); and in the case of wild ruminants (red and fallow deer), an ideal n-6/n-3 ratio; (iv) a higher minerals content, primarily micro-minerals like Zn and the bioavailable form of heme Fe. Additionally, meat from wild game contains ideal levels of B-group vitamins, such as riboflavin, niacin, and B_{12}, as well as vitamin E, which has antioxidant qualities. Overall, because of its sustainable production, which ensures standards for animal welfare, and its nutritional quality, which may support a balanced and healthful diet, wild game meat satisfies consumer demands today. Therefore, wild game meat can diversify the European meat market and is a good substitute for other red meats. (Soriano and Sánchez-García, 2021)

REFERENCES

1. Adzitey, F., & Nurul, H. (2011). Pale soft exudative (PSE) and dark firm dry (DFD) meats: Causes and measures to reduce these incidences-a mini review. *International Food Research Journal, 18*(1).
2. Agencia Española de Seguridad Alimentaria y Nutrición (AESAN). (2012). Informe del Comité Científico de la Agencia Española de Seguridad Alimentaria y Nutrición (AESAN) sobre el riesgo asociado a la presencia de plomo en carne de caza silvestre en España. Available from: https://www.aetox.es/ (accessed on 22 October 2024).
3. Ahmad, R. S., Imran, A., & Hussain, M. B. (2018). Nutritional composition of meat. In *IntechOpen*.
4. Alfaia, C. M. M., Ribeiro, V. S. S., Lourenço, M. R. A., Quaresma, M. A. G., Martins, S. I. V., Portugal, A. P. V., Fontes, C. M. G. A., Bessa, R. J. B., Castro, M. L. F., & Prates, J. A. M. (2006). Fatty acid composition, conjugated linoleic acid isomers and cholesterol in beef from crossbred bullocks intensively produced and from Alentejana purebred bullocks reared according to Carnalentejana-PDO specifications. *Meat Science, 72*, 425–436.
5. Ali Hassan, A., Sandanger, T. M., & Brustad, M. (2012). Selected vitamins and essential elements in meat from semi-domesticated reindeer (*Rangifer tarandus* L.) in mid-and northern Norway: Geographical variations and effect of animal population density. *Nutrients, 4*(7), 724–739.
6. Amici, A., Cifuni, G. F., Contò, M., Esposito, L., & Failla, S. (2015). Hunting area affects chemical and physical characteristics and fatty acid composition of wild boar (Sus scrofa) meat. *Rendiconti Lincei, 26*, S527–S534.
7. Andersen, H. J., Oksbjerg, N., Young, J. F., & Therkildsen, M. (2005). Feeding and meat quality–A future approach. *Meat Science, 70*(3), 543–554.
8. Andreotti, A., Borghesi, F., & Aradis, A. (2016). Lead ammunition residues in the meat of hunted woodcock: A potential health risk to consumers. *Italian Journal of Animal Science, 15*, 22–29.
9. Aranceta, J., & Pérez-Rodrigo, C. (2012). Recommended dietary reference intakes, nutritional goals and dietary guidelines for fat and fatty acids: A systematic review. *British Journal of Nutrition, 107*(S2), S8–S22.
10. Babicz, M., & Kasprzyk, A. (2019). Comparative analysis of the mineral composition in the meat of wild boar and domestic pig. *Italian Journal of Animal Science, 18*, 1013–1020.

11. Badui, S. (2006). Vitaminas y nutrimentos inorgánicos. In S. Badui (Ed.), *Química de los Alimentos* (4th ed., pp. 363–398). México: Pearson.
12. Barbani, R., Bonilauri, P., Sangiorgi, E., & Marliani, G. (2020). Fatty acid composition of wild boar (*Sus scrofa scrofa*) meat compared to commercial hybrid and crossbreed Mora Romagnola swine: A preliminary study. *Rivista Italiana delle Sostanze Grasse, 97*, 7–14.
13. Barron-Hoyos, J. M., Archuleta, A. R., del Refugio Falcon-Villa, M., Canett-Romero, R., Cinco-Moroyoqui, F. J., Romero-Barancini, A. L., & Rueda-Puente, E. O. (2013). Protein quality evaluation of animal food proteins by *in-vitro* methodologies. *Food and Nutrition, 4*, 376–384.
14. Batorska, M., Więcek, J., Kunowska-Slósarz, M., Puppel, K., Slósarz, J., Gołębiewski, M., Kuczyńska, B., Popczyk, B., Rekiel, V. A., & Balcerak, M. (2018). The effect of carcass weight on chemical characteristics and fatty acid composition of Longissimus dorsi and Semimembranosus muscles of European wild boar (*Sus scrofa scrofa*) meat. *Canadian Journal of Animal Science, 98*, 557–564.
15. British Department of Health. (1994). Nutritional aspects of cardiovascular disease. *Report of the Cardiovascular Review Group of the Committee of Medical Aspects of Food Policy*. London: HMSO.
16. Bureš, D., Bartoň, L., Kotrba, R., & Hakl, J. (2015). Quality attributes and composition of meat from red deer (*Cervus elaphus*), fallow deer (*Dama dama*), and Aberdeen Angus and Holstein cattle (*Bos taurus*). *Journal of the Science of Food and Agriculture, 95*, 2299–2306.
17. Cambero, M. I., de la Hoz, L., Sanz, B., & Ordóñez, J. A. (1991). Seasonal variations in lipid composition of Spanish wild rabbit (*Oryctolagus cuniculus*) meat. *Journal of the Science of Food and Agriculture, 56*, 315–362.
18. Chan, W., Brown, J., Lee, S., & Buss, D. H. M. (1995). *Poultry and Game. Fifth Supplement to McCance & Widdowson's the Composition of Foods*. London: The Royal Society of Chemistry and the Ministry of Agriculture Fisheries and Food.
19. Chin, S. F., Liu, W., Storkson, J. M., Ha, Y. L., & Pariza, M. W. (1992). Dietary sources of conjugated dienoic isomers of linoleic acid, a newly recognized class of anticarcinogens. *Journal of Food Composition and Analysis, 5*, 185–197.
20. Chizzolini, R., Zanardi, E., Dorigoni, V., & Ghidini, S. (1999). Calorific value and cholesterol content of normal and low-fat meat and meat products. *Trends in Food Science and Technology, 10*, 119–128.
21. Cobos, A., de la Hoz, L., Cambero, M. I., & Ordóñez, J. A. (1995).

Chemical and fatty acid composition of meat from Spanish wild rabbits and hares. *Zeitschrift für Lebensmittel-Untersuchung und-Forschung, 200*, 182–185.

22. Cordain, L., Watkins, B. A., Florant, G. L., Kelher, M., Rogers, L., & Li, Y. (2002). Fatty acid analysis of wild ruminant tissues: Evolutionary implications for reducing diet-related chronic disease. *European Journal of Clinical Nutrition, 56*, 181–191.
23. Dahlan, I., & Norfarizan Hanoon, N. A. (2008). Chemical composition, palatability and physical characteristics of venison from farmed deer. *Animal Science Journal, 79*, 498–503.
24. Dannenberger, D., Nuernberg, G., Nuernber, K., & Hagemann, E. (2013). The effects of gender, age, and region on macro and micronutrient contents and fatty acid profiles in the muscles of roe deer and wild boar in Mecklenburg-Western Pomerania (Germany). *Meat Science, 94*, 39–46.
25. Daszkiewicz, T., Janiszewski, P., & Wajda, S. (2009). Quality characteristics of meat from wild red deer (*Cervus elaphus* L.) hinds and stags. *Journal of Muscle Foods, 20*, 428–448.
26. de Castro Cardoso Pereira, P. M., & dos Reis Baltazar Vicente, A. F. (2013). Meat nutritional composition and nutritive role in the human diet. *Meat Science, 93*, 586–592.
27. De Smet, S., & Vossen, E. (2016). Meat: The balance between nutrition and health. A review. *Meat Science, 120*, 145–156.
28. Demartini, E., Vecchiato, D., Gaviglio, A., Tempesta, T., & Viganò, R. (2018). Consumer preferences for red deer meat: A discrete choice analysis considering attitudes towards wild game meat and hunting. *Meat Science, 146*, 168–179.
29. Droulez, V., Williams, P. G., Levy, G., Stobaus, T., & Sinclair, A. (2006). Composition of Australian red meat 2002. 2. Fatty acid profile. *Food Australia, 58*, 335–341.
30. Dugan, L. R. (1994). Química de los tejidos animales. Parte 2 Grasas. In J. F. Price & B. S. Schweigert (Eds.), *Ciencia de la Carne y de los Productos Cárnicos* (2nd ed., pp. 93–101). Zaragoza: Acribia.
31. European Food Safety Authority (EFSA). (2012). Lead dietary exposure in the European population. *EFSA Journal, 10*(7), 2831 [59 pp.]. Available from: www.efsa.europa.eu/efsajournal/ (accessed on 22 October 2024).
32. Eze, N. M., Maduabum, F. O., Onyeke, N. G., Anyaegunam, N. J., Ayogu, C. A., Ezeanwu, B. A., & Eseadi, C. (2017). Awareness of food nutritive value and eating practices among Nigerian bank workers: Implications for

nutritional counseling and education. *Medicine, 96*(10).

33. Finstad, G., Wiklund, E., Long, K., Rincker, P. J., Oliveira, A. C. M., & Bechtel, P. J. (2007). Feeding soy or fish meal to Alaskan reindeer (*Rangifer tarandus tarandus*)-Effects on animal performance and meat quality. *Rangifer, 27*, 59–75.

34. Food Standards Australia New Zealand (FSANZ). (2006). Nuttab 2006. Online database of the nutritional composition of Australian foods. Canberra: FSANZ.

35. Frunza, G., Simeanu, D., Pop, C., Lazar, R., Boisteanu, P. C., & Stefan, M. (2019). Contributions on the sensorial, physico-chemical and nutritional characterization of hare meat (*Lepus europaeus pallas*). *Revista de Chimie, 70*, 174–180.

36. Gálik, B., Šmehýl, P., Gašparík, J., Candrák, J., Jahnátek, A., Bíro, D., Rolinec, M., Juráček, M., & Šimko, M. (2018). The effect of age on the fatty acids composition in wild boar (*Sus scrofa*) hunted in the southwest region of Slovakia. *Acta Veterinaria Brno, 87*, 85–90.

37. García Ruiz, A., Mariscal, C., Gonzalez Viñas, M. A., & Soriano, A. (2010). Influence of hunting-season stage and ripening conditions on microbiological, physicochemical and sensory characteristics of venison (*Cervus elaphus*) chorizo sausages. *Italian Journal of Food Science, 22*, 386–394.

38. Gerhard, G. T., Ahmann, A., & Meeuws, K. (2004). Effect of a low-fat diet compared with those of a high-monounsaturated fat diet on body weight, plasma lipids and lipoproteins, and glycemic control in type 2 diabetes. *The American Journal of Clinical Nutrition, 80*, 668–673.

39. Gibier de Chasse-Chasseurs de France. (2020). Available from: http://chasseurdefrance.com/charte-gibier-de-chasse-chasseurs-de-france/ (accessed on 22 October 2024).

40. Givens, I. D., & Gibbs, R. A. (2008). Current intakes of EPA and DHA in European populations and the potential of animal-derived foods to increase them. *Proceedings of the Nutrition Society, 67*, 272–280.

41. Grunert, K. G., Bredahl, L., & Brunsø, K. (2004). Consumer perception of meat quality and implications for product development in the meat sector—A review. *Meat Science, 66*(2), 259–272.

42. Hewison, A. J. M., Angibault, J. M., Boutin, J., Bideau, E., Vincent, J. P., & Sempéré, A. (1996). Annual variation in body composition of roe deer (*Capreolus capreolus*) in moderate environmental conditions. *Canadian Journal of Zoology, 74*, 245–253.

43. Hjeljord, O., & Histøl, T. (1999). Range-body mass interactions of a northern ungulate–A test of hypothesis. *Oecologia, 119*, 326–339.
44. Hocquette, J. F., Gondret, F., Baéza, E., Médale, F., Jurie, C., & Pethick, D. W. (2010). Intramuscular fat content in meat-producing animals: Development, genetic and nutritional control, and identification of putative markers. *Animal, 4*(2), 303–319.
45. Hoffman, L. C., & Wiklund, E. (2006). Game and venison-meat for the modern consumer. *Meat Science, 74*, 197–208.
46. Hoffman, L. C., Kroucamp, M., & Manley, M. (2007). Meat quality characteristics of springbok (*Antidorcas marsupialis*). 2: Chemical composition of springbok meat as influenced by age, gender and production region. *Meat Science, 76*, 762–767.
47. Hoffman, L. C., Muller, M., Schutte, D. W., Calitz, F. J., & Crafford, K. (2005). Consumer expectations, perceptions, and purchasing of South African game meat. *South African Journal of Wildlife Research, 35*, 33–42.
48. Hunt, M. R., Legako, J. F., Dinh, T. T. N., Garmyn, A. J., O'Quinn, T. G., Corbin, C. H., Rathmann, R. J., Brooks, J. C., & Miller, M. F. (2016). Assessment of volatile compounds, neutral and polar lipid fatty acids of four beef muscles from USDA Choice and Select graded carcasses and their relationships with consumer palatability scores and intramuscular fat content. *Meat Science, 116*, 91–101.
49. Hurrell, R., & Egli, I. (2010). Iron bioavailability and dietary reference values. *The American Journal of Clinical Nutrition, 91*(5), 1461S–1467S.
50. Huskisson, E., Maggini, S., & Ruf, M. (2007). The role of vitamins and minerals in energy metabolism and well-being. *Journal of International Medical Research, 35*(3), 277–289.
51. International Agency for Research on Cancer (IARC). (2006). Inorganic and organic lead compounds. In: *IARC Monographs on the Evaluation of Carcinogenic Risks to Humans* (Vol. 87).
52. Jensen, J., Rustad, P. I., Kolnes, A. J., & Lai, Y. C. (2011). The role of skeletal muscle glycogen breakdown for regulation of insulin sensitivity by exercise. *Frontiers in Physiology, 2*, 112.
53. Kamruzzaman, M., Makino, Y., & Oshita, S. (2016). Parsimonious model development for real-time monitoring of moisture in red meat using hyperspectral imaging. *Food Chemistry, 196*, 1084–1091.
54. Karunanayaka, D. S., Jayasena, D. D., & Jo, C. (2016). Prevalence of pale, soft, and exudative (PSE) condition in chicken meat used for commercial

meat processing and its effect on roasted chicken breast. *Journal of Animal Science and Technology, 58*(1), 27.

55. Kasprzyk, A., Stadnik, J., & Stasiak, D. (2019). Technological and nutritional properties of meat from female wild boars (*Sus scrofa L.*) of different carcass weights. *Archives Animal Breeding, 62*, 597–604. DOI: 10.5194/aab-62-597-2019.
56. Kihara, J., Sileshi, G. W., Nziguheba, G., Kinyua, M., Zingore, S., & Sommer, R. (2017). Application of secondary nutrients and micronutrients increases crop yields in sub-Saharan Africa. *Agronomy for Sustainable Development, 37*(4), 25.
57. Króliczewska, B., Miśta, D., Korzeniowska, M., Pecka-Kiełb, E., & Zachwieja, A. (2018). Comparative evaluation of the quality and fatty acid profile of meat from brown hares and domestic rabbits offered the same diet. *Meat Science, 145*, 292–299. DOI: 10.1016/j.meatsci.2018.07.002.
58. Kudrnáčová, E., Bartoň, L., Bureš, D., & Hoffman, L. C. (2018). Carcass and meat characteristics from farm-raised and wild fallow deer (*Dama dama*) and red deer (*Cervus elaphus*): A review. *Meat Science, 141*, 9–27. https://doi.org/10.1016/j.meatsci.2018.02.020.
59. Kwiecińska, K., Kosicka-Gębska, M., Gębski, J., & Gutkowska, K. (2017). Prediction of the conditions for the consumption of game by Polish consumers. *Meat Science, 131*, 28–33. https://doi.org/10.1016/j.meatsci.2017.04.038.
60. Lachowicz, K., Żochowska-Kuiawska, J., Gaiowiecki, L., Sobczak, M., Kotowicz, M., & Żych, A. (2008). Effects of wild boar meat of different seasons of shot: Addition texture of finely ground model pork and beef sausages. *Electronic Journal of Polish Agricultural Universities, 11*, e11.
61. Leaf, A., Kang, J. X., Xiao, Y. F., & Billman, G. E. (1999). n-3 fatty acids in the prevention of cardiac arrhythmias. *Lipids, 34*, S187–S189.
62. Lehnen, T. E., da Silva, M. R., Camacho, A., Marcadenti, A., & Lehnen, A. M. (2015). A review on effects of conjugated linoleic fatty acid (CLA) upon body composition and energetic metabolism. *Journal of the International Society of Sports Nutrition, 12*(1), 36.
63. Lorenzo, J. M., Maggiolino, A., Gallego, L., Pateiro, M., Pérez Serrano, M., Domínguez, R., García, A., Landete-Castillejos, T., & De Palo, P. (2019). Effect of age on nutritional properties of Iberian wild red deer meat. *Journal of the Science of Food and Agriculture, 99*, 1561–1567. https://doi.org/10.1002/jsfa.9334.
64. Lozoff, B., & Georgieff, M. K. (2006). Iron deficiency and brain development. *Seminars in Pediatric Neurology, 13*(3), 158–165.

65. Ludwiczak, A., Składanowska-Baryza, J., & Stanisz, M. (2020). Effect of age and sex on the quality of offal and meat of the wild boar (*Sus scrofa*). *Animals, 10*, 660–670. https://doi.org/10.3390/ani10040660.

66. Mahan, D. C., & Shields, R. G., Jr. (1998). Essential and nonessential amino acid composition of pigs from birth to 145 kilograms of body weight, and comparison to other studies. *Journal of Animal Science, 76*(2), 513–521.

67. Marangoni, F., Corsello, G., Cricelli, C., Ferrara, N., Ghiselli, A., Lucchin, L., & Poli, A. (2015). Role of poultry meat in a balanced diet aimed at maintaining health and wellbeing: An Italian consensus document. *Food and Nutrition Research, 59*(1), 27606.

68. Marescotti, M. E., Caputo, V., Demartini, E., & Gaviglio, A. (2019). Discovering market segments for hunted wild game meat. *Meat Science, 149*, 163–176. https://doi.org/10.1016/j.meatsci.2018.11.019.

69. Marsico, G., Rasulo, A., Dimatteo, S., Tarricone, S., Pinto, F., & Ragni, M. (2007). Pig, F1 (wild boar × pig) and wild boar meat quality. *Italian Journal of Animal Science, 6*, 701–703. https://doi.org/10.4081/ijas.2007.1s.701.

70. McCance, R. A., & Widdowson, E. M. (1960). *The Composition of Foods* (3rd ed.). London: Her Majesty's Stationery Office.

71. Mertin, D., Slamečka, J., Ondruška, Ľ., Zaujec, K., Jurčík, R., & Gašparík, J. (2012). Comparison of meat quality between European brown hare and domestic rabbit. *Slovak Journal of Animal Science, 45*, 89–95.

72. Miller, G. J., Field, R. A., Riley, M. L., & Williams, J. C. (1986). Lipids in wild ruminant animals and steers. *Journal of Food Quality, 9*, 331–343. https://doi.org/10.1111/j.1745-4557.1986.tb00802.x.

73. Moreiras, O., Carbajal, A., Cabrera, L., & Cuadrado, C. (2008). *Tablas de Composición de Alimentos* (15th ed., pp. 104–107). España: Ediciones Pirámide.

74. Nawar, W. W. (1993). Lipidos. In O. R. Fennema (Ed.), *Química de los Alimentos* (2nd ed., pp. 157–274). Zaragoza: Editorial Acribia.

75. Neethling, J., Hoffman, L. C., & Muller, M. (2016). Factors influencing the flavour of game meat: A review. *Meat Science, 113*, 139–153. https://doi.org/10.1016/j.meatsci.2015.11.022.

76. Nuernberg, K., Dannenberger, D., Nuernberg, G., Ender, K., Voigt, J., Scollan, N. D., Wood, J. D., Nute, G. R., & Richardson, R. I. (2005). Effect of a grass-based and a concentrate feeding system on meat quality characteristics and fatty acid composition of longissimus muscle in different cattle breeds. *Livestock Production Science, 94*, 137–147. https://doi.org/10.1016/j.livprodsci.2004.11.036.

77. Papadomichelakis, G., Zoidis, E., Pappas, A. C., & Hadjigeorgiou, I. (2017). Seasonal variations in the fatty acid composition of Greek wild rabbit meat. *Meat Science, 134*, 158–162. https://doi.org/10.1016/j.meatsci.2017.08.001.

78. Pathare, P. B., & Roskilly, A. P. (2016). Quality and energy evaluation in meat cooking. *Food Engineering Reviews, 8*(4), 435–447.

79. Paul, A., & Southgate, D. A. (1978). New food tables. *Nutrition and Food Science, 78*(1), 2, 3.

80. Pedrazzoli, M., Dal Bosco, A., Castellini, C., Ranucci, D., Mattioli, S., Pauselli, M., & Roscini, V. (2017). Effect of age and feeding area on meat quality of wild boars. *Italian Journal of Animal Science, 16*, 353–362. https://doi.org/10.1080/1828051X.2017.1292114.

81. Pereira, P. M., & Vicente, A. F. (2013). Meat nutritional composition and nutritive role in the human diet. *Meat Science, 93*(3), 586–592.

82. Piaskowska, N., Daszkiewicz, T., Kubiak, D., & Janiszewski, P. (2015). The effect of gender on meat (longissimus lumborum muscle) quality characteristics in the fallow deer (*Dama dama* L.). *Italian Journal of Animal Science, 14*, 389–393. https://doi.org/10.4081/ijas.2015.3845.

83. Polak, T., Rajar, A., Gašperlin, L., & Žlender, B. (2008). Cholesterol concentration and fatty acid profile of red deer (*Cervus elaphus*) meat. *Meat Science, 80*, 864–869. https://doi.org/10.1016/j.meatsci.2008.04.005.

84. Poławska, E., Cooper, R. G., Jóźwik, A., & Pomianowski, J. (2013). Meat from alternative species: Nutritive and dietetic value, and its benefit for human health—A review. *CyTA-Journal of Food, 11*, 37–42. https://doi.org/10.1080/19476337.2012.680916.

85. Pollard, J. C., Stevenson-Barry, J. M., & Littlejohn, R. P. (1999). Factors affecting behaviour, bruising and pH in a deer slaughter premises. *Proceedings of the New Zealand Society of Animal Production, 59*, 148–151.

86. Purchas, R. W., Triumf, E. C., & Egelandsdal, B. (2010). Quality characteristics and composition of the longissimus muscle in the short-loin from male and female farmed red deer in New Zealand. *Meat Science, 86*, 505–510. https://doi.org/10.1016/j.meatsci.2010.05.043.

87. Quaresma, M. A. G., Alves, S. P., Trigo-Rodrigues, I., Pereira-Silva, R., Santos, N., Lemos, J. P. C., Barreto, A. S., & Bessa, R. J. B. (2011). Nutritional evaluation of the lipid fraction of feral wild boar (*Sus scrofa scrofa*) meat. *Meat Science, 89*, 457–461. https://doi.org/10.1016/j.meatsci.2011.05.005.

88. Quaresma, M. A. G., Trigo-Rodrigues, I., Alves, S. P., Martins, S. I. V., Barreto, A. S., & Bessa, R. J. B. (2012). Nutritional evaluation of the lipid fraction of Iberian red deer (*Cervus elaphus hispanicus*) tenderloin. *Meat Science, 92*, 519–524. https://doi.org/10.1016/j.meatsci.2012.05.021.

89. Ramanzin, M., Amici, A., Casoli, C., Esposito, L., Lupi, P., Marsico, G., Mattiello, S., Olivieri, O., Ponzetta, M. P., Russo, C., & Marinucci, M. T. (2010). Meat from wild ungulates: Ensuring quality and hygiene of an increasing resource. *Italian Journal of Animal Science, 9*, 318–331. https://doi.org/10.4081/ijas.2010.e61.

90. Ravisankar, P., Reddy, A. A., Nagalakshmi, B., Koushik, O. S., Kumar, B. V., & Anvith, P. S. (2015). The comprehensive review on fat soluble vitamins. *IOSR Journal of Pharmacy, 5*(11), 12–28.

91. Razmaitė, V., & Šiukščius, A. (2019). Seasonal variation in fatty acid composition of wild boar in Lithuania. *Italian Journal of Animal Science, 18*, 350–354. https://doi.org/10.1080/1828051X.2018.1530957.

92. Razmaitė, V., Švirmickas, G. J., & Šiukščius, A. (2012). Effect of weight, sex and hunting period on fatty acid composition of intramuscular and subcutaneous fat from wild boar. *Italian Journal of Animal Science, 11*, 174–179. https://doi.org/10.4081/ijas.2012.e32.

93. Realini, C. E., Duckett, S. K., Brito, G. W., Dalla Rizza, M., & De Mattos, D. (2004). Effect of pasture vs. concentrate feeding with or without antioxidants on carcass characteristics, fatty acid composition, and quality of Uruguayan beef. *Meat Science, 66*, 567–577. https://doi.org/10.1016/S0309-1740(03)00160-8.

94. Roslewska, A., Stanek, M., Janicki, B., Cygan-Szczegielniak, D., Stasiak, K., & Buzala, M. (2016). Effect of sex on the content of elements in meat from wild boars (*Sus scrofa* L.) originating from the province of Podkarpacie (south-eastern Poland). *Journal of Elementology, 21*, 823–832. https://doi.org/10.5601/jelem.2015.20.2.943.

95. Rudel, L. L. (1999). Atherosclerosis and conjugated linoleic acid. *British Journal of Nutrition, 81*, 177–179. https://doi.org/10.1017/S0007114599000367.

96. Russo, C., Balloni, S., Altomonte, I., Martini, M., Nuvoloni, R., Cecchi, F., Pedonese, F., Salari, F., Marilia Sant'ana Da Silva, A., Torracca, B., & Profumo, A. (2017). Fatty acid and microbial profile of the meat (longissimus dorsi muscle) of wild boar (*Sus scrofa scrofa*) hunted in Tuscany. *Italian Journal of Animal Science, 16*, 1–8. https://doi.org/10.1080/1828051X.2016.1261006.

97. Sakomura, N. K., Ekmay, R. D., Mei, S. J., & Coon, C. N. (2015). Lysine,

methionine, phenylalanine, arginine, valine, isoleucine, leucine, and threonine maintenance requirements of broiler breeders. *Poultry Science, 94*(11), 2715–2721.

98. Sales, J., & Kotrba, R. (2013). Meat from wild boar (*Sus scrofa* L.): A review. *Meat Science, 94*, 187–201. https://doi.org/10.1016/j.meatsci.2013.01.012.
99. Sánchez-García, C., Urda, V., Lambarri, M., Prieto, I., Andueza, A., & Villanueva, L. F. (2020). Evaluation of the economics of sport hunting in Spain through regional surveys. *International Journal of Environmental Studies*, 1–15. https://doi.org/10.1080/00207233.2020.1759305.
100. Schley, L., & Roper, T. J. (2003). Diet of wild boar *Sus scrofa* in Western Europe, with particular reference to consumption of agricultural crops. *Mammal Review, 33*, 43–56. https://doi.org/10.1046/j.1365-2907.2003.00010.x.
101. Serrano, M. P., De Palo, P., Maggiolino, A., Pateiro, M., Gallego, L., Domínguez, R., García, A., Landete-Castillejos, T., & Lorenzo, J. M. (2020). Seasonal variations of carcass characteristics, meat quality and nutrition value in Iberian wild red deer. *Spanish Journal of Agriculture Research, 18*. https://doi.org/10.5424/sjar/2020184-16113.
102. Sevillano, J., Moreno-Ortega, A., Amaro, M. A., Arenas, A., Cámara-Martos, F., & Moreno-Rojas, R. (2018). Game meat consumption by hunters and their relatives: A probabilistic approach. *Food Additives & Contaminants: Part A, 35*, 1739–1748. https://doi.org/10.1080/19440049.2018.1488183.
103. Simpson, R. J., & McKie, A. T. (2009). Regulation of intestinal iron absorption: The mucosa takes control? *Cell Metabolism, 10*(2), 84–87.
104. Skewes, O., Morales, R., Mendoza, N., Smulders, F. J. M., & Paulsen, P. (2009). Carcass and meat quality traits of wild boar (*Sus scrofa* s. L.) with 2n=36 karyotype compared to those of phenotypically similar crossbreeds (2n=37 and 2n=38) raised under the same farming conditions: Fatty acid profile and cholesterol. *Meat Science, 83*, 195–200. https://doi.org/10.1016/j.meatsci.2009.04.017.
105. Skobrák, E. B., Bodnár, K., Jónás, E. M., Gundel, J., & Jávor, A. (2011). The comparison analysis of the main chemical composition parameters of wild boar meat and pork. *Scientific Papers Animal Science and Biotechnologies, 44*, 105–112.
106. Skrivanko, M., Hadziosmanovic, M., Cvrtila, Z., Zdolec, N., Filipovic, I., Kozaeinski, L., Florijanèic, T., & Boskovic, I. (2008). The hygiene and quality of hare meat (*Lepus europaeus* Pallas) from Eastern Croatia.

Archiv für Lebensmittelhygiene, 59, 180–184.

107. Soetan, K. O., Olaiya, C. O., & Oyewole, O. E. (2010). The importance of mineral elements for humans, domestic animals and plants: A review. *African Journal of Food Science, 4*(5), 200–222. https://doi.org/10.5897/AJFS.9000287.
108. Soladoye, O. P., Juárez, M. L., Aalhus, J. L., Shand, P., & Estévez, M. (2015). Protein oxidation in processed meat: Mechanisms and potential implications on human health. *Comprehensive Reviews in Food Science and Food Safety, 14*(2), 106–122.
109. Soriano, A., & Sánchez-García, C. (2021). Nutritional composition of game meat from wild species harvested in Europe. In *IntechOpen*.
110. Soriano, A., Montoro, V., Vicente, J., Sánchez-Migallón, B. F., Benítez, S., Utrilla, M. C., & García Ruiz, A. (2016). Influence of evisceration time and carcass ageing conditions on wild venison quality: Preliminary study. *Meat Science, 114*, 130–136. https://doi.org/10.1016/j.meatsci.2015.12.021.
111. Soriano, A., Murillo, P., Perales, M., Sánchez-García, C., Murillo, A., & García Ruiz, A. (2020). Nutritional quality of wild Iberian red deer (*Cervus elaphus hispanicus*) meat: Effects of sex and hunting period. *Meat Science, 168*, 108189. https://doi.org/10.1016/j.meatsci.2020.108189.
112. Stanisz, M., Ludwiczak, A., Składanowska-Baryza, J. A., & Bykowska-Maciejewska, M. (2019). The effect of age and ultimate pH value on selected quality traits of meat from wild boar. *Canadian Journal of Animal Science, 99*, 336–342. https://doi.org/10.1139/cjas-2018-0090.
113. Stanisz, M., Skorupski, M., Ślósarz, P., Bykowska-Maciejewska, M., Składanowska-Baryza, J., Stańczak, Ł., Krokowska-Paluszak, M., & Ludwiczak, A. (2019). The seasonal variation in the quality of venison from wild fallow deer (*Dama dama*): A pilot study. *Meat Science, 150*, 56–64. https://doi.org/10.1016/j.meatsci.2018.12.003.
114. Stasiak, K., Roślewska, A., Stanek, M., Jankowiak, H., Cygan-Szczegielniak, D., & Bocian, M. (2018). Comparison of the fatty acid profile in the meat of pigs and wild boars. *Italian Journal of Food Science, 30*, 707–714. https://doi.org/10.14674/IJFS-1187.
115. Stevenson, J. M., Seman, D. L., & Littlejohn, R. P. (1992). Seasonal variation in venison quality of mature farmed, red deer stags in New Zealand. *Journal of Animal Science, 70*, 1389–1396. https://doi.org/10.2527/1992.7051389x.
116. Suttie, J. M., Goodall, E. D., Pennie, K., & Kay, R. N. B. (1983). Winter food restriction and summer compensation in red deer stags

(*Cervus elaphus*). *British Journal of Nutrition, 50*, 737–747. https://doi.org/10.1079/BJN19830145.

117. Tesařová, S., Ježek, F., Hulánková, R., Plhal, R., Drimaj, J., Steinhauserová, I., & Bořilová, G. (2018). The individual effect of different production systems, age and sex on the chemical composition of wild boar meat. *Acta Veterinaria Brno, 87*, 395–402. https://doi.org/10.2754/avb201887040395.
118. Trocino, A., Birolo, M., Dabbou, S., Gratta, F., Rigo, N., & Xiccato, G. (2018). Effect of age and gender on carcass traits and meat quality of farmed brown hares. *Animal, 12*, 864–871. https://doi.org/10.1017/S1751731117002385.
119. Vicenti, A., Ragni, M., di Summa, A., Marsico, G., & Vonghia, G. (2003). Influence of feeds and rearing system on the productive performances and the chemical and fatty acid composition of hare meat. *Food Science and Technology International, 9*, 279–283. https://doi.org/10.1177/108201303038106.
120. Volpelli, L. A., Valusso, R., & Piasentier, E. (2002). Carcass quality in male fallow deer (*Dama dama*): Effects of age and supplementary feeding. *Meat Science, 60*, 427–432. https://doi.org/10.1016/S0309-1740(01)00156-5.
121. Volpelli, L. A., Valusso, R., Morgante, M., Pittia, P., & Piasentier, E. (2003). Meat quality in male fallow deer (*Dama dama*): Effects of age and supplementary feeding. *Meat Science, 65*, 555–562. https://doi.org/10.1016/S0309-1740(02)00248-6.
122. Wiklund, E., Farouk, M., & Finstad, G. (2014). Venison: Meat from red deer (*Cervus elaphus*) and reindeer (*Rangifer tarandus tarandus*). *Animal Frontiers, 4*, 55–61. https://doi.org/10.2527/af.2014-0034.
123. Wiklund, E., Johansson, L., & Malmfors, G. (2003). Sensory meat quality, ultimate pH values, blood parameters and carcass characteristics in reindeer (*Rangifer tarandus* L.) grazed on natural pastures or fed a commercial feed mixture. *Food Quality and Preference, 14*, 573–581. https://doi.org/10.1016/S0950-3293(02)00151-9.
124. Wiklund, E., Manley, T., Littlejohn, R., & Stevenson-Barry, J. (2003). Fatty acid composition and sensory quality of *Musculus longissimus* and carcass parameters in red deer (*Cervus elaphus*) grazed on natural pasture or fed a commercial feed mixture. *Journal of the Science of Food and Agriculture, 83*, 419–424. https://doi.org/10.1002/jsfa.1384.
125. Williams, P. (2007). Nutritional composition of red meat. *Nutrition & Dietetics, 64*(Suppl. 4), S113–S119. https://doi.org/10.1111/j.1747-0080.2007.00197.x.
126. Williams, P., Droulez, V., Levy, G., & Stobaus, T. (2007). Composition of

Australian red meat 2002. 3. Nutrient profile. *Food Australia, 59*, 331–341.

127. Winkelmayer, R., Stangl, P. V., & Paulsen, P. (2011). Assurance of food safety along the game meat production chain: Inspection of meat from wild game and education of official veterinarians and 'trained persons' in Austria. In P. Paulsen, A. Bauer, M. Vodnansky, R. Winkelmayer, & F. J. M. Smulders (Eds.), *Game Meat Hygiene in Focus: Microbiology, Epidemiology, Risk Analysis and Quality Assurance* (1st ed., pp. 245–258). The Netherlands: Wageningen Academic Publishers.
128. Wood, J. D., Enser, M., Fisher, A. V., Nute, G. R., Sheard, P. R., Richardson, R. I., Hughes, S. I., & Whittington, F. M. (2008). Fat deposition, fatty acid composition and meat quality: A review. *Meat Science, 78*(4), 343–358.
129. Wood, J. D., Richardson, R. I., Nute, G. R., Fisher, A. V., Campo, M. M., Kasapidou, E., Sheard, P. R., & Enser, M. (2003). Effects of fatty acids on meat quality: A review. *Meat Science, 66*, 21–32. https://doi.org/10.1016/S0309-1740(03)00022-6.
130. Woods, V. B., & Fearon, A. M. (2009). Dietary sources of unsaturated fatty acids for animals and their transfer into meat, milk and eggs: A review. *Livestock Science, 126*(1), 1–20.
131. World Health Organization, & Food and Agriculture Organization. (2003). *Diet, Nutrition and the Prevention of Chronic Diseases* (Technical Report Series No. 916). Geneva: World Health Organization.
132. Wyness, L., Weichselbaum, E., O'Connor, A., Williams, E. B., Benelam, B., Riley, H., & Stanner, S. (2011). Red meat in the diet: An update. *Nutrition Bulletin, 36*(1), 34–77.
133. Yu, T. Y., Morton, J. D., Clerens, S., & Dyer, J. M. (2017). Cooking-induced protein modifications in meat. *Comprehensive Reviews in Food Science and Food Safety, 16*(1), 141–159.
134. Żochowska-Kujawska, J., Lachowicz, K., Sobczak, M., & Bienkiewicz, G. (2010). Utility for production of massaged products of selected wild boar muscles originating from wetlands and an arable area. *Meat Science, 85*, 461–466. https://doi.org/10.1016/j.meatsci.2010.02.016.
135. Zomborszky, Z., Szentmihály, G., Sarudi, I., Horn, P., & Szabó, C. S. (1996). Nutrient composition of muscles in deer and boar. *Journal of Food Science, 61*, 625–627. https://doi.org/10.1111/j.1365-2621.1996.tb13172.x.

Chapter 3

Nutritional Composition of Plant-Based Meat Alternatives

Contents

Introduction.. 88

3.1. Nutrients in Plant-Based Meats and Meat... 89

3.2. Plant-Based Meat Versus Conventional Meat: A Nutritional Stance . 100

3.3. Nutritional Challenges in Plant-Based Meat Products...................... 103

3.4. Current and Potential Solutions to Improve Plant-Based Meat Nutrition 113

References.. 118

INTRODUCTION

The Food and Agriculture Organization (FAO) projects that by 2050, there will be nearly 10 billion people on Earth, and to feed them, food production must rise by 60%. Nevertheless, higher output does not ensure food security. The amount of food being produced now is sufficient to feed everyone, but a third of it is lost or wasted, and what is left over is either unfit for consumption or inaccessible. In 2022, 2 billion people lacked access to wholesome food, and 735 million people experienced chronic hunger.

The way forward is to reduce waste and switch to a healthier diet. The macro- and micronutrient composition of a healthy diet is balanced. In recent years, plant-based alternative proteins have gained attention as a viable option. As an alternative to goods derived from animals, "plant-based foods" are defined as foods prepared entirely of plant-based ingredients. This covers dairy, seafood, eggs, and plant-based meat. Similar to food products derived from animals, plant-based foods also provide complex carbohydrates and fiber, as well as vital nutrients like protein, fat, vitamins, and minerals. These foods mimic the look, flavor, and cooking qualities of conventional animal-based foods.

Vegetables, whole grains, and legumes are examples of plant-based whole foods that have been shown in numerous studies to have positive health effects and to reduce the risk of diet-related disease and death. The majority of research has found that eating a diet high in plant-based foods has advantages for both the environment and human health. Furthermore, these diets are good for the environment and for people.

The market for plant-based meat in India is expected to grow at a rate of 44% annually from 2022 to 2030, from 12.6. USD million to 228 million, thanks to these findings and consumer trends. A recent study indicated that 72% of consumers who purchased plant-based meat planned to do so again, citing nutrition, health, and protein content as important factors. Given the growing popularity of plant-based meat, it is crucial to comprehend its nutritional aspects. Claims about the relationship between plant-based meat, nutrition, and health need to be supported by evidence.

According to preliminary research, switching from traditional meat to plant-based alternatives may lower the risk of heart disease, enhance gut health, and assist in maintaining a healthy weight. However, it is important to address the nutritional challenge posed by low-quality protein and the presence of anti-nutritional factors linked to plant-based meat ingredients. Reducing antinutritional factors and improving the nutrients' bioavailability and digestibility in plant-based meat are important considerations.

This perspective chapter concentrates on the nutritional aspects of plant-based meat products due to the extensive and in-depth nature of the subject matter. The first parts of this chapter will review their nutritional value and macro- and micronutrient composition in comparison to their counterparts derived from animals.

The following sections concentrate on the nutritional difficulties associated with plant-based meat, highlighting the factors that are counterproductive to nutrition and the scientific ways to address them. The use of "phytase," a phosphatase enzyme, will be covered in this context as a potential remedy for the problems posed by anti-nutrients in plant-based meat substitutes.

3.1. NUTRIENTS IN PLANT-BASED MEATS AND MEAT

Innovative plant-based meat substitutes, like the Beyond Burger® and Impossible™ Burger, are gaining popularity among consumers and have drawn significant funding, media attention, and scientific study. Other food companies have produced their own versions of these products as a result of their success. The market for plant-based meat is anticipated to reach a valuation of over $30 billion by 2026, owing to its rapid growth. Meat substitutes are promoted as being healthier and less harmful to the environment than red meat because they are designed to taste and feel like red meat. The new generation of plant-based meat substitutes uses plant proteins (like soy, pea, potato, rice, wheat, and/or mycoprotein), fats (like canola, coconut, soybean, and/or sunflower oil), and other unique ingredients (like soy leghemoglobin, red-colored vegetable extracts, and/or flavoring agents) to mimic the sensory experience and macronutrient content of meat. Furthermore, a growing number of naturally occurring vitamins and minerals (such as zinc, iron, and B vitamins) are added to plant-based meat substitutes.

This allows plant-based meat substitutes to closely resemble meat's Nutrition Facts panels (Figure 3.1). Additionally, plant-based meats align with recommendations for dietary transitions toward reduced meat consumption and increased plant-based diets, particularly in Western civilization, and may allay concerns about the effects of red meat on human health and climate change. Furthermore, those who are omnivores but want to consume fewer animal products—flexitarians—are the target market for the innovative meat substitutes.

Figure 3.1. *Nutrition Facts panels of 4 oz. (113 grams) of novel plant-based meat alternatives and ground beef.*

Source: van Vliet S, Kronberg SL and Provenza FD (2020) Plant-Based Meats, Human Health, and Climate Change. Front. Sustain. Food Syst. 4:128. doi: 10.3389/fsufs.2020.00128.

3.1.1. Omnivory or Optionality?

Whether plant-based alternatives to animal foods can sufficiently meet nutritional needs is a central question in discussions of plant-based food replacements. Being omnivores, humans typically find it easier to obtain certain nutrients from plants, but consuming animal products usually meets our needs for other nutrients more effectively. For instance, plant-based diets are far more likely to meet our needs for magnesium and vitamin C than animal products. Furthermore, plant-based diets frequently have higher levels of vitamin E, potassium, thiamin, manganese, and folate. Additionally, a variety of phytochemicals found in plant foods play significant regulatory roles in human health. Furthermore, a large body of *in vitro* and *in vivo* experimental data indicates that certain harmful effects of compounds (such as heterocyclic amines, nitroso compounds, malondialdehyde, and advanced glycation end products) present in cooked red meat may be counteracted by plant compounds.

More research is required to firmly establish this hypothesis, but these findings may provide a mechanistic explanation—though not the only one—for why high-quality omnivorous diets—also rich in plants—do not exhibit the usual associations between red meat consumption and unfavorable health outcomes that are frequently observed in population studies of individuals consuming a typical Standard American/Western Diet.

However, minerals like iron and zinc, long-chain polyunsaturated fatty acids (PUFAs) like eicosapentaenoic acid (EPA) and docosahexaenoic acid (DHA), and vitamins A (retinol), B_{12} (adenosyl- and hydroxocobalamin), D (cholecalciferol), and K_2 (menaquinone-4) are more easily obtained from animal sources than from plant sources. These nutrients are crucial for the growth and regeneration of tissue. Although plant foods frequently include the precursors of these nutrients, a significant portion of the population has poor enzymatic conversion of plant-precursors into forms that the body can use. For instance, depending on genetic variability between individuals, the conversion of carotenoids (provitamin A) to retinol (vitamin A) ranges from ~3.5 to 28%. This illustrates how "poor converters" cannot obtain enough retinol when relying solely on plant foods. Retinol deficiency is particularly common in developing countries, especially in young children and women who are childbearing. These individuals mainly rely on eating fruits and vegetables that contain provitamin A (mainly β-carotene) to meet their vitamin A requirements, but many of them do not get enough of it.

Naturally, genetic variations pertaining to nutrient metabolism also account for the fact that while some people thrive on plant-based diets, others adhering to a vegan or vegetarian diet report health issues linked to nutrient deficiencies. In the US, for instance, the number of former vegans and vegetarians is five times higher than the number of current vegans and vegetarians. of them, 53% said they followed the diet for less than a year. Intraindividual variations in nutrient metabolism make it extremely unlikely that everyone can thrive on a plant-based diet, even though a variety of factors, such as social factors and food options, contribute to the challenges associated with sticking to plant-based diets. More research is required to confirm this theory, but it is probably true for people who follow "carnivorous" diets, which consist primarily or only of animal products.

Although it is debatable whether human omnivory truly translates to true optionality for either plant- or animal-based diets, widespread recommendations to replace traditional animal foods with plant-based foods to meet the majority of our nutritional needs have arisen due to worries about the detrimental effects of animal foods on human and environmental health. Modern food technologies have also made it possible to produce plant-based foods that can match the macronutrient, vitamin, and mineral content of animal foods by using isolated plant proteins, bioengineered ingredients, and/or synthetic vitamins and minerals. This has led to a shift away from animal foods and toward plant-based alternatives.

Furthermore, consumers may be drawn to novel plant-based meats that resemble meat in appearance, texture, and flavor because they may be able to more fully satiate people's "intrinsic desire" for meat. For instance, the intrinsic motivational salience—the desire for meat—was retained in vegetarians on a brain level comparable to omnivores, despite their subjective dislike of animal products. Notably, this "subconscious motivation for eating meat" was noted even following a single overnight fast—far from being a condition akin to starvation. In the section that follows, we address the question of whether plant-based meat substitutes can satisfy the nutritional needs that are typically satisfied by consuming animal products, considering how similar novel plant-based meat substitutes are to meat.

3.1.2. Protein

For adults, the recommended daily allowance (RDA) for protein is 0.8 g protein/kg body weight (or about 56 g for a 70 kg person). This amount, however, ought to be considered the bare minimum to avert deficiency in early adulthood rather than an amount that encourages ideal health. Additionally, the RDA for protein is deemed inadequate for middle-aged and older adults (over 50 years of age) as well as for adults who aim to optimize cellular adaptations through consistent physical activity or exercise.

Because animal foods, like meat, provide dietary protein at a modest calorie load and are thought to be of higher protein quality than plant sources, they are frequently suggested as a way to meet protein needs. The two main metrics used to assess the quality of dietary protein sources are the digestible indispensable amino acid score (DIAAS) and the protein digestibility-corrected amino acid score (PDCAAS). Animal proteins (> 0.9) typically have higher scores than plant proteins (ranging from 0.4 to 0.9). The presence of "anti-nutrients" such as trypsin inhibitors and phytotates, which obstruct protein digestion and absorption, contributes to the lower PDCAAS/DIAAS of plant sources. The use of soy, pea, and other plant protein isolates and/or concentrates makes novel meat substitutes advantageous. These pure protein sources have similar PDCAAS/DIAAS to most animal proteins, including meat, because they contain fewer anti-nutritional factors.

Comparing isolated plant proteins to animal sources, one could anticipate a comparable anabolic response based on their respective PDCAAS/DIAAS. However, when compared to isonitrogenous amounts of animal proteins with similar PDCAAS/DIAAS, several studies have shown that purified plant proteins have a lower skeletal muscle anabolic response. Leucine is one of the essential amino acids that cannot be produced *in vivo* and must be obtained through diet; lysine and methionine are particularly low in plant proteins. As an example, the most popular plant-based protein sources for innovative plant-based substitutes, soy, and pea protein, are both significantly lower in methionine and slightly lower in lysine than beef (Figure 3.2).

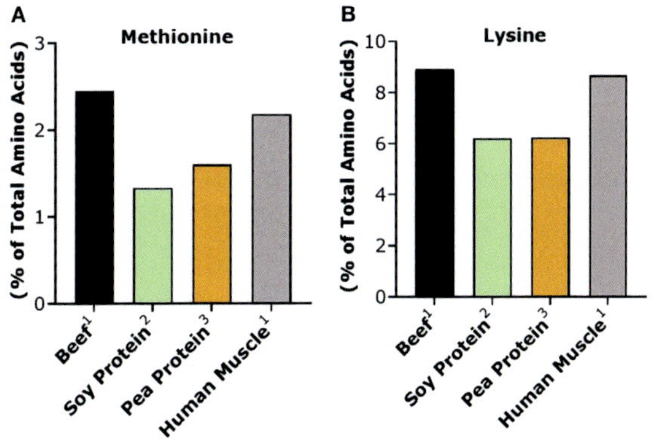

Figure 3.2. *Methionine (A) and lysine (B) content of beef and plant proteins commonly used in plant-based meat alternatives.*

Source: van Vliet S, Kronberg SL and Provenza FD (2020) Plant-Based Meats, Human Health, and Climate Change. Front. Sustain. Food Syst. 4:128. doi: 10.3389/fsufs.2020.00128.

One way to address an unbalanced amino acid profile is to combine plant proteins that are higher in methionine and lower in lysine (like soy, potato, and pea) with isolated plant proteins that are lower in lysine and higher in methionine (like wheat, rice, hemp, and maize) in a single product. Another option is to add crystalline amino acids to the product. It is unlikely that the amount of isolated rice protein in a popular pea-based alternative—which is complimentary to pea protein—will be enough to raise the product's methionine content, given the ingredients' listed order. Like soy, potatoes are high in lysine and low in methionine. In a similar vein, a popular soy-based substitute contains very little potato protein (less than 2% of the total ingredients). Although we have previously proposed that combining various complementary plant sources is a viable method to enhance the anabolic response of skeletal muscle to ingested plant protein, no research has yet established whether or not this raises the anabolic response of muscle from plant sources to the level of animal proteins. Interestingly, when compared to a leucine-matched amount of whey protein, consuming complementary plant proteins still resulted in 30–40% lower circulating essential amino acid availability. These findings imply that, even when plant-based sources of amino acids are combined to create a comprehensive amino acid profile, the anabolic potential may still be lower than that of animal-based protein.

One possible countermeasure to the reduced anabolic effects of plant vs. Eat more plant proteins instead of animal proteins. Anabolic responses in the muscles are comparable when 40 g of soy protein is consumed to 20 g of whey protein. To enhance the muscle mass gains induced by resistance exercise, a high dose of rice protein (48 g) added to an omnivorous diet is equally as effective as a protein-matched amount of whey protein. In contrast, research indicates that skeletal muscle gains induced by training are superior when lower doses of animal vs. proteins from plants (from 17.5 to 25 g). However, it is reasonable to assume that consuming plant-based meat substitutes as a component of an omnivorous diet won't have a detrimental effect on skeletal muscle mass or protein requirements.

3.1.3. Vitamin B_{12}

Among the class of substances known as corrinoids, the cobalamin (vitamin B_{12}) are the most well-known. Red blood cell synthesis, myelin formation, DNA synthesis, and the preservation of central nervous system function are all aided by cobalamin, also known as vitamin B_{12}. Vitamin B_{12} must be consumed by humans through food or supplementation. Evolutionary pressures most likely led to preferential absorption of B_{12} in the small intestine as a result of regular animal consumption, even though gut bacteria in our large intestine still produce some corrinoids, including trace amounts of active forms of cobalamin. Because of this, there is now probably a 1.5. Ma dependence on exogenous B_{12}. In fact, the receptors required for absorbing vitamin B_{12} are exclusively located in the small intestine of contemporary humans, upstream from the location where bacteria produce corrinoid compounds.

The most common forms of biologically active B_{12} are hydroxocobalamin in dairy products and eggs, and adenosyl cobalamin in animal flesh. These active forms of vitamin B_{12} bioaccumulate in animal products primarily through the consumption of soil and excrement by non-ruminant herbivores, microbial synthesis in the guts of ruminants, and phytoplankton by aquatic animals. A prevalent misconception is that most cattle get extra B_{12}, and that people are just taking B_{12} supplements through a "middle-man." The truth is that B_{12} is rarely given to cattle, probably because of its high cost and limited benefits for production, even though there is some evidence that high-producing dairy cows require more B_{12} than their microbes produce.

The majority of B_{12} in plants is found in biologically inactive corrinoids, or B_{12} analogs, which may compete with the transport of biologically active B_{12} and exacerbate a B_{12} deficiency. Limited amounts of B_{12} are also present in some plant foods, such as mushrooms and fermented vegetables. Several plant-based foods, especially those designed to take the place of animal foods (vegan spreads, cereals, nondairy milks, and plant-based meat substitutes), are frequently supplemented with extra B_{12} to prevent deficiencies. Commonly used in these products as an additional form of B_{12}, cyanocobalamin is reasonably inexpensive to produce and exhibits stability when exposed to heat.

The synthetic form of vitamin B_{12} known as cyanocobalamin is typically found in trace amounts in human tissue, especially in smokers. All forms of vitamin B_{12}, including the synthetic form cyanocobalamin and the naturally occurring forms adenosylcobalamin and hydroxycobalamin, are absorbed similarly well. However, using cyanocobalamin to meet B_{12} requirements may raise concerns because its tissue retention rates and consequent metabolic activity are lower than those of the naturally occurring forms. Furthermore, because B_{12} in animal diets is protein-bound, it is somewhat shielded from light degradation. However, supplementing with cyanocobalamin can also help adults and children with their vitamin B_{12} deficiency, which is another reason it's recommended to fortify meat substitutes with B_{12}. This is particularly crucial for older people (even omnivores), vegans, and vegetarians, as they frequently have low B_{12} status and must rely on food fortification and/or supplements to meet their B_{12} needs. It's crucial to emphasize that less than 25% of plant-based meat substitutes contain vitamin B_{12}.

3.1.4. Iron

Dietary iron is found in foods that are animal-based, especially red meat, as heme iron, and in foods that are plant-based, especially grains, pulses, green leafy vegetables, and some fruits. Omnivores frequently have higher serum ferritin levels (a measure of iron status), which can be explained by the fact that heme iron is 5–10 times more bioavailable than non-heme iron. Phytates, polyphenols, and calcium—plant and dairy products—are among the plant compounds that inhibit the absorption of iron, especially non-heme iron.

While the majority of plant-based foods primarily consist of non-heme iron, a popular soy-based alternative contains heme iron that has been purified from yeast that has been genetically modified to express the leghemoglobin protein, which is typically found in the root nodules of soy plants (i.e., soy leukemia).

In a human epithelial cell culture model, leghemoglobin's iron uptake exhibited comparable bioavailability to that of bovine hemoglobin, although its amino acid sequence differs significantly from that of its animal hemoglobin counterparts. Crucially, preliminary research on the safety of soy leghemoglobin derived from yeast indicates minimal genotoxicity and immunogenicity in *in vitro* and short-term (28-day) *in vivo* animal experiments. Future research should validate the long-term safety and bioavailability of consuming soy leghemoglobin derived from yeast *in vivo* in humans, especially in children where soy allergy is more prevalent. A recent consumer survey indicates that although there is some consumer concern about the use of genetically modified ingredients in food products, soy leghemoglobin is present in a popular soy-based meat alternative, but this does not seem to be a barrier to consumption or the product's perceived healthfulness.

A popular meat substitute made from peas has non-heme iron that is found in peas naturally. Even though non-heme iron has a lower bioavailability than heme iron, it can nevertheless be a significant dietary source of iron. Ironically, meat, and vitamin C are the primary boosters of non-heme iron absorption. For this reason, including meat in plant-based diets increases the amount of non-heme iron that is absorbed from plants. As a chelator in the stomach, vitamin C also improves the absorption of iron. Notably, a popular soy-based meat substitute on the market is enhanced with sodium ascorbate (vitamin C), which likely mitigates some of the phytates (found in soy protein) that inhibit iron absorption.

Even with greater awareness, iron deficiency is still one of the most prevalent nutritional deficiencies in both developed and developing nations. Older people, children, and teenage girls are the demographic groups most at risk for iron deficiency. Conversely, there is growing evidence that a high heme iron intake contributes to the development of cardiovascular disease (CVD). Heme iron intake and CVD are particularly associated with US cohorts, but the relationship is not as strong in cohorts outside of the US. It's interesting to note that the first cohort of the NHANES, which was studied in the 1970s when red meat consumption was higher than it is today, did not show any correlation between high heme iron intake and risk of CVD even in the US. The variation observed among the studies implies that the background diet, which includes red meat consumption, could be a significant moderating factor. Specifically, the harmful consequences of consuming red meat could be prolonged if meat is a part of the Standard American Diet, which is high in processed foods and insufficiently balanced with whole food plant sources. For instance, heme iron absorption is inhibited by polyphenols, phytates, calcium, and fibers; this could account for the observation in certain epidemiolocal studies that the risk of heme iron intake and CVD vanishes when diet quality is significantly adjusted (i.e., diets rich in whole plant foods as well).

3.1.5. Zinc

Zinc deficiency can be problematic in both developed and developing nations, much like iron deficiency can be, and people who limit their intake of animal products typically have lower zinc levels. Anti-nutrients like lectins, phytates, and certain fibers can reduce the amount of zinc absorbed from plant sources. As with iron, eating plant-based zinc can enhance the absorption of zinc when combined with animal-based diets. Although there isn't much zinc in soy protein, a popular soy-based substitute is fortified with zinc gluconate to raise the amount to the same as in beef. Even so, the absorption of zinc from fortified plant foods is less than that of beef, even with an equivalent zinc content. We point out that sufficient zinc levels may be obtained by a carefully thought-out vegan diet high in legumes, nuts, seeds, and other plant foods high in zinc. It's important to keep in mind that supplementing to meet zinc (and iron) needs may decrease the absorption of other minerals like copper, which raises the need for other foods. By eating foods high in copper from plants, such as nuts, seeds, and leafy greens, the latter can be lessened.

3.1.6. Essential Fatty Acids

Essential fatty acids, such as ω-6 linoleic acid (C18:2, LNA) and ω-3 fatty acid alpha-linolenic acid (C18:3, ALA), are not synthesized *in vivo* by humans and are only available through diet. The long-chain polyunsaturated fatty acids (LCPUFA) docosahexaenoic acid (C22:6 n-3, DHA) and eicosapentaenoic acid (C20:5 n-3, EPA) are derived from ALA. While DHA and EPA are only present in animal foods and some types of algae, ALA, and LNA are frequently found in plant foods and are also present in animal foods in trace amounts.

Although ALA can be converted through a series of elongation and desaturation steps to DHA and EPA, the conversion rate is often less than 1%. Furthermore, the presence of co-factors like iron, zinc, selenium, and vitamin B_6, which are less bioavailable from plant-based diets, affects this conversion efficiency. For these reasons, compared to omnivores, vegetarians may have lower DHA and EPA levels.

The significance of DHA and EPA for immunomodulation, cardiovascular health, vision, and cognitive function has been thoroughly investigated. Low circulating levels of DHA are linked to accelerated brain aging. DHA makes up 30–40% of total fatty acids and is a major component of the brain phospholipid membrane. However, research indicates that the human brain only needs 5 mg of DHA daily; therefore, 1200 mg of ALA is thought to be able to meet these minimal needs, even though this isn't thought to be the best amount for overall health. Although there are no official daily recommended intakes for DHA and EPA, many studies show that combined intakes of these two important nutrients, ranging from 250 to 1000 mg/day, improve various health parameters, including cognitive function. As a result, various health organizations frequently recommend these amounts.

Significant amounts of the ω-3 fatty acid ALA can be found in some vegetable oils, including camelina oil (38% ALA), perilla oil (60% ALA), chia seed oil (64% ALA), and flax seed oil (53% ALA). However, consumption of the latter two oils is primarily limited to Asian and Nordic countries, respectively. While a modest intake of these oils can provide the body with the necessary amount of ALA to ensure minimum DHA requirements, most vegetable oils consumed in industrialized countries come from ω-6 LNA-rich seed oils, like canola, soybean, corn, and sunflower oil, which contain less than 10% ALA. For example, the primary oils in the innovative plant-based meat substitutes, canola oil, and sunflower oil, only have 1% and 10% of ALA, respectively. Since ALA and DHA have already low rates of conversion, plant-based meat substitutes are unlikely to contribute significantly to the diet in terms of very long-chain PUFAs.

An additional possible problem is that ω-6 fatty acids, like LNA, directly compete with ALA for the enzymes that are involved in the processes of elongation and desaturation, which makes it harder to extract DHA and EPA from ALA. This is especially problematic because the modern Western diet's increased consumption of high-LNA seed oils has led to an ω-6-to-ω-3 fatty acid ratio of 16:1, whereas historical intakes place this ratio closer to 1:1. The elevated ratio of ω-6 to ω-3 fatty acids is thought to be a significant contributing factor to the rising prevalence of metabolic disorders and overall death from all causes in Western nations.

The reduction of inflammation observed in experiments involving the substitution of ω-3-rich ALA oils for ω-6-rich LNA oils provides a mechanistic explanation for the potential cardioprotective effects of ALA oil consumption. Therefore, an appropriate way to improve the novel plant-based meat substitutes would be to use high ALA oils, which are rich in ω-3, rather than high LNA oils, which are rich in ω-6 fatty acids. The fact that high ALA oils are more likely to oxidize lipids is a significant factor, which may be why high LNA oils are usually used in plant-based meat substitutes. That being said, producers of plant-based meat substitutes that take into account the use of high ALA oils in their products should investigate the addition of natural antioxidants and the entrapment of high ALA oils with isolated plant proteins, as these strategies may increase their healthfulness.

It is frequently claimed that foods derived from land animals, like beef, only contain trace amounts of ω-3 fatty acids, making them an unvalued dietary source of these vital fatty acids. This idea, however, ignores the abundance of the ω-3 fatty acid docosapentaenoic acid (C22:5, DPA) found in beef, especially in beef raised on pasture, which causes *in vivo* conversion to raise platelet EPA and DHA levels. Although offal meat can also provide significant amounts of DHA (e.g., 100 g of grass-fed beef liver contains 80 mg of DHA), eating organ meat is becoming less common in Western diets, with the majority of the ω-3 fatty acids DHA and EPA coming from marine sources.

3.1.7. Secondary Nutrients

Although we have discussed several critical individual nutrients so far, foods in their unprocessed state are far more complex than the contents of essential fatty acids, amino acids, vitamins, and minerals would imply. Thousands to hundreds of thousands of biochemicals found in food sources are crucial to human metabolism. Even though many of these nutrients are regarded as conditionally or non-essential depending on life stages and are frequently overlooked in discussions of the nutritional needs of humans, it is still important to consider how they may affect human metabolism.

For instance, creatine's capacity to improve athletic performance has been the subject of much research; however, creatine also has significant effects on cognition. Since creatine can only be found in animal products, vegetarians, and vegans frequently have lower body stores of the vitamin. However, when vegetarians were given additional creatine, their performance on memory tests significantly improved. Analogously, animal foods are the only sources of the antioxidants carnosine, taurine, and anserine (nearly exclusively). In humans, higher anserine and carnosine intake protects neurocognition.

Taurine is an amino acid that is almost exclusively present in animal foods. While very small amounts of taurine can be found in some plant foods, such as grains, legumes, and cereals (a thousand times less than in animal foods), these amounts are not enough to meet human needs. It's commonly believed that since taurine can be produced *in vivo* from methionine and cysteine by the enzyme cysteinesulfinic acid decarboxylase (CSD), one can satisfy their taurine needs by consuming plant proteins high in methionine and cysteine, which are present in sufficient amounts in a variety of grains and legumes. The human body can produce taurine from methionine and cysteine thanks to CSD levels, but these levels are not high enough to sustain tissue concentrations over time. Taurine affects almost every important organ in the body and is essential for the proper functioning of the brain, eyes, mitochondria, skeletal muscle cell differentiation, and cardiovascular system. More research is required to fully comprehend the effects of these variations in secondary nutrients on both short- and long-term health when comparing plant-based meat substitutes to meat.

3.1.8. Fortifying Foods to Mimic the Natural Food Source

Natural whole foods are incredibly complex, and the reductionist method of attempting to "mimic" whole food sources (meat or other foods) by combining multiple isolated nutrients probably underestimates the complexity and health benefits of consuming whole foods. This is a recurring concern. Specifically, supplementing a low-meat diet with zinc and other meat-derived minerals did not produce the same zinc status as if these minerals were included in the diet as part of the meat's natural matrix. Furthermore, a recent large population-based study found that adequate intakes of zinc, copper, and vitamins A and D when derived from foods, but not from supplements, were associated with decreased risk of cardiovascular disease and all-cause mortality. In a similar vein, diets high in carotenoid content are linked to a lower risk of cardiovascular disease, retinopathies, and several cancers. However, the findings of epidemiological and interventional research indicate that taking supplements containing carotenoid and/or vitamin A may even increase the risk of certain subpopulations of cancer or cardiovascular disease. Analogous conclusions have been drawn from studies on calcium, which indicate that supplementing with the mineral may raise the risk of cardiovascular disease but not eating calcium-rich foods. Lastly, a systematic review of RCTs involving close to 300,000 participants revealed similar results for vitamin C and selenium supplements, which have no beneficial effects on mortality. Therefore, it would not be the best idea to just consume these nutrients outside of their natural food matrices to promote health. Therefore, regardless of a person's diet, getting nutrients from whole food sources rather than supplements is emphasized (van Vliet et al., 2020)

3.2. PLANT-BASED MEAT VERSUS CONVENTIONAL MEAT: A NUTRITIONAL STANCE

Calorie intake is not the only aspect of nutrition. The nutritional quality of any food product is determined by the quantity of macronutrients and micronutrients as well as their capacity to meet consumer dietary requirements. This also applies to plant-based meat; in fact, the rise of this product category can be attributed to the growing need for protein and the ensuing innovations in the food sector. Diet-related illnesses and deaths have generally been linked to underconsumption of plant-based foods and overconsumption of red and processed meat as well as foods high in sugar, salt, and saturated fat. Conversely, plant-based meat has a higher fiber content and typically less saturated fat than conventional meat while still meeting many of the same nutritional needs. Plant-based meat substitutes can have their digestibility and nutrient content improved with the right processing method selection. A comparison of the various macro- and micronutrients and energy that plant-based meat and conventional meat provide is given in the ensuing subsections.

Among the macronutrients, protein content (percentage) is the most important factor in plant-based meat substitutes. This is pertinent because researchers studying substitute proteins and amino acids are looking into ways to meet the FAO's meat regulations. According to a recent report by the Good Food Institute (GFI) Europe, most plant-based meat products in the European market, which includes the UK, Sweden, Germany, and the Netherlands, have a slightly lower protein content per 100 g than conventional meat. Nutritious protein sources, however, are defined not only by the weight of the protein but also by the percentage of calories (at least 20%) that come from the protein, as opposed to the fat or carbs in a food item. Therefore, based on the energy that protein content in plant-based foods provides, the report concluded that these foods are more similar to conventional meat.

As a function of their protein source and product format, different plant-based meat products have different proportions of essential amino acids. For example, only plant-based patties and meatballs had a lower protein content than their conventional meat counterparts, according to a nutritional label analysis study done on 269 plant-based meat products sold on the Italian market across various formats (steaks, patties, meatballs, cutlets, and cured meats). Cereals, legumes, and a combination of plant protein sources were present in the other plant-based meat products that had higher protein content than their animal-derived counterparts, which may have contributed to their higher protein content. Compared to a combination of legumes and vegetables (10.1 g/100 g) or cereals and vegetables (5.4 g/100 g), legumes (15.0 g/100 g) offer a significantly higher protein content. When evaluating nutrition, it's also important to consider the amount and source of fat in food, and plant-based foods are frequently linked to lower fat content foods. According to a recent study, plant-based patties have an average total fat content of 30% and lower saturated fat levels (30%) than animal-based patties (40%). In general, plant-based meat substitutes have about the same amount of carbohydrates as traditional meat. It does, however, differ based on the kind of product and its ingredients. Additionally, certain plant-based meat substitutes might include extra carbohydrates as flavorings, binders, or fillers.

The literature provides strong evidence that consuming a high-fiber diet lowers the risk of cardiovascular disease and death. Consuming a lot of fiber is linked to the development of a healthy gut and microbiota as well as a reduction in inflammation. In contrast to traditional meat, plant-based meat is thought to be high in dietary fiber. The tiny amount of fiber in meat products derived from animals is supplied by additional plant-based ingredients. Fiber is a must for a diet that is balanced. An increased consumption of fiber has been linked to a decreased risk of major illnesses such as cardiovascular disease, coronary artery disease, pancreatic cancer, and all-cause mortality. Compared to their traditional animal meat counterparts, which have higher fat contents, particularly saturated fat, plant-based meat products typically have fewer calories per serving in terms of energy. This is especially true if the plant-based meat has less fat and is made of leaner ingredients. The fiber from plant sources in plant-based meat may help with satiety and fullness without adding a lot of calories (Shankaran and Kumari, 2024). Comparative information about the nutritional makeup of traditional meat and plant-based meat is displayed in Table 3.1.

Table 3.1. *Median Values for Nutrients per 100 g of Plant-Based Meat Versus Conventional Meat*

Product (per 100 g)	Product	Calories (kcal)	Saturated Fat (g)	Protein (g)	Fiber (g)
Sausages	Conventional meat	277 (17)	6.8 (1.4)	15 (1.3)	1.0 (0.8)
		279 (42.5)	7.5 (1.3)	20 (3.3)	0.9 (0.85)
	Plant-based	149 (15.5)	2.5 (2.23)	8.6 (6.55)	6.4 (2.25)
Beef patties	Conventional meat	247.5 (35.25)	7.6 (0.68)	20 (1.48)	0.5 (0.28)
		255.5 (57.75)	7.5 (3.78)	21 (3.9)	0.5 (0.23)
	Plant-based	212.5 (48.5)	1.6 (0.8)	16.6 (1.38)	6.1 (1.5)
Chicken nuggets	Conventional meat	262.5 (17)	2.1 (0.88)	14.0 (1.15)	1.6 (0.48)
	Plant-based	253 (23.25)	0.9 (0.13)	12.4 (0.85)	4.9 (1.78)
Mince	Conventional meat	195.5 (33.75)	4.5 (1.83)	20.8 (3.88)	0.0 (0.2)
	Plant-based	141 (30.25)	0.9 (2.48)	17.4 (3.48)	5.3 (1.05)
Meatballs	Conventional meat	216.5 (47)	5.4 (2.45)	20.5 (3.38)	0.5 (0.60)
	Plant-based	177 (31.75)	1.3 (0.55)	15.6 (2.38)	5.1 (1.53)

Source: Ishwarya Shankaran P, Kumari P. Nutritional Analysis of Plant-Based Meat: Current Advances and Future Potential. Applied Sciences. 2024; 14(10):4154. https://doi.org/10.3390/app14104154.

Depending on the components and fortification used, plant-based meat substitutes' micronutrient contents can differ greatly from those of traditional animal meat. Vitamin C, vitamin A, folate, and minerals like magnesium and potassium can all be found in plant-based meats that are made from legumes, grains, vegetables, and seeds. Meats derived from animals, particularly red meats like beef, are rich in heme iron, which the body absorbs more easily than non-heme iron from plant-based sources. To improve their nutritional value, some plant-based meat substitutes are fortified with iron. Meat, fish, eggs, and dairy products are among the main sources of vitamin B_{12}. Unless they are fortified with vitamin B_{12}, plant-based meat substitutes usually do not naturally contain this nutrient. Zinc can be obtained from both plant- and animal-based sources, although zinc from plant sources may have a lower bioavailability because of anti-nutritional elements like phytates that can prevent the mineral from being absorbed.

3.3. NUTRITIONAL CHALLENGES IN PLANT-BASED MEAT PRODUCTS

The primary factor limiting plant-based meat's nutritional quality in comparison to conventional meat is its lower protein quality, which is brought on by an incomplete amino acid profile—a result of the absence of one or more essential amino acids that the body is unable to synthesis on its own. Complete proteins are those made from animal sources, such as meat, dairy, and eggs, which typically contain adequate amounts of each essential amino acid. The digestibility and bioavailability of the amino acids present in plant-based meat can be lowered by an incomplete amino acid profile and the presence of anti-nutritional components. Mineral and other nutrient absorption is hampered by antinutritional factors. These difficulties and the resulting ramifications are expounded upon in the following sections.

3.3.1. Protein Quality

Any food product's level of protein determines how nutritious it is. Assuming normal physical activity, protein quality is defined as the minimum intake of high-quality protein that contains all the essential amino acids and supports the maintenance of an appropriate body composition while promoting growth at a normal rate for age. The Food and Drug Administration (FDA) states that an adult's daily diet must include 2000 calories and 50 grams of protein. However, a person's unique needs for protein will vary depending on their age, sex, degree of activity, and other variables. For the average sedentary adult, a dietary allowance of 0.8 g/kg of body weight is advised to prevent protein deficiency.

The amount of essential amino acids in protein and its digestibility are typically used to determine the quality of the protein. The digestible indispensable amino acid score, or DIAAS, is a metric that the FAO has recommended be used to assess the quality of proteins. It can be calculated using the following formula:

$$\text{DIAAS}(\%) = 100 \times \frac{\text{mg of digestible dietary indispensable amino acidin 1 g of the dietary protein}}{\text{mg of the same dietary indispensable amino acid in 1 g of the reference protein}} \quad (1)$$

Despite being produced from protein-rich ingredients like soy, wheat gluten, and pea, plant-based meat substitutes frequently don't have a well-balanced mix of vital amino acids. In actuality, plant-based meat substitutes contain less alanine, glycine, and methionine and more glutamic acid and cysteine. This might be the result of these amino acids being lacking in the corresponding plant protein sources. Legumes are deficient in sulfur-containing essential amino acids like cysteine and methionine, while cereals are deficient in lysine. Because plant proteins lack one or more essential amino acids, they are therefore nutritionally incomplete. The essential amino acid content of proteins derived from various plant and animal sources is compared in Table 3.2. Lowering muscle protein synthesis linked to plant-based diet consumption may result from a deficiency of plant proteins in essential amino acids. Given that older people are more susceptible to muscle atrophy, this is a serious concern. To ensure that you are getting enough of each amino acid, it is advised to eat a wide range of plant-based proteins.

Table 3.2. *Comparison of the Essential Amino Acid Content Between Selected Sources of Plant Proteins and Animal Proteins (Values Expressed in g/100 g of Protein)*

Essential Amino Acid/ Protein Source	Histidine	Lysine	Isoleucine	Leucine	Methionine	Phenylalanine	Threonine	Tryptophan	Valine
Finger millet	2.37 ± 0.46	2.83 ± 0.34	3.70 ± 0.44	8.86 ± 0.54	2.74 ± 0.27	5.70 ± 1.27	3.8 ± 0.45	0.91 ± 0.30	5.65 ± 0.44
Pearl millet	2.15 ± 0.37	3.19 ± 0.49	3.45 ± 0.74	8.52 ± 0.86	2.11 ± 0.50	4.82 ± 1.18	3.55 ± 0.40	1.33 ± 0.30	4.79 ± 1.04
Sorghum	2.07 ± 0.20	2.31 ± 0.40	3.45 ± 0.63	12.03 ± 1.51	1.52 ± 0.50	5.10 ± 0.50	2.96 ± 0.17	1.03 ± 0.21	4.51 ± 0.71
Little millet	2.35 ± 0.18	2.42 ± 0.10	4.14 ± 0.08	8.08 ± 0.06	2.21 ± 0.10	6.14 ± 0.10	4.24 ± 0.12	1.35 ± 0.10	5.31 ± 0.16
Kodo millet	2.14 ± 0.07	1.42 ± 0.17	4.55 ± 0.22	11.96 ± 1.65	2.69 ± 0.16	6.27 ± 0.34	3.89 ± 0.16	1.32 ± 0.19	5.49 ± 0.23
Amaranth seed (pale brown)	1.98 ± 0.50	5.50 ± 0.35	2.85 ± 0.04	4.94 ± 0.17	1.95 ± 0.12	4.75 ± 0.41	2.99 ± 0.21	1.69 ± 0.10	4.30 ± 0.27
Quinoa	2.98	5.55	3.75	6.08	2.24	4.35	3.01	1.25	4.55
Bengal gram	2.39 ± 0.34	6.06 ± 0.52	4.25 ± 0.26	6.91 ± 0.65	1.12 ± 0.31	5.97 ± 1.06	3.24 ± 0.34	1.09 ± 0.25	4.09 ± 0.41
Black gram	2.64 ± 0.11	6.22 ± 0.21	3.75 ± 0.92	7.93 ± 0.45	1.31 ± 0.35	5.68 ± 0.91	2.99 ± 0.19	1.07 ± 0.12	4.61 ± 1.06
Field bean, black	3.21	6.79	4.57	8.91	1.36	5.88	4.12	0.73	5.24

Field bean, brown	3.23	6.75	4.59	8.88	1.38	5.72	3.97	0.78	5.16
Field bean, white	2.82 ± 0.76	6.13 ± 0.68	4.41 ± 0.36	8.48 ± 0.66	1.40 ± 0.18	5.76 ± 0.32	3.97 ± 0.31	0.89 ± 0.07	4.96 ± 0.36
Green gram	2.55 ± 0.26	6.09 ± 1.06	4.07 ± 1.11	7.90 ± 0.99	1.05 ± 0.23	6.20 ± 0.61	3.36 ± 0.60	1.24 ± 0.42	5.21 ± 1.28
Lentil	1.93 ± 0.32	6.12 ± 0.71	3.74 ± 0.91	7.10 ± 0.86	0.54 ± 0.19	5.10 ± 0.73	3.31 ± 0.24	0.81 ± 0.07	5.02 ± 0.39
Cowpea, brown	2.93 ± 0.42	6.67 ± 1.26	4.10 ± 0.58	7.49 ± 1.23	1.38 ± 0.16	5.47 ± 0.49	3.80 ± 0.61	1.05 ± 0.03	4.87 ± 0.59
Cowpea, white	3.25	7.14	4.40	7.96	1.53	5.63	4.10	0.92	5.31
Black gram	2.64 ± 0.11	6.22 ± 0.21	4.34 ± 0.23	7.40 ± 0.31	1.16 ± 0.16	6.26 ± 0.70	3.55 ± 0.31	0.95 ± 0.07	4.58 ± 0.51
Chicken (poultry, leg, skinless)	4.47 ± 0.17	7.26 ± 0.28	2.76 ± 0.18	7.84 ± 0.17	6.04 ± 0.32	7.07 ± 0.21	5.66 ± 0.29	1.16 ± 0.22	4.92 ± 0.22
Goat, liver	4.23 ± 0.47	8.39 ± 0.98	3.04 ± 0.42	6.02 ± 0.25	6.27 ± 0.18	5.64 ± 1.19	6.58 ± 0.68	1.22 ± 0.11	4.53 ± 0.99
Beef, round (leg)	5.07 ± 0.36	8.09 ± 0.22	3.67 ± 0.50	6.43 ± 0.16	5.41 ± 0.17	7.00 ± 0.85	5.10 ± 0.13	1.42 ± 0.16	4.67 ± 0.18
Pork, shoulder	4.83 ± 0.29	8.39 ± 0.12	3.86 ± 0.46	5.50 ± 0.13	5.32 ± 0.44	6.51 ± 0.58	4.39 ± 0.43	1.32 ± 0.11	4.41 ± 0.29
Whole egg	4.65 ± 0.07	8.37 ± 0.19	2.15 ± 0.07	7.95 ± 0.18	6.72 ± 0.27	6.29 ± 0.32	6.34 ± 0.23	1.59 ± 0.15	5.82 ± 0.20

Egg white	4.54 ± 0.17	8.35 ± 0.20	2.25 ± 0.14	7.94 ± 0.25	6.24 ± 0.33	5.28 ± 0.56	6.90 ± 0.21	1.50 ± 0.11	6.89 ± 0.19
Egg yolk	5.69 ± 0.09	8.58 ± 0.29	2.19 ± 0.08	7.24 ± 0.23	6.17 ± 0.22	7.08 ± 0.38	6.14 ± 0.10	1.47 ± 0.12	5.66 ± 0.33
Cow milk	2.14 ± 0.13	8.59 ± 0.20	6.20 ± 0.55	10.66 ± 0.41	2.51 ± 0.12	5.09 ± 0.42	4.81 ± 0.07	1.46 ± 0.11	6.40 ± 0.24
Whey	1.53 ± 0.03	7.37 ± 0.18	5.30 ± 0.04	8.52 ± 0.15	1.69 ± 0.03	2.59 ± 0.02	5.47 ± 0.11	1.63 ± 0.02	4.97 ± 0.05

Source: Ishwarya Shankaran P, Kumari P. Nutritional Analysis of Plant-Based Meat: Current Advances and Future Potential. Applied Sciences. 2024; 14(10):4154. https://doi.org/10.3390/app14104154.

3.3.2. Anti-Nutritional Factors

Another significant nutritional issue with plant-based meat is the high concentration of anti-nutritional components, such as lectins, oxalates, tannins, saponins, and enzyme inhibitors (Table 3.3). "Anti-nutrients" are substances that are either synthetic or naturally occurring by plants as a defense mechanism. These substances hinder the body's ability to absorb nutrients and decrease intake, digestion, and utilization of nutrients. Reduced nutrient bioavailability is the main health effect of anti-nutritional factors linked to plant-based meat. Furthermore, when anti-nutrients are added to plant-based meat product substitutes, they may impart an unwanted flavor, like bitterness. In addition to anti-nutritional elements, plant sources could also include indigestible components like fibers or non-starch polysaccharides that prevent enzymes from accessing proteins and decrease protein digestibility.

Table 3.3. *Plant-Based Sources and Effects of Anti-Nutritional Factors*

Anti-Nutrients	Source	Interaction With Nutrients
Phytic acid	Cereals, legumes, nuts, and seeds that are used to produce extruded plant-based meat products	It can bind to amino acids and minerals such as iron, zinc, and calcium, making them less bioavailable for absorption by the body.
Lectins	Legumes, grains, and vegetables belonging to the group of plants in the Solanaceae family (ex. tomatoes, eggplant)	Some lectins may interfere with nutrient absorption and cause digestive issues.
Oxalates	Spinach, beet greens, and nuts	Oxalates can bind to calcium and other minerals, potentially reducing their absorption.
Tannins	Tea, coffee, and certain plant foods like legumes, nuts, and some fruits.	Tannins can interfere with iron absorption and may cause digestive discomfort.
Protease inhibitors	Legumes and grains	Protease inhibitors can interfere with protein digestion.
Saponins	Legumes, whole grains, and some vegetables	Saponins may interfere with nutrient absorption and cause digestive issues.

Source: Ishwarya Shankaran P, Kumari P. Nutritional Analysis of Plant-Based Meat: Current Advances and Future Potential. Applied Sciences. 2024; 14(10):4154. https://doi.org/10.3390/app14104154.

Phytic acid is the most important anti-nutritional factor for human nutrition out of all of them. Plants, soils, legumes, cereals, seeds, and nuts all contain this compound (Table 3.4). It is the main form of phosphorus storage in plant tissues and is chemically defined as the hexaphosphoric ester of the hexahydric cyclic alcohol meso-inositol. Because of its special structure—which consists of 12 interchangeable hydrogen ions and a strong negative charge—it can bind to amino acids and minerals (such as iron, zinc, magnesium, and calcium) to form an insoluble complex called phytate, which prevents the body from absorbing proteins and micronutrients. Numerous studies using single-meal isotopes have demonstrated the impact of phytate on the absorption and bioavailability of both intrinsic and added iron. Even at very low intestinal pH, phytate has low solubility and functions over a wide pH range.

Table 3.4. *Examples of High-Phytate Plant Materials, as a Percentage of Dry Weight*

Food	Phytic Acid
Almonds	0.4–9.4%
Beans	0.6–2.4%
Brazil nuts	0.3–6.3%
Hazelnuts	0.2–0.9%
Lentils	0.3–1.5%
Maize, corn	0.7–2.2%
Peanuts	0.2–4.5%
Peas	0.2–1.2%
Rice	0.1–1.1%
Rice bran	2.6–8.7%
Sesame seeds	1.4–5.4%
Soybeans	1.0–2.2%
Tofu	0.1–2.9%
Walnuts	0.2–6.7%
Wheat	0.4–1.4%
Wheat bran	2.1–7.3%
Wheat germ	1.1–3.9%

Source: Ishwarya Shankaran P, Kumari P. Nutritional Analysis of Plant-Based Meat: Current Advances and Future Potential. Applied Sciences. 2024; 14(10):4154. https://doi.org/10.3390/app14104154.

Furthermore, phytate can affect how well lipids and carbohydrates are absorbed. Phytate–carbohydrate complexes reduce the solubility, digestibility, and absorption of glucose by inhibiting amylase activity and decreasing the breakdown of carbohydrates. Lower lipid availability and metabolism are the result of metallic soaps produced in the gut lumen by lipid–phytate complex formation. Therefore, to increase the nutritional content of plant-based meat, phytates must be removed. To lower the amount of phytates in food, there are several methods for processing and preparation. Research and development (Table 3.5) has been concentrating on creating techniques and solutions like extrusion, soaking, germination, cooking, fermentation, and the addition of isolated phytase during the past 10 years. To some extent, all techniques aid in the breakdown of phytates in food. On the other hand, enzymatic techniques—like adding phytase and fermenting—have shown to be more beneficial than physical extraction techniques. The bioavailability of added and native minerals can be increased and total dephytinization can be achieved by using added or native phytase during food processing or consumption. Therefore, phytase may improve the benefits of fortification and the absorption of natural minerals and amino acids found in plant-based meat.

Table 3.5. *Methods Used to Reduce Phytic Acid in Food Material*

Method	Variables Analyzed	Food Source	Phytic Acid Reduction (%)	Conditions	Study
Phytase addition	Phytase activity	Dark rye bread	99% IP6 degraded	Sourdough fermented 10–12 h	Engelsen et al., 2007
Phytase addition	Phytic acid content	Whole wheat flour	35.3–69.2	N/D	Akhter et al., 2012
Phytase addition	Phytic acid reduction	Infant cereals	68.6–93.4	Cereal mix roasted in oven at 120°C for 30 min. 60 g roasted flour suspended in 124 mL deionized water. Phytase added to the suspension 20 min incubation at 55°C and pH 5.5 under gentle agitation.	Sanz-Penella at al., 2012

Phytase addition	Phytase activity	Bread with bran	90%	N/D	Watson W et al., 2014
Phytase addition	Phytic acid content	70:30 blend of pea and oat protein	32%	15% dry matter; pH = 6.7. The suspension was heated to 40°C. 1.5% of phytase (Phyzyme® XP) added to the suspension 4-h incubation at 40°C under minimum agitator speed.	Kaleda et al., 2020
Phytase addition	Phytase activity	Sorghum flour	65% IP6 degraded	Grains (four varieties) washed and dried in an oven at 50°C for 4 h. Grains milled in a knife mill and sieved in a 20-mesh sieve. Flour stored at −18°C	Rebellato et al., 2020
Soaking	Phytase activity	Grains and seeds	10–60	100 g whole grains or seeds soaked in 500 mL water in a covered beaker for 16 h in the dark at 25°C	Egli et al., 2002
Soaking	Heat pretreatment	Brown rice	14–28	Soaking in demineralized water for 24 h at 25°C	Liang, Han, Nout, and Hamer, 2008
Autoclave	Autoclaving time, pH	Wheat and rice bran	Wheat: max 95 Rice: max 95.6	Autoclave at 121°C for 0.5, 1.0, or 1.5 h, pH levels of 3.5–6.5	Özkaya, Turksoy, Özkaya, and Duman, 2017
Germination	Time of malting	Millet	72 h: 23.9 96 h: 45.3	Germination for 72 or 96 h	Coulibaly et al., 2011, Makokha et al., 2002

Germination	Time (steeping and sprouting)	Brown rice	Steeping: <20	Steeping: 24 h	Liang et al., 2008
			Sprouting: 60	Sprouting: 72 h	
			Red kidney beans: 35.90		
Extrusion	Type of beans	Fava and red kidney beans	Fava beans: 26.7	156°C, 100 rpm, 385 g/min	Alonso et al., 2000
			Red kidney beans: 21.4		
Extrusion	Moisture-	Bran of wheat, rice, barley, oat	Wheat: 64.4	115°C and moisture content of 20%	Kaur et al., 2015
	Temperature		Barley: 63.5		
			Oat: 26.4		
Extrusion	Moisture-Temperature	Whole rice grain	25	160°C and moisture content of 16.5%	Albarracín et al., 2015
Fermentation	Type of fermented microorganism	Whole wheat dough	70	Sourdough fermentation or lactic acid (pH 5.5)	Leenhardt, Levrat-Verny, Chanliaud, and Rémésy, 2005
Fermentation	Use starter enrichment (0/1/2/3 cycles)	Brown rice	56–96	Brown rice and demineralized water were fermented naturally (24 h at 30°C)	Liang et al., 2008

Fermen-tation	Type of flour	Bran-enriched bread	Whole-wheat bread: 44	*Bifidobacterium* strains	Sanz-Penella, Tamayo-Ramos, Sanz, and Haros, 2009
		Whole–wheat flour	Bran-enriched bread: 67		

Source: Ishwarya Shankaran P, Kumari P. Nutritional Analysis of Plant-Based Meat: Current Advances and Future Potential. Applied Sciences. 2024; 14(10):4154. https://doi.org/10.3390/app14104154.

3.4. CURRENT AND POTENTIAL SOLUTIONS TO IMPROVE PLANT-BASED MEAT NUTRITION

Our analyses indicate that the most effective approach would be to reduce the phytate content and enhance the quality of the protein in plant-based meat in a sustainable manner.

3.4.1. Balanced Amino Acid Composition for Better Meat Protein Quality

Finding a clever way to combine two or more protein sources has frequently been the method used to address the deficiency of one or more essential amino acids in plant proteins and produce plant-based meat with a balanced amino acid composition. For example, adding protein from rice (a cereal low in lysine) and chickpeas (a legume low in cysteine, methionine, and tyrosine) to a protein blend can help raise its PDCAAS (protein digestibility corrected amino acid score). As an alternative, producers of plant-based meat may add the missing essential amino acids to their products.

The preferred technique for determining the amount of protein in a person's diet is the protein digestibility-corrected amino acid score (PDCAAS), according to the FAO/WHO. The technique consists of comparing the concentration of the amino acid in a reference (scoring) pattern with the concentration of the test protein's first limiting essential amino acid. The preschool-age child's essential amino acid requirements are used to determine this scoring pattern. The test protein's true fecal digestibility is taken into account when calculating the final chemical score (Shankaran and Kumari, 2024).

3.4.2. Phytase Treatment to Reduce Phytate Content in Plant-Based Meat

The enzyme that breaks down phytate is called phytose, and it is naturally present in microbes, plants, and animals. Myoinositol (1,2,3,4,5,6)-hexakisphosphate phosphohydrolase is the chemical name for it. To create free inorganic P and break the phytate complex, which releases minerals and amino acids, phytolase sequesters orthophosphate groups from the inositol ring of phytic acid (Figure 3.3). Phytase derived from microorganisms is the most stable type of phytase used in food processing, with extensive industrial research having been done on it under a variety of pH and temperature conditions. Phytase has been used to improve the quality and digestibility of amino acids as well as the bioavailability of iron, zinc, and calcium in foods based on cereals and legumes (Shankaran and Kumari, 2024).

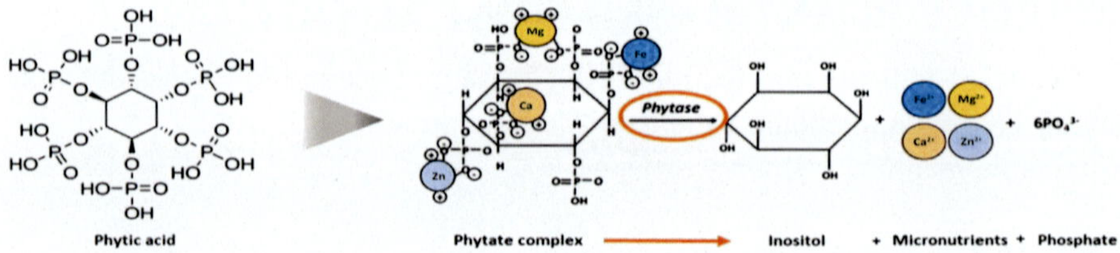

Figure 3.3. *Phytic acid, phytate complex, and phytase enzymatic reaction representation.*

Source: Ishwarya Shankaran P, Kumari P. Nutritional Analysis of Plant-Based Meat: Current Advances and Future Potential. Applied Sciences. 2024; 14(10):4154. https://doi.org/10.3390/app14104154.

Numerous studies have evaluated the impact of phytase on the absorption of micronutrients. Eight studies ranging from 1.0 to 11.6 times greater absorption of iron are summarized in Table 3.6. These studies demonstrated statistically significant improvements in absorption. Three studies specifically looked at how phytase, which breaks down phytate during food processing, affects iron bioavailability in humans among the iron absorption studies that were included in the review. Phytase served as an active enzyme component in five investigations. One processing study employed Aspergillus niger as the type of microorganism. While endogenous phytase was used in the final study, phytase from Aspergillus niger was utilized in four of the studies that added the active enzyme to meals. Six studies were examined with regard to zinc absorption, and each of them revealed statistically significant increases in zinc absorption (between 1.4 and 2.0 times higher absorption). Four employed Aspergillus niger-derived phytase or cereals to aid in food processing, and two used the phytase as an ingredient in fortified porridges.

Table 3.6. *Summary of Clinical Trials on Phytase Effects with Statistically Significant Improvements in Iron and Zinc Absorption*

Study	Study Subjects		Sample Size	Test Meal	Phytase	Outcome
	Adults/Women/Children/Infants	Details				
Iron Absorption Studies						
Layrisse et al., 2000	Adults	15 to 50 years	14	Bread with white wheat flour	*A. niger*; ~300 FTU/100 g flour	$p<0.05$; Fe absorption increased 1.7- to 2-fold
Davidsson et al., 2001	Women	20 to 28 years	10	Pea formula	Not disclosed	$p<0.0001$; Fe absorption increased 1.6-fold
Hurrell et al., 2003	Adults	21 to 38 years	6 to 11 per study	Cereal porridges (wheat, rice, oat, maize, sorghum)	*Aspergillus niger*	$p<0.0001$ to <0.05; Fe absorption increased 1.3- to 11.6-fold
Zhang et al., 2007	Adults	19 to 35 years	20	Oat drink	Exogenous phytase	$p = 0.003$; Fe absorption increased 1.8-fold
Troesch et al., 2009	Women	NA	16 to 18	Maize porridge	Genetically modified culture of *Aspergillus niger*; ~190 FTU/serving	$p<0.05$; Fe absorption increased 1.2- to 1.8-fold
Cercamondi et al., 2013	Children	19 to 36 months	16 to 18	Millet porridge	Aspergillus niger; 400 FTU/serving	$p<0.001$ to <0.05; Fe absorption increased 1.6- to 2.8-fold

Koréissi-Dembélé et al., 2013	Women	18 to 30 years	16	Fonio wheat flour porridge	Endogenous wheat phytase; dose not disclosed	$p<0.0001$; Fe absorption increased 3.2-fold
Monnard et al., 2017	Women	18 to 45 years	14	Whole-grain maize flour-based meal	*A. niger*; 190 FTU/portion	$p<0.001$; Fe absorption increased 1.9-fold
Zinc Absorption						
Davidsson et al., 2004	Infants	61 to 191 days	9	Soy formula	*Aspergillus niger*	$p = 0.03$; Zn absorption increased 1.4-fold
Egli et al., 2004	Adults	Not disclosed	9	Wheat and soy porridge enriched with	Endogenous wheat phytase	$p = 0.005$; Zn absorption increased 1.5 fold
Kim et al., 2007	Women	22 to 75 years	7 to 10	High- vs. low-phytate diet of common Korean food	*Aspergillus niger*; 10 FTU/100 g brown rice; 20 FTU/100 g soybean curd residue; 3 months	$p<0.001$; Zn absorption increased 1.7 to 1.9-fold
Brnic et al., 2010	Adults	18 to 45 years	9	Maize porridge	Genetically modified culture of Aspergillus niger; 190 FTU/portion	$p<0.001$; Zn absorption increased 1.9-fold

Brnic et al., 2017	Children	12 to 24 months	35	Millet-based porridge	Genetically modified culture of *Aspergillus niger*; 20.5. FTU/portion	$p<0.0001$; Fe absorption increased 1.7-fold
Zyba et al., 2019	Children	18 to 23 months	26	Millet-based porridge	*A. niger*—exogenous phytase; 588 FTUs/portion	$p<0.001$; Zn absorption increased 1.9-fold

Note: The activity of the phytase is measured as the amount of enzyme that liberates 1 mmol of inorganic phosphorous per minute and is expressed in phytase units (FTUs).

Source: Ishwarya Shankaran P, Kumari P. Nutritional Analysis of Plant-Based Meat: Current Advances and Future Potential. Applied Sciences. 2024; 14(10):4154. https://doi.org/10.3390/app14104154.

A denatured protein's bioavailability and digestibility may be improved by phytotase. According to studies, phytase can raise the amount of soluble protein in foods like bread and defatted soymilk by 41% and 28%, respectively. A food product's protein digestibility rises as its soluble/solubility increases. The effects of lactic acid bacteria fermentation and enzymatic phytase treatment on the digestibility, protein quality, and reduction of phytic acid in fava bean flour were investigated in a study that used simulated *in vitro* digestion. Protein solubility was enhanced at lower pH levels when phytase treatment dramatically reduced phytic acid levels by up to 89%. This increase in solubility improved the proteins from fava beans' digestibility and released more free amino nitrogen during the first stage of digestion.

Phytase enhanced iron absorption by one-to when added to plant-derived protein products like soy-protein isolate, soy formula, pea formula, and oat drink. According to a recent study, adding phytase to a pea-protein blend reduced phytate by 32%, enhancing the blend's nutritional qualities. Since the addition of protease did not increase the release of volatile compounds or their precursors, its effects on the extrusion process and the physiochemical, textural, and sensorial properties of extrudates were statistically insignificant. Thus, adding phytase to plant protein improves its nutritional qualities without compromising its intended flavor (Shankaran and Kumari, 2024).

REFERENCES

1. Akhter, S., Saeed, A., Irfan, M., & Malik, K. A. (2012). *In vitro* dephytinization and bioavailability of essential minerals in several wheat varieties. *Journal of Cereal Science, 56*, 741–746.
2. Albarracín, M., González, R., & Drago, S. (2015). Soaking and extrusion effects on physicochemical parameters, phytic acid, nutrient content and mineral bio-accessibility of whole rice grain. *International Journal of Food Science and Nutrition, 66*, 210–215.
3. Alonso, R., Aguirre, A., & Marzo, F. (2000). Effect of extrusion and traditional processing methods on antinutrients and *in vitro* digestibility of protein and starch in faba and kidney bean. *Food Chemistry, 68*, 159–165.
4. Asgar, M. A., Fazilah, A., Huda, N., Bhat, R., & Karim, A. A. (2010). Nonmeat protein alternatives as meat extenders and meat analogs. *Institute of Food Technologists, 9*, 513–529.
5. Aune, D., Keum, N., Giovannucci, E., Fadnes, L. T., Boffetta, P., Greenwood, D. C., Tonstad, S., Vatten, L. J., Riboli, E., & Norat, T. (2016). Whole grain consumption and risk of cardiovascular disease, cancer, and all cause and cause specific mortality: Systematic review and dose-response meta-analysis of prospective studies. *BMJ, 353*, i2716.
6. Berghout, J. A. M., Venema, P., Boom, R. M., & Van der Goot, A. J. (2015). Comparing functional properties of concentrated protein isolates with freeze-dried protein isolates from lupin seeds. *Food Hydrocolloids, 51*, 346–354.
7. Bohn, L., Meyer, A. S., & Rasmussen, S. K. (2008). Phytate: Impact on environment and human nutrition. A challenge for molecular breeding. *Journal of Zhejiang University Science B, 9*, 165–191.
8. Bohrer, B. M. (2019). An investigation of the formulation and nutritional composition of modern meat analogue products. *Food Science and Human Wellness, 8*, 320–329.
9. Bottin, J. H., Swann, J. R., Cropp, E., Chambers, E. S., Ford, H. E., Ghatei, M. A., & Frost, G. S. (2016). Mycoprotein reduces energy intake and postprandial insulin release without altering glucagon-like peptide-1 and peptide tyrosine-tyrosine concentrations in healthy overweight and obese adults: A randomised-controlled trial. *British Journal of Nutrition, 116*, 360–374.
10. Brnić, M., Hurrell, R. F., Songré-Ouattara, L. T., Diawara, B., Kalmogho-Zan, A., Tapsoba, C., Zeder, C., & Wegmüller, R. (2017). Effect of phytase on zinc absorption from a millet-based porridge fed to young Burkinabe

children. *European Journal of Clinical Nutrition, 71*, 137–141.

11. Bryant, C. J. (2019). We cannot keep meating like this: Attitudes towards vegetarian and vegan diets in the United Kingdom. *Sustainability, 11*(23).
12. Cercamondi, C. I., Egli, I. M., Mitchikpe, E., Tossou, F., Hessou, J., Zeder, C., Hounhouigan, J. D., & Hurrell, R. F. (2013). Iron bioavailability from a lipid-based complementary food fortificant mixed with millet porridge can be optimized by adding phytase and ascorbic acid but not by using a mixture of ferrous sulfate and sodium iron EDTA. *Journal of Nutrition, 143*, 1233–1239.
13. Chang, C., Tu, S., Ghosh, S., & Nickerson, M. T. (2015). Effect of pH on the interrelationships between the physicochemical, interfacial and emulsifying properties for pea, soy, lentil and canola protein isolates. *Food Research International, 77*, 360–367. 10.1016/J.FOODRES.2015.08.012.
14. Chiang, J. H., Tay, W., Ong, D. S. M., Liebl, D., Ng, C. P., & Henry, C. J. (2021). Physicochemical, textural and structural characteristics of wheat gluten-soy protein composited meat analogues prepared with the mechanical elongation method. *Food Structure, 28*. 10.1016/J.FOOSTR.2021.100183.
15. Coffey, A. A., Lillywhite, R., & Oyebode, O. (2023). Meat versus meat alternatives: Which is better for the environment and health? A nutritional and environmental analysis of animal-based products compared with their plant-based alternatives. *Journal of Human Nutrition and Dietetics, 36*, 2147–2156.
16. Costa-Catala, J., Toro-Funes, N., Comas-Basté, O., Hernández-Macias, S., Sánchez-Pérez, S., Latorre-Moratalla, M. L., Veciana-Nogués, M. T., Castell-Garralda, V., & Vidal-Carou, M. C. (2023). Comparative assessment of the nutritional profile of meat products and their plant-based analogues. *Nutrients, 15*, 2807.
17. Coulibaly, A., Kouakou, B., & Chen, J. (2011). Phytic acid in cereal grains: Structure, healthy or harmful ways to reduce phytic acid in cereal grains and their effects on nutritional quality. *American Journal of Plant Nutrition and Fertilization Technology, 1*, 1–22.
18. Crimarco, A., Springfield, S., Petlura, C., Streaty, T., Cunanan, K., Lee, J., Fielding-Singh, P., Carter, M. M., Topf, M. A., Wastyk, H. C., et al. (2020). A randomized crossover trial on the effect of plant-based compared with animal-based meat on trimethylamine-N-oxide and cardiovascular disease risk factors in generally healthy adults: Study with appetizing plant food—Meat eating alternative trial (swap-meat). *American Journal of Clinical Nutrition, 112*, 1188–1199.

19. Cutroneo, S., Angelino, D., Tedeschi, T., Pellegrini, N., & Martini, D. (2022). Nutritional quality of meat analogues: Results from the food labelling of Italian products (FLIP) project. *Frontiers in Nutrition, 9*, 852831.
20. Davidsson, L., Dimitriou, T., Walczyk, T., & Hurrell, R. F. (2001). Iron absorption from experimental infant formulas based on pea (*Pisum sativum*)-protein isolate: The effect of phytic acid and ascorbic acid. *British Journal of Nutrition, 85*, 59–63.
21. Davidsson, L., Ziegler, E. E., Kastenmayer, P., van Dael, P., & Barclay, D. (2004). Dephytinisation of soybean protein isolate with low native phytic acid content has limited impact on mineral and trace element absorption in healthy infants. *British Journal of Nutrition, 91*, 287–294.
22. Day, L., Cakebread, J. A., & Loveday, S. M. (2022). Food proteins from animals and plants: Differences in the nutritional and functional properties. *Trends in Food Science & Technology, 119*, 428–442.
23. de Boer, J., & Aiking, H. (2019). Strategies towards healthy and sustainable protein consumption: A transition framework at the levels of diets, dishes, and dish ingredients. *Food Quality and Preference, 73*, 171–181.
24. De Marchi, M., Costa, A., Pozza, M., Goi, A., & Manuelian, C. L. (2021). Detailed characterization of plant-based burgers. *Scientific Reports, 11*, 2049.
25. Dekkers, B. L., Boom, R. M., & van der Goot, A. J. (2018). Structuring processes for meat analogues. *Trends in Food Science & Technology, 81*, 25–36.
26. Dekkers, B. L., Nikiforidis, C. V., & van der Goot, A. J. (2016). Shear-induced fibrous structure formation from a pectin/SPI blend. *Innovative Food Science and Emerging Technologies, 36*, 193–200. https://doi.org/10.1016/j.ifset.2016.07.003.
27. Dersjant-Li, Y., Awati, A., Schulze, H., & Partridge, G. (2014). Phytase in non-ruminant animal nutrition: A critical review on phytase activities in the gastrointestinal tract and influencing factors. *Journal of the Science of Food and Agriculture, 95*, 878–896. https://doi.org/10.1002/jsfa.6608.
28. Duodu, K. G., Taylor, J. R., Belton, P. S., & Hamaker, B. R. (2003). Factors affecting sorghum protein digestibility. *Journal of Cereal Science, 38*, 117–131.
29. Egli, I., Davidsson, L., Juillerat, M. A., Barclay, D., & Hurrell, R. F. (2006). The influence of soaking and germination on the phytase activity and phytic acid content of grains and seeds potentially useful for complementary feeding. *Journal of Food Science, 67*, 3484–3488.

30. Egli, I., Davidsson, L., Zeder, C., Walczyk, T., & Hurrell, R. F. (2004). Dephytinization of a complementary food based on wheat and soy increases zinc, but not copper, apparent absorption in adults. *Journal of Nutrition, 134*, 1077–1080.

31. English, L. K., Ard, J. D., Bailey, R. L., Bates, M., Bazzano, L. A., Boushey, C. J., Brown, C., Butera, G., Callahan, E. H., de Jesus, J., et al. (2021). Evaluation of dietary patterns and all-cause mortality. *JAMA Network Open, 4*, e2122277.

32. Farsi, D. N., Gallegos, J. L., Koutsidis, G., Nelson, A., Finnigan, T. J., Cheung, W., Muñoz-Muñoz, J. L., Commane, D. M. (2023). Substituting meat for mycoprotein reduces genotoxicity and increases the abundance of beneficial microbes in the gut: Mycomeat, a randomized crossover control trial. *European Journal of Nutrition, 62*, 1479–1492.

33. Feizollahi, E., Mirmahdi, R. S., Zoghi, A., Zijlstra, R. T., Roopesh, M. S., & Vasanthan, T. (2021). Review of the beneficial and anti-nutritional qualities of phytic acid, and procedures for removing it from food products. *Food Research International, 143*, 110284.

34. Fischer, M., Egli, I., Aeberli, I., Zimmermann, M., & Hurrell, R. (2015). Strategies to enhance iron absorption from tef-injera in young women. *European Journal of Nutrition and Food Safety, 5*, 1169–1170.

35. Geerts, M. E., Dekkers, B. L., van der Padt, A., & van der Goot, A. J. (2018). Aqueous fractionation processes of soy protein for fibrous structure formation. *Innovative Food Science and Emerging Technologies, 45*, 313–319.

36. Gemede, H. F., & Ratta, N. (2014). Antinutritional factors in plant foods: Potential health benefits and adverse effects. *International Journal of Nutrition and Food Science, 3*, 284–289.

37. Gibbs, J., & Leung, G. K. (2023). The effect of plant-based and mycoprotein-based meat substitute consumption on cardiometabolic risk factors: A systematic review and meta-analysis of controlled intervention trials. *Dietetics, 2*, 104–122.

38. Gorissen, S. H., Crombag, J. J., Senden, J. M., Waterval, W. A., Bierau, J., Verdijk, L. B., & van Loon, L. J. (2018). Protein content and amino acid composition of commercially available plant-based protein isolates. *Amino Acids, 50*, 1685–1695.

39. Grabowska, K. J., Zhu, S., Dekkers, B. L., De Ruijter, N. C. A., Gieteling, J., & Van Der Goot, A. J. (2016). Shear-induced structuring as a tool to make anisotropic materials using soy protein concentrate. *Journal of Food Engineering, 188*, 77–86. https://doi.org/10.1016/j.jfoodeng.2016.05.010.

40. Graça, J., Godinho, C. A., & Truninger, M. (2019). Reducing meat consumption and following plant-based diets: Current evidence and future directions to inform integrated transitions. *Trends in Food Science & Technology, 91*, 380–390.

41. Greiner, R., & Konietzny, U. (2006). Phytase for food application. *Food Technology and Biotechnology, 44*, 125–140.

42. Gupta, R. K., Gangoliya, S. S., & Singh, N. K. (2015). Reduction of phytic acid and enhancement of bioavailable micronutrients in food grains. *Journal of Food Science and Technology, 52*, 676–684.

43. Haque, M. A., Timilsena, Y. P., & Adhikari, B. (2016). Food proteins, structure, and function. In *Reference Module in Food Science* (pp. 1–7). Elsevier. https://doi.org/10.1016/B978-0-08-100596-5.03057-2.

44. Heusala, H., Sinkko, T., Sözer, N., Hytönen, E., Mogensen, L., & Knudsen, M. T. (2020). Carbon footprint and land use of oat and faba bean protein concentrates using a life cycle assessment approach. *Journal of Cleaner Production, 242*, 118376. https://doi.org/10.1016/j.jclepro.2019.118376.

45. Hurrell, R. F., Juillerat, M. A., Reddy, M. B., Lynch, S. R., Dassenko, S. A., & Cook, J. D. (1992). Soy protein, phytate, and iron absorption in humans. *American Journal of Clinical Nutrition, 56*, 573–578.

46. Hurrell, R. F., Reddy, M. B., Juillerat, M. A., & Cook, J. D. (2003). Degradation of phytic acid in cereal porridges improves iron absorption by human subjects. *American Journal of Clinical Nutrition, 77*, 1213–1219.

47. Ishwarya Shankaran, P., & Kumari, P. (2024). Nutritional analysis of plant-based meat: Current advances and future potential. *Applied Sciences, 14*(10), 4154. https://doi.org/10.3390/app14104154.

48. Joye, I. (2019). Protein digestibility of cereal products. *Foods, 8*, 199.

49. Kaleda, A., Talvistu, K., Tamm, M., Viirma, M., Rosend, J., Tanilas, K., Kriisa, M., Part, N., & Tammik, M. L. (2020). Impact of fermentation and phytase treatment of pea-oat protein blend on physicochemical, sensory, and nutritional properties of extruded meat analogs. *Foods, 9*, 1059.

50. Kalhoff, H., & Kersting, M. (2017). Programming long-term health: Nutrition and diet in infants aged 6 months to 1 year. In J. M. Saavedra & A. M. Dattilo (Eds.), *Early Nutrition and Long-Term Health: Mechanisms, Consequences, and Opportunities* (pp. 499–535). Woodhead Publishing.

51. Kaur, S., Sharma, S., Singh, B., & Dar, B. N. (2015). Effect of extrusion variables (temperature, moisture) on the antinutrient components of cereal brans. *Journal of Food Science and Technology, 52*, 1670–1676.

52. Kim, J., Paik, H. Y., Joung, H., Woodhouse, L. R., Li, S., & King, J. C.

(2007). Effect of dietary phytate on zinc homeostasis in young and elderly Korean women. *Journal of the American College of Nutrition, 26*, 1–9.

53. Kore'issi-Dembe'le,' Y., Fanou-Fogny, N., Moretti, D., Schuth, S., Dossa, R. A. M., Egli, I., Zimmermann, M. B., & Brouwer, I. D. (2013). Dephytinisation with intrinsic wheat phytase and iron fortification significantly increase iron absorption from fonio (*Digitaria exilis*) meals in West African women. *PLoS ONE, 8*, e70613.
54. Kumar, V., Sinha, A. K., Makkar, H. P. S., & Becker, K. (2010). Dietary roles of phytate and phytase in human nutrition: A review. *Food Chemistry, 120*, 945–959.
55. Ladjal-Ettoumi, Y., Boudries, H., Chibane, M., & Romero, A. (2016). Pea, chickpea and lentil protein isolates: Physicochemical characterization and emulsifying properties. *Food Biophysics, 11*, 43–51.
56. Layrisse, M., Gracia-Casal, M. N., Solano, L., Baron, M. A., Arguello, F., Li, O., Ramirez, J., Leets, I., & Tropper, E. (2000). Iron bioavailability in humans from breakfasts enriched with iron bis-glycine chelate, phytates and polyphenols. *Journal of Nutrition, 130*, 2195–2199.
57. Leenhardt, F., Levrat-Verny, M. A., Chanliaud, E., & Rémésy, C. (2005). Moderate decrease of pH by sourdough fermentation is sufficient to reduce phytate content of whole wheat flour through endogenous phytase activity. *Journal of Agricultural and Food Chemistry, 53*, 98–102.
58. Liang, J., Han, B. Z., Nout, M. J., & Hamer, R. J. (2008). Effects of soaking, germination and fermentation on phytic acid, total and *in vitro* soluble zinc in brown rice. *Food Chemistry, 110*, 821–828.
59. Longvah, T., Ananthan, R., Bhaskarachary, K., & Venkaiah, K. (2017). *Indian Food Composition Table* (p. 578). National Institute of Nutrition.
60. Lu, F., Kuhnle, G. K., & Cheng, Q. (2018). The effect of common spices and meat type on the formation of heterocyclic amines and polycyclic aromatic hydrocarbons in deep-fried meatballs. *Food Control, 92*, 399–411. https://doi.org/10.1016/j.foodcont.2018.05.018.
61. Mahgoub, S. E. O., & Elhag, S. A. (1998). Effect of milling, soaking, malting, heat-treatment and fermentation on phytate level of four Sudanese sorghum cultivars. *Food Chemistry, 61*, 77–80.
62. Makokha, A., Oniang'o, R., Njoroge, S., & Kamar, O. (2002). Effect of traditional fermentation and malting on phytic acid and mineral availability from sorghum (*Sorghum bicolor*) and finger millet (*Eleusine coracana*) grain varieties grown in Kenya. *Food & Nutrition Bulletin, 23*, 241–255.
63. Marica, B., Wegmüller, R., Zeder, C., Gabriela, S., & Hurrell, R. F.

(2010). Influence of phytase, EDTA, and polyphenols on zinc absorption in adults from porridges fortified with zinc sulfate or zinc oxide. *Journal of Nutrition, 144*, 1467–1473.

64. Monnard, A., Moretti, D., Zeder, C., Steingötter, A., & Zimmermann, M. B. (2017). The effect of lipids, a lipid-rich ready-to-use therapeutic food, or a phytase on iron absorption from maize-based meals fortified with micronutrient powders. *American Journal of Clinical Nutrition, 105*, 1521–1527.
65. Moretti, D., Biebinger, R., Bruins, M. J., Hoeft, B., & Kraemer, K. (2013). Bioavailability of iron, zinc, folic acid, and vitamin A from fortified maize. *Annals of the New York Academy of Sciences, 1312*, 54–65.
66. Nielsen, M. M., Damstrup, M. L., Thomsen, A. D., Rasmussen, S. K., & Hansen, Å. (2007). Phytase activity and degradation of phytic acid during rye bread making. *European Food Research and Technology, 225*, 173–181.
67. Özkaya, B., Turksoy, S., Özkaya, H., & Duman, B. (2017). Dephytinization of wheat and rice brans by hydrothermal autoclaving process and the evaluation of consequences for dietary fiber content, antioxidant activity and phenolics. *Innovative Food Science and Emerging Technologies, 39*, 209–215.
68. Pinckaers, P. J., Trommelen, J., Snijders, T., & van Loon, L. J. (2021). The anabolic response to plant-based protein ingestion. *Sports Medicine, 51*, 59–74.
69. Pramitha, J. L., Rana, S., Aggarwal, P. R., Ravikesavan, R., Joel, A. J., & Muthamilarasan, M. (2021). Diverse role of phytic acid in plants and approaches to develop low-phytate grains to enhance bioavailability of micronutrients. *Advances in Genetics, 107*, 89–120.
70. Rebellato, A. P., Orlando, E. A., Theodoropoulos, V. C. T., Greiner, R., & Pallone, J. A. L. (2020). Effect of phytase treatment of sorghum flour, an alternative for gluten-free foods and bioaccessibility of essential minerals. *Journal of Food Science and Technology, 57*, 3474–3481.
71. Rimbach, G., Pallauf, J., Moehring, J., Kraemer, K., & Minihane, A. M. (2008). Effect of dietary phytate and microbial phytase on mineral and trace element bioavailability: A literature review. *Current Topics in Nutraceutical Research, 6*, 131–144.
72. Rosa-Sibakov, N., Re, M., Karsma, A., Laitila, A., & Nordlund, E. (2018). Phytic acid reduction by bioprocessing as a tool to improve the *in vitro* digestibility of faba bean protein. *Journal of Agricultural and Food Chemistry, 66*, 10394–10399.

73. Saito, T., Kohno, M., Tsumura, K., Kugimiya, W., & Kito, M. (2001). Novel method using phytase for separating soybean β-conglycinin and glycinin. *Bioscience, Biotechnology, and Biochemistry, 65*, 884–887.
74. Sánchez-Velázquez, O. A., Ribéreau, S., Mondor, M., Cuevas-Rodríguez, E. O., Arcand, Y., & Hernández-Álvarez, A. J. (2021). Impact of processing on the *in vitro* protein quality, bioactive compounds, and antioxidant potential of 10 selected pulses. *Legume Science, 3*, e88.
75. Sandberg, A. S., & Andersson, H. (1988). Effect of dietary phytase on the digestion of phytate in the stomach and small intestine of humans. *Journal of Nutrition, 118*, 469–473.
76. Sanz-Penella, J. M., Frontela, C., Ros, G., Martinez, C., Monedero, V., & Haros, M. (2012). Application of bifidobacterial phytases in infant cereals: Effect on phytate contents and mineral dialyzability. *Journal of Agricultural and Food Chemistry, 60*, 11787–11792.
77. Sanz-Penella, J. M., Tamayo-Ramos, J. A., Sanz, Y., & Haros, M. (2009). Phytate reduction in bran-enriched bread by phytase-producing bifidobacteria. *Journal of Agricultural and Food Chemistry, 57*, 10239–10244.
78. Schlemmer, U., Frølich, W., Prieto, R. M., & Grases, F. (2009). Phytate in foods and significance for humans: Food sources, intake, processing, bioavailability, protective role and analysis. *Molecular Nutrition & Food Research, 53*, S330–S375.
79. Shaghaghian, S., McClements, D. J., Khalesi, M., Garcia-Vaquero, M., & Mirzapour-Kouhdasht, A. (2022). Digestibility and bioavailability of plant-based proteins intended for use in meat analogues: A review. *Trends in Food Science & Technology, 129*, 646–656.
80. Singh, B., & Satyanarayana, T. (2008). Phytase production by *Sporotrichum thermophile* in a cost-effective cane molasses medium in submerged fermentation and its application in bread. *Journal of Applied Microbiology, 105*, 1858–1865.
81. Thavarajah, P., Thavarajah, D., & Vandenberg, A. (2009). Low phytic acid lentils (*Lens culinaris* L.): A potential solution for increased micronutrient bioavailability. *Journal of Agricultural and Food Chemistry, 57*, 9044–9049.
82. Toribio-Mateas, M. A., Bester, A., & Klimenko, N. (2021). Impact of plant-based meat alternatives on the gut microbiota of consumers: A real-world study. *Foods, 10*, 2040.
83. Troesch, B., Egli, I., Zeder, C., Hurrell, R. F., de Pee, S., & Zimmermann, M. B. (2009). Optimization of a phytase-containing micronutrient powder

with low amounts of highly bioavailable iron for in-home fortification of complementary foods. *American Journal of Clinical Nutrition, 89*, 539–544.

84. Troesch, B., Jing, H., Laillou, A., & Fowler, A. (2013). Absorption studies show that phytase from *Aspergillus niger* significantly increases iron and zinc bioavailability from phytate-rich foods. *Food & Nutrition Bulletin, 34*, S90–S101.
85. Van Vliet, S., Burd, N. A., & van Loon, L. J. (2015). The skeletal muscle anabolic response to plant-versus animal-based protein consumption. *Journal of Nutrition, 145*, 1981–1991.
86. van Vliet, S., Kronberg, S. L., & Provenza, F. D. (2020). Plant-based meats, human health, and climate change. *Frontiers in Sustainable Food Systems, 4*, 128.
87. Varadan, R., Solomatin, S., Holzschietinger, C., Cohn, E., Klapholz-Brown, A., Shiu, J. W. Y., & Fraser, R. (2015). Ground meat replicas. *WO2015153666A1*.
88. Vashishth, A., Ram, S., & Beniwal, V. (2017). Cereal phytases and their importance in improvement of micronutrients bioavailability. *3 Biotech, 7*, 42.
89. Veronese, N., Solmi, M., Caruso, M. G., Giannelli, G., Osella, A. R., Evangelou, E., Maggi, S., Fontana, L., Stubbs, B., & Tzoulaki, I. (2018). Dietary fiber and health outcomes: An umbrella review of systematic reviews and meta-analyses. *American Journal of Clinical Nutrition, 107*, 436–444.
90. Wang, H., Chen, Y., Hua, Y., Kong, X., & Zhang, C. (2014). Effects of phytase-assisted processing method on physicochemical and functional properties of soy protein isolate. *Journal of Agricultural and Food Chemistry, 62*, 10989–10997.
91. Watson, R. R., Preedy, V. R., & Zibadi, S. (2014). *Wheat and Rice in Disease Prevention and Health: Benefits, Risks and Mechanisms of Whole Grains in Health Promotion*. Academic Press.
92. Wu, Y. F., Baek, H. H., & Cadwallader, K. R. (2000). Development of a meat-like process flavoring from soybean-based enzyme-hydrolyzed vegetable protein (E-HVP). *Journal of Food Science, 65*, 1220–1227.
93. Yang, Y., Churchward-Venne, T. A., Burd, N. A., Breen, L., Tarnopolsky, M. A., & Phillips, S. M. (2012). Myofibrillar protein synthesis following ingestion of soy protein isolate at rest and after resistance exercise in elderly men. *Nutrition & Metabolism, 9*, 57.

94. Zahari, I., Östbring, K., Purhagen, J. K., & Rayner, M. (2022). Plant-based meat analogues from alternative protein: A systematic literature review. *Foods, 11*, 2870.

95. Zargarzadeh, N., Mousavi, S. M., Santos, H. O., Aune, D., Hasani-Ranjbar, S., Larijani, B., & Esmaillzadeh, A. (2023). Legume consumption and risk of all-cause and cause-specific mortality: A systematic review and dose–response meta-analysis of prospective studies. *Advances in Nutrition, 14*, 64–76.

96. Zhang, C., Guan, X., Yu, S., Zhou, J., & Chen, J. (2022). Production of meat alternatives using live cells, cultures and plant proteins. *Current Opinion in Food Science, 43*, 43–52. https://doi.org/10.1016/j.cofs.2021.11.002.

97. Zhang, F., Fan, D., Huang, J., & Zuo, T. (2022). The gut microbiome: Linking dietary fiber to inflammatory diseases. *Medical Microecology, 14*, 100070.

98. Zhang, H., Onning, G., Oste, R., Gramatkovski, E., & Hulthén, L. (2007). Improved iron bioavailability in an oat-based beverage: The combined effect of citric acid addition, dephytinization and iron supplementation. *European Journal of Nutrition, 46*, 95–102.

99. Zhang, L., Langlois, E., Williams, K., Tejera, N., Omieljaniuk, M., Finglas, P., & Traka, M. H. (2024). A comparative analysis of nutritional quality, amino acid profile, and nutritional supplementations in plant-based products and their animal-based counterparts in the UK. *Food Chemistry, 448*, 139059.

100. Zyba, S. J., Wegmüller, R., Woodhouse, L. R., Ceesay, K., Prentice, A. M., Brown, K. H., & Wessells, K. R. (2019). Effect of exogenous phytase added to small-quantity lipid-based nutrient supplements (SQ-LNS) on the fractional and total absorption of zinc from a millet-based porridge consumed with SQ-LNS in young Gambian children: A randomized controlled trial. *American Journal of Clinical Nutrition, 110*, 1465–1475.

Chapter 4

Protein Quality and Digestibility

Contents

Introduction 130

4.1. Role of Plant Protein on the Quality and Structure of Meat Analogs 130

4.2. Effect of Different Plant Proteins on the Quality and Structure of Meat Analog 135

4.3. Digestibility and Bioavailability of Proteins 146

4.4. Different Sources of Plant-Proteins in Meat Analogues 147

4.5. Effect of Main Development Processes on Digestibility and Bioavailability of Meat Analogues 153

References 160

INTRODUCTION

When assessing the nutritional value of meat, digestibility, and protein quality are critical components to consider. Since meat contains all nine essential amino acids in proportions that closely match human dietary needs, it is regarded as a high-quality protein. This is especially true of meat derived from animal sources like fish, poultry, and beef. As a result, it serves as a complete protein source, offering the building blocks required for tissue healing, muscle growth, and general body upkeep. The high quality protein content of meat is largely due to its bioavailability, or the ability of the body to effectively absorb and use the protein after it has been consumed. Another important factor in determining the quality of protein is digestibility or the degree to which the protein is absorbed and broken down by the body. With digestibility values frequently surpassing 90%, meat proteins are known for their high absorption and utilization by the body. Cooking methods are one factor that can affect meat's digestibility; some techniques improve meat's digestibility by dissolving protein structures. On the other hand, overcooking at high temperatures can make meat proteins less soluble by forming compounds that the body finds more difficult to break down.

Meat is a great source of essential nutrients because of its high protein quality and ease of digestion, especially for those looking for an efficient way to meet their protein needs. It is important for human nutrition because it contains important micronutrients like iron, zinc, and B vitamins in addition to protein. Although plant-based proteins can also be beneficial to a healthy diet, meat is a better option for individuals who want to maximize their protein intake because plant-based proteins frequently lack one or more essential amino acids and have a lower digestibility.

4.1. ROLE OF PLANT PROTEIN ON THE QUALITY AND STRUCTURE OF MEAT ANALOGS

Meat consumption was essential to human evolution and is one of the main dietary components of people all over the world. The USA and Australia have the highest rates of meat consumption per capita, and the most popular meats worldwide are chicken, beef, mutton, and hog.

Over the past 20 years, the world's demand for meat has grown by 58% as a result of rising population and swift economic expansion. Iron, vitamins, proteins, and other essential nutrients are supplied to the human diet through the consumption of meat and meat products. Still, it's clear that this source of protein has a big impact on the environment. In this context, it is commonly acknowledged that the production of livestock negatively affects water footprint, water pollution, and water shortages in addition to greenhouse gas emissions (Farchi et al., 2017). Additionally, some people are concerned about the negative effects of eating meat on human health, including the higher risk of zoonoses, chronic illnesses, and health issues related to air pollution (Marinova and Bogueva, 2019). Because there is a great demand on the local and international supply chains for animal meat, many consumers have attempted to replace more non-meat foods in their diets to offset the amount of animal muscle-based foods. The comparatively low carbon conversion efficiency linked to the production of animal meat, the detrimental health effects of meat consumption, the environmental costs associated with animal husbandry, and the effects on animal welfare are the main factors driving the demand for meat substitutes (Sanchez-Sabate and Sabate, 2019). To meet the needs of health-conscious consumers and lessen their impact on climate change, the food industry, food organizations, and food policymakers that are involved in the sustainable production and consumption of foods anticipate a shift toward more carbon-efficient sustainable food value addition. Meat substitutes, also referred to as meat analogs, are food items that are becoming more and more common in the global market due to their inherent value, although they may be considered alternatives to animal meat.

Products that aim to imitate animal products in terms of appearance, flavor, aroma, functionality, and cooking experience are called plant-based (PB) meat analogs. Two well-known companies that provide alternatives to PB meat are Beyond Meat and Gardein. Meat analogs or substitutes can be divided into three categories: cell-based, like *in vitro* or cultured meat, fermentation-based mycoproteins, and PB meat substitutes (wheat protein or related, soy, pea, or seed proteins). It is well known that PB protein, which is intended to alter the chewing texture of meat, offers extra health advantages. Likewise, mycelium protein is thought to be healthful and has a chewing texture similar to meat. Furthermore, low-cost alternative proteins targeted as meat analogs are currently being developed for the improvement of sensory, nutritional, and textural properties due to the high costs associated with processing meat, taking climate and other environmental factors into consideration. To create meat substitutes, whole plant proteins or combinations of plant proteins with other non-animal proteins, like micro-algae, insect, and fungal proteins, may be used. As a result, PB meat alternatives, analogs, or replacements account for a sizeable share of this market's growth and expansion. The standard PB meat analog products have attracted a lot of funding for research as well as interest from consumers and advertising. Research on meat analogs has advanced recently as a result of the increased focus on alternative protein products in global food research. Research on meat alternatives has advanced in recent years as a result of food scientists' growing commitment to alternative protein sources. These meat substitutes are frequently marketed as being more environmentally friendly and sustainable than conventional diets derived from animals. They are said to be higher in dietary fiber and a few other beneficial compounds than meat.

According to UBS of Switzerland, the market for alternatives to PB meat is expected to grow from $46,6 billion in 2018 to $85 billion in 2030 and $30,9 billion by 2026. This new market is probably in a good position to grow thanks to future innovations. Burger King and Impossible jointly launched PB beefless Impossible Whoppers in May 2019. They added vegetarian Impossible Pork and Impossible Sausage to their menu in January 2020. In August 2019, Kentucky Fried Chicken (KFC) started serving boneless chicken wings. In a similar vein, Beyond Meat and Light life processed fewer chicken nuggets in August 2019. Ground beef, vegetarian burger patties, and meat substitutes have all been offered by the largest supermarket chain in the US, Kroger, under the Simple Truth Emerge brand. Several other fast-food chains are currently evaluating non-meat dishes, and it appears that this new market potential will be sustained for the foreseeable future. The goal of producing meat analogs is to improve their physical attributes so that they are more palatable to consumers and have similar nutritional benefits to meat derived from animals.

There are now steaks, pieces, strips, patties, sausages, chicken-like blocks, nuggets, and ground beef-like products available as PB meat analogs. Meat-like flavor is necessary for PB meat alternatives to be successful. To persuade regular consumers of animal-based meat to change their dietary habits and consume less animal-based meat, taste—especially mouth feel—is crucial. There's no doubt that additional variables, like the availability of meal sides, could play a role. This result was highlighted in the consumer study report for the Protein Foods, Environment, Technology, and Society program (PROFETAS). On the other hand, whole-muscle meat imitators can be created technologically through cell culture; however, recreating whole-muscle meat from plant proteins is difficult. Developing PB meat analogs with a delicate texture that closely resembles myofilaments in terms of softness and juiciness is also difficult, but not impossible. Therefore, the majority of PB analog product development research has gone into rebuilding the products. These meatless mimicking analog products are divided into two categories: (i) coarse-particle meatless products, which include emulsified foods like frankfurters and bologna substitutes, and (ii) fine-particle meatless products, which include meatless burgers, patties, sausages, meatballs, and imitation chicken nuggets. Protein is essential to the creation of products that resemble PB meat. To create and replicate the texture, appearance, flavor, taste, juiciness, and mouth feel of PB meat products, a variety of functional ingredients are required. To create fibrous structures that resemble meat, several techniques such as thermo-extrusion, shear, spinning, and cross-linking are necessary. By using these techniques, protein-carbohydrate polymer interactions can be enhanced by unfolding and denaturing native protein structures. In addition, extra vitamins and minerals are added to give the product a nutritional value equivalent to that of animal-based meat, and red pigments are added to give the product a meat-like appearance.

Even with their remarkable early success, PB meat alternatives continue to face challenges. Two additional potential barriers that have not received enough attention are food safety and nutrition. Furthermore, there exist technological obstacles for plant-based meat substitutes that attempt to replicate the taste and consistency of meat derived from animals. The use of widely accessible but non-allergic plant proteins, the addition of a variety of ingredients and additives to produce a meat-like consumer preference, the potential for detrimental chemical changes to heat-sensitive compounds, and potential microbiological contamination, for example, all require extensive investigation. The purpose of this chapter is to address these issues by going over the functional quality and value of the primary protein sources, different types of plant-based protein, the impact of different plant proteins on the structure and quality of meat analogs, the bioavailability of meat analogs, consumer perception, and issues. This chapter will also cover the various kinds of protein sources that aren't animal-based (like beef, pork, mutton, chicken, and duck) as well as the kinds of PB products that are currently on the market. Producing meat-like substitutes from plant proteins is thought to require a deep understanding of processing.

4.1.1. Plant Protein-Based Product Types

4.1.1.1. Texturized Vegetable Protein (TVP)

TVPs are structured materials made from water and sources of PB protein that have been processed to transform them from powdered forms. In the market, there are two varieties of TVP. The fibrous structure of high moisture TVP (HM-TVP), which resembles muscle meat. The most widely used high-moisture texturization techniques that enable the creation of anisotropic fiber structures from single or combinations of plant proteins are extrusion and shear cell technology. Extrusion is the most widely used low-moisture texturization technique. By altering machinery and process parameters, TVP's textures, flavors, and forms can be adjusted to satisfy a range of vegan or hybrid food products. Making meat alternative products like sausages, nuggets, and patties is the main application of TVP. Meat extenders such as LM-TVP are used to improve the meat's hardness, juiciness, color stability, and viscoelasticity. The most widely used TVPs are made from combinations of soy, wheat, or pea proteins. Chewy and juicy flesh meat analogs are created when a porous structured LM-TVP is hydrated. Improving the structural capacity of protein sources is gaining more attention in an effort to replicate meat-like components. Additionally, to create a wide range of products that aim for versatility and superior meat-like qualities, food scientists are experimenting with various plant protein kinds or mixers to create TVP with a range of textures, flavors, and nutritional contents (Figure 4.1).

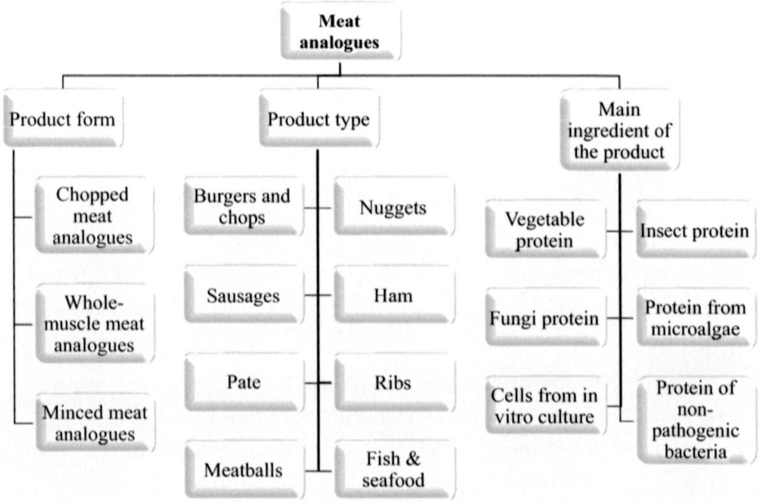

Figure 4.1. *Types of meat analogues available on the market.*

Source: Kołodziejczak K, Onopiuk A, Szpicer A, Poltorak A. Meat Analogues in the Perspective of Recent Scientific Research: A Review. Foods. *2022; 11(1):105. https://doi.org/10.3390/foods11010105.*

In a variety of PB meat products, including steak, sliced meats, sausages, and hamburger patties, TVP is widely used as a meat extender or to simulate flesh. Bakhsh and associates. (2021b) discovered that TVP's qualities were comparable to those of beef and pork hamburgers when it was used as the primary ingredient. However, as this previous study pointed out, one disadvantage of using TVP and texturized soy protein isolate is that the surface of PB patties made from those two components appeared granular. Abdullah Sharima et al. (2018) created a chicken nugget-like dish using TVP and chickpea flour, which was recommended as a potential PB product.

4.1.1.2. Meat Analogs (MA)

A meat analog (MA) is made up of the following components: 50–80% water, 10–25% TVP, 4–20% non-textured proteins, 3–10% flavorings, 0–15% fat, 1–2% binding agents, and 0–0–5% coloring agents. This blend of ingredients results in meat that has extremely pleasing sensory characteristics. Textured proteins can be used to replace muscle meat in two different ways. These two methods involve combining textured proteins with muscle meat, such as meat extension, and completely substituting all meat with textured proteins to prepare meals that are vegan or vegetarian. While cooking alters the sensory qualities of meat extenders—such as appearance, texture, or mouthfeel—it also enhances the functional and sensory qualities of food products when combined with meat. However, the goal of meat analogs is to replicate the overall look, feel, taste, and texture of meat when it is hydrated and cooked without the addition of any meat-based ingredients. Minor ingredients or chemicals may be used to improve the final texture or aid in texturizing the basic components. Flavorings and colorings are added because the texture, flavor, and appearance of a product affect consumer acceptance. The final products' color, texture, and aromas are all influenced by the amount of fat used in PB meat products.

Meat analogs are defined by many researchers as products that are made directly from whole muscle meat, either with or without the addition of other ingredients, and most frequently by extrusion. Meat analogs, according to some researchers, are restructured meat products such as sausages, edible mushroom sausages, emulsion-type sausages, and minced meat substitutes made from an edible mushroom fungus that is used to make burger patties. This suggests that the fungus has the productive potential to produce nutrient-dense meals. The production of meat analogs composed solely of faba bean protein concentrate obtained by a dry-fractionation process was enhanced by the application of response surface methodology, which also yielded promising outcomes.

4.2. EFFECT OF DIFFERENT PLANT PROTEINS ON THE QUALITY AND STRUCTURE OF MEAT ANALOG

For the development of meat alternatives, various plant protein sources, including legumes, oil seeds, cereals, whole grains, and combinations of different plant proteins (oil seeds-whole grains and legume-cereals), could be utilized in place of meat. Researchers are currently considering lowering the use of wheat protein and soy protein because of their propensity to induce celiac disease and allergies, respectively. Food scientists are currently looking into other PB raw materials, such as legumes, to make meat substitutes because of their high nutritional value—which includes vitamins, minerals, dietary fiber, essential amino acids, and low-calorie count—and low-calorie content. Many scientists are working on plant protein-based meat analogs (PB meat analogs) that use extrusion, shear cell technology, and freeze structuring. These include emulsion-type sausages, minced meat substitutes, northern Thai style sausages, and high-moisture plant protein-based meat analogs. Our lab has created a blend of various plant proteins with other non-plant (non-animal) proteins, such as chickpea-mushroom based nuggets, minced meat based on mushrooms and legumes, high sauce soy in protein-based mushrooms meat analogs, and sausages based on chickpeas and mushrooms. Using this as a foundation, upcoming bench research will concentrate on methods to enhance the capabilities of these plant-based proteins for health benefits—like enhanced prebiotic properties and management of glycemic control—that meat cannot offer.

Hybrid food products could be a good way to reduce the amount of meat consumed and its harmful effects on the environment while also supplying meals that are nutrient-dense and well-balanced. They do this by substituting a certain percentage of animal-based meat with plant-based proteins. The strategy of replacing plant-based PB proteins with animal meat protein is still hindered by the unpredictable structure and processing characteristics of isolated plant proteins (Figure 4.2).

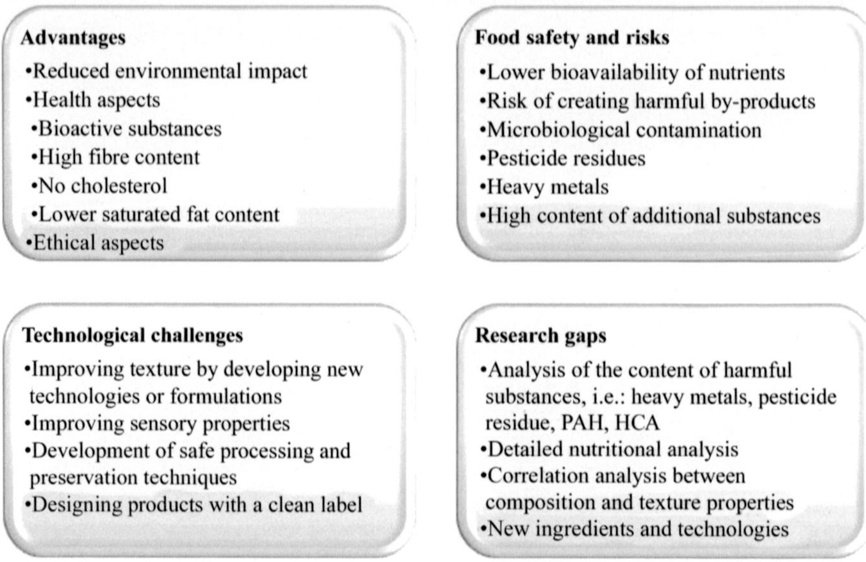

Figure 4.2. *Advantages, risks, technological challenges, and research gaps associated with meat analogues (PAH- polycyclic aromatic hydrocarbons; HCA- heterocyclic aromatic amines).*

Source: Kolodziejczak K, Onopiuk A, Szpicer A, Poltorak A. Meat Analogues in the Perspective of Recent Scientific Research: A Review. Foods. *2022; 11(1):105. https://doi.org/10.3390/foods11010105.*

4.2.1. Effect of Legume Protein on the Quality and Structure of MA

To reduce the amount of meat consumed and its detrimental effects on the environment, hybrid products—which replace some of the meat with plant proteins—might be a good option for producing well-balanced, nutrient-dense meals. The viability of utilizing two different types of pea protein components and high-moisture extrusion of beef mince with a fat content of 7 or 17% to produce mixed animal-plant-based MA. Using a twin-screw extruder with a long cooling die (5 × 20 × 250 mm), animal-plant-based MA was created. The screws were 40:1 in length to diameter, with an 11 mm diameter. Hybrid extrudates were made by milling texturized pea protein concentrate (PPC) or pea protein isolate (PPI) in beef at a 1:1 ratio. The developed extrudates were kept at –20°C and vacuum-packed. Both the properties of the ingredients and the extrusion processing variables (especially temperature) affected variations in the hybrid extrudates' appearance, texture, and sensory attributes. Beef and PPI-containing extrudates were soft and multilayered, whereas beef and PPC-containing extrudates were firmer and less fibrous. Extrudates with beef and PPC had an umami flavor and a meaty smell, while those with beef and PPI tasted and smelled more like peas (Table 4.1).

Table 4.1. *Composition of Meat Analogues in Scientific Research*

Product Type	Composition	Tested Parameters	References
Meat analogue	Soy protein concentrate, wheat gluten, vegetable oil, pumpkin powder, wheat starch, salt	Protein content, moisture, pH, color, textural properties, sensory properties, microscopy, protein solubility	(Chiang et al., 2019; Jin et al., 2021)
Chicken analogue	Wheat gluten, soy protein isolate, water, soybean oil, wheat starch	Protein content, moisture, pH, textural properties analysis, scanning electron microscopy, total amino acids, protein solubility	(Chiang et al., 2021; Joye, 2019).
Meat analogue	Soy protein isolate, wheat gluten, water, salt	pH, maximum swelling, water holding capacity, ionic strength	(Cornet et al., 2021; House et al., 2010)

Product Type	Composition	Tested Parameters	References
Meat analogue	Pea protein isolates, defatted soy protein isolates, oat protein	Protein content, texture profile analysis, water absorption, oil absorption, sensory analysis	(De Angelis et al., 2020; Berrazaga et al., 2019)
Meat analogue	Protein concentrate from faba beans	Specific mechanical energy (SME), differential scanning calorimetry (DSC), color, cooking yield, water binding capacity, oil binding capacity, textural properties, sensory evaluation	(do Carmo et al., 2021; Jung et al., 2022)
Meat analogue	Soy protein isolate, wheat gluten, and natural flavor powder	Volatile compounds profile, scanning electron microscopy, low-field nuclear magnetic resonance (NMR), fourier transform-infrared (FT-IR) spectroscopy	(Guo et al., 2020; Cornet et al., 2022)
Meat analogue	Rapeseed protein concentrate, soy protein concentrate, wheat gluten, water, salt	Macrostructure, color, tensile strength, scanning electron microscopy, confocal laser scanning microscopy, X-ray microtomography	(Jia et al., 2021; Kang et al., 2022)
High moisture extruted (HME) protein	Soy protein isolate, whey protein concentrate	Texture, cryo-imaging, micro-CT, rheological measurements, scanning electron microscopy	(Wittek et al., 2021; Kapravelou et al., 2013)

Source: Kołodziejczak K, Onopiuk A, Szpicer A, Poltorak A. Meat Analogues in the Perspective of Recent Scientific Research: A Review. Foods. *2022; 11(1):105. https://doi.org/10.3390/foods11010105.*

A pea protein-based MA that uses high moisture extrusion to use *Haematococcus pluvialis* residue (HPR). High moisture levels of 50 g/100 g were achieved through the extrusion of mixed raw materials with HPR concentrations ranging from 10 g/100 g to 40 g/100 g. Because of the reddish hue of HPR, the dried red meat color was the analog of the extrudate meat color. By releasing the layers and fibrous structures, HPR may significantly improve the textural characteristics of MA. At an HPR concentration of 1.28±0.05, the meat analogs with the highest fibrous degree (10 g/100 g db) were shown. The fibrous degree of MAs was 1.15 ± 0.03 after increasing the HPR concentration to 40 g/100 g db. This was significantly higher than the extrudates without HPR (1.08 ± 0.05). The addition of HPR increased the moisture content and β-sheet structure of extrudates, which in turn produced MA with enhanced texture. All things considered, HPR may be used as a raw material for high moisture extrudate meat analogs, improving the physical qualities of meat substitutes.

The viability of using a short concentric die and wet extrusion to produce minced meat substitutes (MMS) from faba bean protein concentrate (FBPC) and oat fiber concentrate (OFC). The findings imply that the mechanical qualities of the raw and fried MMS, such as their chewiness and Young's modulus, were either higher or equivalent to those of the raw or fried beef minced meat (BMM). While still being versatile, raw, and fried MMS was found to be significantly stiffer than raw and fried BMM. Fried MMS had greater gumminess, chewiness, and hardness than fried BMM in terms of texture attributes. Mechanically speaking, MMS with a high fiber-to-protein ratio were the closest to BMM after frying. Because of the reconfiguration of covalent crosslinks and myofibrillar structures brought about by the denaturation of meat proteins, the BMM exhibits a higher tolerance to compression (high springiness). However, frying MMS results in a reduction in the strength of polymeric crosslinks, such as hydrogen bonds and van der Waals interactions. Even though frying accelerated MMS browning, MMS color changes were correlated with the kinds and content of OFC. A higher concentration of OFC was linked to MMS which was lighter, redder, and yellower. Fried BMM, on the other hand, was comparatively darker and redder than any MMS.

PPI was studied by Dinani et al. (2021) and is thought to be a sustainable and greener substitute for soy for its possible use in the handling of MA. When making MAs, ingredients are combined in a ratio of 40% (50% PPI and 50% WG) with 60% deionized water. This process shows that the fibrous and mechanical properties of the processed MA were enhanced by l-lysine (CYS) and l-ascorbic acid (AA). However, processing MA at a low temperature yields a weak, less fibrous MA, while processing MA at a higher temperature yields a strong, multi-layered MA. On the other hand, the fibrous and anisotropic MA generated by SPI-WG mixes is similar and occurs over a wider shearing temperature range (110 to 140°C). At low shearing temperatures (120°C), blends containing SPI yield materials that are three times stronger than blends containing pea protein. However, at higher shearing temperatures (140°C), the strength of SPI-WG and PPI-WG mixes is comparable. The soy and pea blends' strengths clearly decreased when they were processed at 130°C.

4.2.2. Effect of Mixture of Cereal and Legume Proteinson the Quality and Structure of MA

Efficiently extruded a blend of pea and oat proteins into a highly fibrous MA with a range of physicochemical, textural, and sensory properties. The pea and oat mix powders were made using fermentation and protease.

The FAO's minimum recommendations for adult essential amino acid intake are met by this pea-oat protein mixture; however, it also contains 1.5% phytic acid, a known anti-nutrient. Firmness, fibrousness, and wetness were more influenced by extrusion conditions, while water holding capacity, water solubility index, bitterness, cohesiveness, and springiness were more influenced by blend composition. Because the lipid and carbohydrate contents promoted phase separation, the extrudates containing 70% oat exhibited the highest levels of fibrousness. A few samples even featured fibers that smelled neutral and looked like chicken. However, wetness, and water holding capacity were also below average, suggesting that combining with other materials could yield even bigger improvements. In general, oat-pea protein blends work well as an alternative to soy and WG protein when creating meat analogs that have a mild flavor and high nutritional content.

Dry fractionated pea protein (PPDF), along with PPI, SPI, and oat protein (OP), were combined to create mixed alimentary (MA) formulations. Utilizing PPDF, it's feasible to manufacture MA with a protein content of 55 grams per 100 grams, making it a suitable choice for developing meat alternatives due to its high starch, fat, and mineral content. Owing to the inherent carbohydrates in the dry fractionated PPDF, the resulting MA exhibited a more sponge-like structure and lower hardness (ranging from 13.55 to 18.33. N) compared to MA made from PPI and SPI (27 N). Furthermore, the water absorption capacity of PPDF-OP and PPDF-PPI combinations was significantly lower than that of PPI-OP and SPI-OP mixtures. The pea protein-based extrudates demonstrated a high capacity for oil absorption, which could be advantageous in incorporating oil into the final product. Sensory analysis revealed that protein isolate extrudates possessed stronger neutral sensory characteristics, while PPDF-OP had a more potent aroma and taste profile. In conclusion, employing dry-fractionated proteins facilitates the efficient production of safe and eco-friendly plant-based meat alternatives.

A range of PB composites with distinct textural qualities were created by employing a special freeze-structuring method to create PB meat analogs. To create MAs measuring 5 cm x 3 cm x 2 cm, different ratios of pea protein and wheat protein were combined: 17:0, 13:4, 8.5:8.5, 4:13, and 0:17. The composite with a 3:1 protein blend of wheat and peas had the best texture profile, weighing approximately 12,335 g in terms of hardness, 7217 g in terms of chewiness, and 0.82 in terms of springiness. The degree of protein cross-linking and the viscoelastic properties of analogs were found to have a significant influence on the textural features and structure of the materials, which were found to be related to their viscoelastic capabilities. The hardness, chewiness, and viscoelastic properties of the meat analogs appear to be enhanced by the addition of pea protein. The dough made with pea protein formed the strongest gel, according to the dough's rheological analysis, while the dough made with WP showed the opposite tendency. It has been demonstrated that the degree of the protein molecule gel network significantly affects the texture behavior of meat substitutes.

The potential use of pea protein in combination with varying amounts of chickpea flour (CF) to create a meaty texture. The lowest water retention capacity (55.5%) and highest bulk density (272 g/L) were observed in a total of 20% CF. The results of the textural profile analysis showed that at 10 and 20% CF, the optimal values for hardness, springiness, and chewiness were 674 to 1024 g, 72.1 to 80.7%, and 400 to 439, respectively. Protein interactions produced an effective network with 91.3% springiness without the addition of CF. Hardness dropped from 38 to 84% and chewiness from 73 to 92% when more than 20% CF was added. By adding CF at a rate of 10 to 20%, the ideal protein-to-starch ratio was reached, enabling the development of product attributes that could potentially mimic the texture of meat.

4.2.3. Effect of Mixture of Different Types of Legume Protein on the Quality and Structure of MA

The potential to use faba bean protein concentrate (FBPC) in high-moisture extrusion (HME) to produce MA. MAs produced at temperatures between 130 and 140°C with a water-to-product feed rate ratio of 4 and a feed rate of 11 rpm (1.10. Kg/h) exhibited good mouth feel and elasticity, attributes associated with this product category. Additionally, the specific mechanical energy, or SME, varied between 145.53 and 172 kJ/kg. Higher protein denaturation was caused by high extrusion moisture content, and this was reflected in higher water binding capacity (WBC). Furthermore, a correlation was found between the WBC and the textural characteristics of MAs. The various browning indices of the MA indicate that browning was facilitated by low moisture content and higher temperatures. It was also linked to a tiny quantity of easily accessible reducing sugars without interacting with the more common oligosaccharides.

Ferawati et al. (2021) looked at the efficacy of protein isolates and concentrates from faba beans and yellow peas as meat substitutes. Utilizing protein isolates from faba beans and yellow peas, high moisture meat alternatives

(HMMAs) with a fibrous multi-layer structure were extruded. Higher hardness (p) was found in the HMMA made of faba beans and yellow pae. <0.05) than the control sample (commercial soy-based HMMA), but similar hardness with cooked chicken and beef, depend on the extruder conditions. There was no significant difference ($p > 0.05$) in springiness between the control sample and the HMMA. The control group had less chewiness ($p < 0.05$) compared to all other HMMA, the reference chicken, and the reference beef. The chewiness of the HMMA made from yellow peas was comparable to that of cooked chicken and beef.

4.2.4. Effect of Oils Seeds/Whole Grain Protein on the Quality and Structure of MA

There are several ways to incorporate 3% black quinoa into bologna-style sausage (BS), either as whole seeds or as a fiber-rich fraction from the wet milling process. When BS is processed with 3% black quinoa or by-products high in fiber, it exhibits improved technological and nutritional qualities. The developed sausages showed reduced water activity, lower lipid oxidations, and better emulsion stability when compared to the control sausages. According to color analysis, 3% of the fiber-rich quinoa by-products differed in color from the control sausages ($p < 0.05$). The human eye is also capable of detecting this color difference. Because of their color, customers found the sausages more palatable when a portion of fiber-rich black quinoa was added. The addition of quinoa caused the residual nitrite levels to rise, indicating that less nitrite may be required and supporting the development of more natural products.

4.2.5. Effect of Mixture of Oil Seeds and Legume Protein on the Quality and Structure of MA

To create novel extrudates derived from plants, Zahari et al. (2021) mixed yellow pea isolates (YPI) and rapeseed protein concentrate (RPC) 50:50 in a twin-screw extruder for high moisture extrusion cooking. It was discovered that the moisture content had a greater effect on the texture of the extrudates than the screw speed. With a low-speed screw, the extrudates had a higher hardness at the same moisture content. Both extrudates and blended protein powder showed good qualities in terms of color, texture, amino acid content, and nutritional composition. The appearance, functionality, and texture of the extrudates, in addition to their nutritional value, may help make a powder made from rapeseed and pea protein more acceptable to consumers. The challenge of guaranteeing a sufficient and sustainable supply of proteins is addressed in part by this research. It evaluated RPC's potential as a novel protein source for MAs using shear cell technology with RPC alone and RPC-WG blends. To create fibrous materials,

RPC may be used even in the absence of WG. Furthermore, RPC produces more noticeable fibers than SPC does under the same circumstances. Compared to SPC at 40°C, RPC offers more potential for producing a wide variety of fibrous products when paired with WG. This could be a result of RPC particles melting at 140°C or higher. The fibrous structure of RPC was improved by adding WG at 140°C, but the RPC-WG product's color changed to a whiter appearance when more WG was added. Images from confocal laser scanning microscopy showed that the protein behaved as a continuous phase and the RPC intrinsic polysaccharides as a dispersed phase. One could argue that the RPC is an additional source of SPC for MAs.

4.2.6. Effect of Cereals and Mixture of Cerelas-Oils Seed Protein on the Quality and Structure of MA

The effects of WG concentrations using a high moisture extrusion technique on the distribution of moisture in MA, secondary protein structure, microstructure, and retention of volatile compounds. The microstructure, water molecules, internal protein binding capacity, and protein secondary structure are all altered by the WG and water contents, which also affect the flavor characteristics of the MAs. Retention of volatile flavor components was higher in the MA prepared with a higher WG and less water. The influence of WG and moisture content in raw materials on the microstructure, water distribution, protein structure, and interfactial force of MAs can regulate the rate at which volatile flavor components are retained while offering a diverse array of food products to cater to different consumer demands.

The physicochemical characteristics play a critical role in controlling the fibrous structure of MA based on the type of extrusion (cold or hot), the conditions of the extrusion, including feed rate, barrel temperature, screw speed, moisture content, and raw material particle size and addition of WG. But springiness was only marginally impacted by screw speed. Using constant screw speed and identical formulations, all HMMAs outperformed low moisture meat analogs (LMMAs) in terms of integrity index, cutting strength, and stability of springiness. The nitrogen solubility index, however, was lower. Based on the physicochemical characteristics, it is hypothesized that the increased cross-link production in HMMAs happens in the cooling die stage.

The impact of the SPC to WG ratio on the physicochemical characteristics of extruded MAs at constant mass. Instrumental and sensory assessments indicated that MA containing 30% WG had the highest levels of chewiness, hardness, fibrous structure, and texturization. In the MA with 20 and 30% WG content, large fibrous structures connected by many smaller threads were seen. It was found that the extrusion conditions denatured the proteins, facilitating greater crosslinking and aiding in the formation of fibrous structures. Physicochemical, textural, and structural properties of various MAs at 52% moisture using WG and SPI at various ratios (100:0, 80:20, 60:40, and 40:60% w/w dry protein basis). The MAs made with 100% WG had the lowest levels of chewiness and hardness. Yet, the appropriate concentration of WG had no effect on the ultrastructure. Although the lysine content of MAs rose in tandem with SPI concentrations, it remained significantly lower than that of steamed chicken. The use of WG and WG-SPI has been suggested as a way to produce a workable dough for the mechanical elongation method of processing MAs.

The impact of substituting gluten on the rheological and textural characteristics of a meat substitute based on SPI. Compared to the control sample, the meat analog's viscosity and degree of flexibility were higher after the hydrocolloids were added. The textural features, such as hardness, cohesiveness, springiness, and gumminess, were enhanced by adding more guar gum and κ-carrageenan; however, xanthan gum did not show these same properties. In comparison to the control sample, 1% of xanthan gum had higher levels of hardness and gumminess. The rheological and textural analysis suggested that the gluten in the MAs might be replaced with either 6% κ-carrageenan + 1% xanthan gum or 6% guar gum + 1% xanthan gum.

4.2.7. Effect of Cereal Protein on the Quality and Structure of MA

The effects of rice flour (RF) on low-fat emulsified sausage's sensory qualities, cooking loss, modulus of storage, and physicochemical characteristics. The addition of RF at 2, 4, and 6% increased ($p<0.05$) the emulsified sausage's moisture content when compared to the control. When RF was used to emulsify sausage, the results showed significantly ($p <0.05$) higher emulsion stability and lower cooking loss as a result of a slower rate of moisture release than in control samples. Compared to control samples and other RF concentrations, the hardness of low-fat sausage values increased ($p <0.05$) with the addition of 6% RF. An analysis of sensory data showed that adding 4% rice flour increased consumer acceptance. However, up to 2% of rice flour could be added to the formulation to successfully improve the textural and sensory quality. We conclude that adding RF to emulsified sausage with reduced fat content may be a promising way to enhance its quality attributes and sensory aspects. The viability of using rice protein isolate (RPI) and SPI to create meat substitutes. Due to SPI's and RPI's respective high and low water attractions, the substitution of SPI with RPI reduces SME associated with textured rice proteins. Furthermore, supplementation raised the true and apparent densities but decreased the porosity and expansion ratio. Amino acid analysis revealed that methionine, which was absent in SPI-based MA, was augmented in MA made by combining SPI and RPI. This result clearly demonstrated that SPI-RPI-based MAs exhibit superior nutritional quality compared to SPI- and RPI-based MAs alone. Due to gluten's disulfide cross-linking and hydrophobicity, the addition of RPI reduced the total soluble matter of MAs. Because fewer nutrients are lost during rehydration, MA with a low moisture content is preferred. Consequently, RPI may be substituted for SPI as an ingredient in the processing of MAs.

4.2.8. Effect of Other Protein Sources on the Quality and Structure of MA

TVP, which is intended to replace meat, is created by converting soy protein through the extrusion process. The addition of *Xanthosoma sagitifolium* powder (XSP) is expected to improve the functional features of the TVP matrix. The effects of water, mixing time, and XSP on TVP characteristics. The result shows that TVP's water holding capacity, oil binding capacity, texture, and protein solubility are all influenced ($p < 0.05$) by the quantity of XSP, mixing time, and water addition. XSP has an increasing effect on WHC up to a 20% addition; an increasing effect occurs over a 20% addition. The addition of XSP to the texture resulted in a decrease in protein solubility and random OHC. The texture of the MAs increased and their ability to hold water, bind oil, and dissolve proteins decreased when the mixing times were extended from 6 to 18 minutes. The addition of water significantly ($p < 0.05$) decreased the solubility of proteins, WHC, OHC, and mouthfeel.

Mishal et al. (2022) created a meat alternative (MA) that has sensory and nutritional attributes similar to animal-based meat, potentially addressing the current meat scarcity and meat demand. To evaluate the quality of the meat patty, a control meat patty was made without jackfruit. There was an improvement in the meat mimic's look, feel, taste, and acceptance when 22–28% unripe jackfruit pulp was added. A change in the amount of jackfruit pulp affected the MA's binding, juiciness, hardness, and softness as well as its chewiness. The consumer panel recommended processing MA using 26% unripe jackfruit pulp. The ideal concentration of the plant-based meat substitute is 26% for jackfruit pulp, 26% for TSP, 17.14% for mushrooms, 4.3% for wheat gluten, 2.86% for chickpea protein flour, 4.33% for refined flour, and 8.57% for coconut oil. The processed MA exhibited a moisture content of approximately 48.73%, a fat content of 2.46%, a pH of 6.80, a cooking yield of 90.25%, a reduction in diameter of 38.16%, and a reduction in thickness of 11.83%.

4.3. DIGESTIBILITY AND BIOAVAILABILITY OF PROTEINS

Proteins are naturally occurring polymers made up of peptide bonds tying amino acids together. The native structure of a protein molecule is determined by the kind, quantity, and arrangement of amino acids along a polypeptide chain. This structure affects how well a protein is absorbed and digested in the human digestive system. Furthermore, the process of protein digestion and absorption is contingent upon various other factors, including the protein's denaturation state, aggregation state, and the effects of the food matrix. Protein digestion mostly takes place in the small intestine because of the presence of many proteases, such as trypsin and chymotrypsin, but it starts in the stomach because of the presence of pepsin in the gastric fluids. The ease with which proteases can access the peptide bonds within polypeptide chains determines the rate and degree of protein digestion.

The ability of an ingested protein to be broken down into amino acids by the proteases in the gastrointestinal tract (GIT) is known as digestibility. The percentage of an ingested protein's amino acids that are actually absorbed by the body is referred to as bioavailability (Stein et al., 2007). In actual use, these terms ought to correspond with particular amino acids because certain amino acids are easier to release and absorb from proteins than others.

Dietary proteins are vital for growth and physiological processes because they are the main source of nitrogen in the human diet (Sa´et al., 2020). The gastrointestinal tract's epithelium cells may absorb the free amino acids and small peptides produced by the breakdown of proteins. Nutritional deficiencies may result if a protein is not completely absorbed and digested in the upper gastrointestinal tract, leaving the body without access to certain amino acids (Xie et al., 2022). Undigested proteins may also negatively impact the gut microbiome in the colon, which may have unfavorable consequences for overall health.

Many researchers have recently concentrated on the digestibility and/or bioavailability of the proteins in meat substitutes due to their significance in human nutrition. Zhou and colleagues examined the gastrointestinal fate and *in vitro* digestibility of meat derived from plants versus meat derived from animals (Zhou et al., 2021). In comparison to the plant proteins in soy-based meat substitutes, these authors discovered that the digestion of the meat proteins in ground beef occurred more quickly in the gastric phase but more slowly in the small intestine phase. In a different study, Xie, and colleagues found that, in the intestinal phase, the digestibility of the proteins in plant-based meat analogs was slower than that of real meat, but in the gastric phase, they found different results for beef and pork meat (Xie et al., 2022). The various protein types utilized, the processing and cooking techniques applied, and the compositions of the food matrices encircling the proteins could all be contributing factors to the variations in digestion behavior seen in these studies (Xie et al., 2022; Zhou et al., 2021). It is generally recognized that the type of protein, the effects of the food matrix, and the conditions of manufacture affect the digestibility and bioavailability of proteins (Reynaud et al., 2021).

Due to the presence of antinutritional factors, indigestible cell walls, and unique protein structures in plants, plant proteins have a lower digestibility (75–80%) than animal proteins (90–95%) (Sa et al., 2020). Plant proteins can be processed to improve their nutritional profile by taking into account the factors that decrease their digestibility. To increase the digestibility and bioavailability of plant proteins, for example, antinutrients can be eliminated or chelated, plant cell walls can be broken down, or protein structure can be altered through pretreatments and processing. The relative digestibility of various meat products and their plant-based equivalents, as well as the reasons behind these variations, obviously require further study. It will be necessary to conduct this research using both *in vitro* and *in vivo* techniques, including cell culture models, animal feeding studies, and simulated gastrointestinal tract models.

4.4. DIFFERENT SOURCES OF PLANT-PROTEINS IN MEAT ANALOGUES

The market for alternative proteins, particularly meat alternatives, is expanding quickly as a result of rising consumer demand (Kyriakopoulou et al., 2019). It has been possible to make meat substitutes that resemble hamburgers, sausages, nuggets, meatballs, and whole muscle products using a variety of plant-based protein sources. Although extrusion technology is frequently used to create these products, other techniques like 3D printing and shear cells are also being used (McClements and Grossmann, 2021a; 2021b). Usually, the goal of these products is to imitate the look, feel, flavor, and mouthfeel of actual meat products. Nonetheless, their composition must align with the nutritional characteristics of authentic meat products, including amino acid composition, bioavailability, and digestibility, to prevent unfavorable health outcomes. It has previously been demonstrated that many plant protein varieties possess the technological, functional, and nutritional qualities needed to displace meat proteins. For example, meat analogs that mimic many of the desirable physicochemical, sensory, and nutritional qualities of real meat have been created using soybean and wheat proteins (Ishaq et al., 2022). However, not much research has been done on the digestibility and bioavailability of these plant proteins, especially in meat substitutes. Despite this, there are numerous other types of plant proteins that can also be utilized for this purpose. As a result, we highlight the current knowledge about the nutritional characteristics of plant proteins in this section and concentrate on the various sources of plant proteins that can be utilized to create meat substitutes.

4.4.1. Cereals

Plant proteins found in cereals like wheat, rice, hemp, corn, barley, and oats are a significant part of the human diet. Wheat gluten's viscoelastic qualities make it especially useful for achieving structures and textures resembling meat. The formation of a three-dimensional network of proteins, primarily gliadins and glutenins, bound together by hydrophobic interactions, hydrogen bonds, and intramolecular disulfide bonds, is responsible for the viscoelastic properties. These networks are created in the hydrated gluten during the shear cell or extrusion processes, resulting in the formation of fibrils that resemble the texture and structure of meat products in certain ways (Ahmad et al., 2022; Kyriakopoulou et al., 2021). The degree of gluten cross-linking can be changed under reducing conditions, high temperatures or hydrostatic pressures, pH, and ionic strength changes, protein subunit isolation, the use of cross-linking enzymes (particularly transglutaminase), and interactions between the proteins and polyphenols can all alter the structural organization of the protein molecules.

The protein molecules' bioavailability and digestibility may also be impacted by all of these variables. It should be mentioned that people who have allergies or sensitivity to gluten may be at risk for health problems if they consume gluten-containing products. By extruding cereal proteins through electrospinning syringes, fibrous structures can also be induced in them. To form individual fibers or fibrous networks, zein, the primary storage protein in corn, can be dissolved in ethanol and then extruded using mechanical elongation and electrospinning techniques. Plant-based meat substitutes that mimic chicken have been made by combining soy protein networks with zein fibers made this way (Kyriakopoulou et al., 2021). The texture of the meat analogs is improved when the zein is added in a fibrous form. Gums can be added to gels containing fibrous zein to increase their water-holding capacity (Mattice and Marangoni, 2020).

The proportion of an amino acid's profile that is actually absorbed by the body and its levels of essential amino acids, in particular, determine the nutritional quality of food proteins (Raja et al., 2022). The digestible indispensable amino acid score (DIAAS) or the protein digestibility corrected amino acid score (PDCAAS) are commonly used to assess the quality of proteins. These parameters are determined by the protein's amino acid profile and the percentage of essential amino acids that the body actually absorbs (Joye, 2019). A protein is said to contain all of the essential amino acids needed in a human diet if its DIAAS or PDCAAS score is 100 or higher; a score of less than 100 suggests that the protein is missing one or more essential amino acids. The DIAAS or PDCAAS values of numerous significant plant protein sources are less than 100. For example, legumes are low in methionine, while cereals are low in lysine. Therefore, it's critical to understand the amino acid composition, digestibility, and bioavailability of proteins to estimate protein quality. Measuring the portion of amino acids that are not absorbed in a particular gastrointestinal tract segment is a common method of determining their bioavailability. Because amino acids passing through the colon may be metabolized by microorganisms in the large intestine, it is preferable to measure the concentration of non-absorbed amino acids at the end of the small intestine (ileum) rather than in the feces for more accurate estimates of bioavailability (Han et al., 2019).

Research has demonstrated that the ability of proteins to reach peptide bonds is influenced by their three-dimensional conformation, which in turn affects the digestibility of proteins. For example, it has been reported that the amount of β-sheet and β-turn structure in proteins is negatively correlated with protein digestibility. The viscoelastic properties of wheat flour are largely determined by these secondary structures, but their high hydrophobicity makes them indigestible.

By preventing the breakdown of proteins, antinutritional components found in plants may also lower the bioavailability of amino acids (Vanga et al., 2020). Common components of PBPs include dietary fibers and phytotate, which decrease protein bioavailability and digestibility by forming stable complexes with metal ions, which are enzyme cofactors, and proteins (Joye, 2019). Phytate builds up in the aleurone layer of most cereals, including rice, wheat, and millet. This can be avoided by milling the grain (Joye, 2019). Heating and shearing are two food processing techniques that can deactivate phytate and phytases. By limiting the accessibility of proteases to the proteins, dietary fibers may slow down the digestion and absorption of proteins in the stomach and small intestine (Joye, 2019). For example, they could create hydrogels that trap proteins, make complexes with proteins or enzymes, or make the gastrointestinal fluids more viscous. Researchers have also looked into the possibility of using pseudo-cereal proteins to create meat substitutes. Quinoa is a pseudo-cereal that has a high percentage of essential amino acids (39%) and a relatively high protein content (12–20%). Specifically, it has high concentrations of both lysine (6.5%) and methionine (2.6%), making it a useful supplement for legume proteins deficient in methionine. Quinoa has been used as a binder, gelling agent, meat extender, and fat substitute in meat products. Researchers Shokry and colleagues demonstrated that using quinoa flour instead of soy flour enhanced the cooking and sensory qualities of beef burgers, as well as their ability to retain water and oil (Shokry, 2016). The desired functional attributes of quinoa proteins make them suitable for use in the formulation of meat substitutes. For example, in meat analogs, quinoa protein has been used in place of wheat protein to create gluten-free products. It has been demonstrated that the quinoa protein possesses foaming, emulsifying, and water-holding qualities comparable to soy protein. It has been effectively used in meat substitutes, such as plant-based sausages, as a binder, gelling, and thickening agent.

4.4.2. Legumes

Common legumes include peas, lentils, lupines, chickpeas, soybeans, fava beans, mung beans, and other bean varieties; they provide 20–40% (dry weight) of the protein in diets. Because these legume proteins have functional qualities like gelling, water holding, binding, and emulsifying, they are frequently used to create meat substitutes. The two most prevalent globulin legume proteins, vicilin, and legumemin, are found in varying amounts in various legume species. These storage proteins are soluble in diluted aqueous salt solutions but insoluble in pure water. The type and ratio of globulin proteins (2S, 7S, 11S, and 15S) as well as the presence of non-protein substances like carbohydrates affect the functional characteristics of these proteins (Kyriakopoulou et al., 2021). A wide range of meat substitute products, such as minced, emulsified, and whole muscle ones, have been made using legume proteins. Because these proteins form crosslinked protein networks, they improve the chewiness and hardness of meat analogs while also increasing their protein content.

Because of their plentiful supply, high protein content, excellent protein quality, and desired functional characteristics, soybean proteins make a good source of proteins for use in plant-based meat substitutes. Protein isolates (90% protein), protein concentrates (70% protein), and defatted soy flours (50% protein) can all be made from them. The physicochemical, sensory, and nutritional qualities needed for a given application determine which ingredient is used. Research on the bioavailability of soy protein indicates that amino acids are excreted from the body through urea and that soy protein is highly degraded. Therefore, skeletal muscle synthesis requires less soy protein (Berrazaga et al., 2019). Leucine, isoleucine, and valine added to soy protein fortification may enhance skeletal muscle protein synthesis, increase peripheral availability of dietary amino acids, and lessen elderly splanchnic extraction of amino acids.

The use of legume proteins to create meat substitutes may have various drawbacks. The fact that some people have soy allergies is a disadvantage of using soy proteins in the formulation of these products. Legumes may also include antinutritional elements that reduce the digestibility of proteins and heighten the loss of nitrogen in excrement. Certain polyphenols, for example, can interact with and inactivate proteases, which lowers the digestibility of proteins. The possibility of legumes including compounds like raffinose that cause gastrointestinal distress presents another difficulty when utilizing them as functional ingredients. Humans cannot hydrolyze and absorb this sugar because their guts lack the enzyme α-galactosidase. The human intestine tract's microflora ferments this and other carbohydrates anaerobically, which can cause gastrointestinal symptoms like diarrhea. By employing the proper pre-treatments and processing techniques (such as soaking, dehulling, heating, extrusion, and enzymatic hydrolysis) to eliminate or inactivate the antinutritional factors and enhance protein digestibility and bioavailability, these difficulties are frequently resolved.

4.4.3. Oilseeds

Other oilseeds, such as cotton, sunflower, and rapeseed, are also used to extract plant proteins; these extractions typically occur as a byproduct of the edible oil processing sector. Affordable, plentiful, and frequently possessing favorable functional characteristics are the oilseed proteins. According to reports, sunflower proteins have similar qualities to soybean proteins in terms of water holding, emulsification, gelling, and foaming. Typically, pressure treatment and n-hexane extraction are used to break down the cell walls of sunflower seeds, increasing their digestibility by stimulating enzyme activity. According to Ditta and King (2017), on the other hand, excessively high temperatures and pressures can decrease the bioavailability of sunflower protein and its essential amino acids. Through the use of effective extraction techniques, the potentially harmful effects of polyphenols and antinutritional factors in defatted oilseeds can be offset. Although using high-intensity ultrasound can counteract this effect, the phenolic compounds in extracted sunflower proteins can lessen the hardness of gels. If properly processed (usually by high temperature and pressure extrusion methods), proteins from canola and rapeseed can also display acceptable functional characteristics. Cruciferin (globulin) and napin (albumin) make up the majority of the proteins found in rapeseed and canola oil. They can form cohesive gels and show good foaming and emulsifying properties when exposed to polysaccharides and salts. To create meat analogs with desired textural qualities, they can be utilized.

With a good essential amino acid profile and high digestibility, hemp seed protein—which makes up 25–30% of the seed—is a promising option for applications involving meat substitutes. It is usually utilized as a defatted hemp seed protein meal (HPM), which has a protein content of about 44%, or as a hemp seed protein isolate (HPI), which has a protein content of about 84%. The functional characteristics of HPM and HPI are impacted by the extraction and isolation techniques employed. When HPI and HPM's amino acid compositions were compared, HPI's Arg/Lys ratio was higher. Because hemp proteins have good functional qualities, they can be used as suitable replacements for soy proteins in meat analogs made by extrusion with high moisture content. For whole, meal, and dehulled hemp seed, the percentage protein digestibility and PDCAAS values are reported to be approximately 84–66% and 50%, 91–38% and 47%, and 84–52% and 64%, respectively. Lysine was found to be the primary limiting amino acid in all compositions. The hull is removed, which raises the PDCAAS score and protein digestibility. These findings show that the PDCAAS of hemp proteins is on par with or higher than that of some cereals, nuts, and pulses.

4.4.4. Nuts

When creating meat substitutes, some types of nuts can also be a good source of protein. For example, peanut protein concentrates foam, gel, and bind well at high temperatures. Fibrous filaments can be created by high-moisture extrusion of peanut-defatted flour, and these filaments can be used to make meat substitutes with a protein content of 55% that resemble minced or emulsion. The main protein component of walnut protein is glutelins (70%), which is accompanied by globulins (18%), albumins (7%), and prolamins (5%). Walnut protein has good digestibility and a reasonable essential amino acid profile. Peanuts have a higher protein content than other nuts. The two primary globulins, arachin, and conarachin, make up approximately 87% of the total protein in peanuts, making globulins the most prevalent type of protein in the plant (Arya et al., 2016). Because roasting causes protein aggregation or structural alterations, peanut proteins are frequently not very digestible. Freitas et al.(2012) found that cashew nuts had no limiting amino acid, but baru almond, peanuts, and Brazil nuts contained lysine as the limiting amino acid. While the protein digestibility of these nuts is relatively high, there are significant differences in their PDCAAS, ranging from 57% for baru almonds to 90% for cashew nuts.

4.4.5. Chia Seed

To create meat substitutes, chia seeds have also been utilized as a source of plant proteins. Their protein content ranges from 19 to 23%, and their proteins have good functional qualities like the ability to hold water, emulsify, and gel. This protein source is limited in lysine (3.7%), but abundant in glutamic acid (12%), arginine (11%), and aspartic acid (6.1%). To enhance the nutritional value and sustainability of meat products, chia seed proteins have been utilized in place of animal fat. Because chia seeds have a high water and oil holding capacity, they cook with minimal loss of flavor and texture, making them a suitable source of protein for sausages and meat substitutes. Chia protein also has the benefit of not containing gluten, which is beneficial for people who have celiac disease.

Olivos-Lugo et al. (2010) looked into the functional and nutritional characteristics of chia seed proteins. The reason behind the high digestibility of a chia seed protein isolate was the denaturation of the protein during the isolation procedure. To encourage protein unfolding, this procedure included two treatments: an alkaline (pH 12, 1.5 hours) and a thermal (90°C for 10 min). This enhanced the proteins' accessibility to proteases and facilitated more digestion. As per the authors, chia seeds are a good alternative to meat because of their high protein content, which is comparable to that of lentils, peas, and chickpeas.

The primary goals of adding chia seeds to meats derived from plants and animals are to increase their fiber content and decrease their fat content. Further investigation is required regarding the amino acid composition, protein digestibility, and bioavailability of chia seeds as a meat alternative.

4.5. EFFECT OF MAIN DEVELOPMENT PROCESSES ON DIGESTIBILITY AND BIOAVAILABILITY OF MEAT ANALOGUES

Extrusion is the primary method used to create plant-based meat substitutes, but other techniques like shear cell, 3D printing, electrospinning, freeze structuring, cell culturing, and wetspinning technologies are also being investigated (Figure 4.3). All methods of producing analogs of fibrous meat can be broadly categorized into two approaches: top-down and bottom-up. Force fields are more effectively used to create microstructural protein fibers in top-down methods including freeze structuring, wet-spinning, electrospinning, extrusion, phase separation, and cell culturing. On the other hand, bigger structures that are more akin to the meat structure are created by assembling manufactured microstructure blocks using bottom-up techniques like 3D printing.

PBPs are exposed to harsh conditions during the common plant-based meat production method (extrusion technique, sometimes after shear cells). The protein becomes denatured as a result of the applied heat treatment and shear stress, which change intra- and intermolecular bonds. On occasion, it can strengthen disulfide and hydrogen bonds, increase hydrophobic interactions, and form reversible protein networks. As a result, the product's digestibility and bioavailability alter, and the protein structure denatures (Cornet et al., 2022). Furthermore, distinct reactions occur at different temperatures in different plant protein sources. For example, heating wheat protein inside extruders produces glutenin polymers through disulfide bonding, giving the product viscoelastic and nutritional qualities. However, the disulfide bonding in soy protein results in a decrease in viscosity and the inactivation of its antinutritional elements.

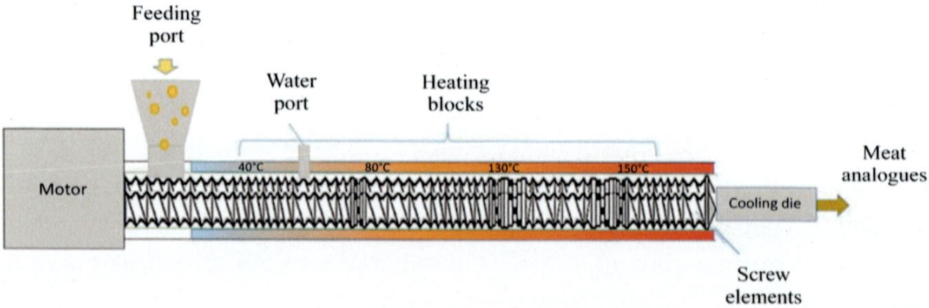

Figure 4.3. *Illustration of high-moisture extrusion (HME) with twin-screw extruder.*

Source: Zahari I, Östbring K, Purhagen JK, Rayner M. Plant-Based Meat Analogues from Alternative Protein: A Systematic Literature Review. Foods. *2022; 11(18):2870. https://doi.org/10.3390/foods11182870.*

Overall, it appears that the key to producing meat analogs with the best bioavailability, digestibility, and functional properties is to choose the most appropriate primary process for each plant protein source and optimize its parameters.

4.5.1. Green Processing Methods

The new green technologies have had a big impact on the processing of PBPs intended for the production of meat analogs. The public's growing interest in chemical-free products and green technologies have spurred processing method innovation. These innovative methods include, but are not limited to, enzymatic, sonication, microwave, electric field (PEF), high-pressure process (HPP), and ohmic heating (OH) technologies. Together with environmental friendliness, green processes also consider health and safety issues. We investigate how alternative protein digestibility and bioavailability are affected by green processing techniques. Protein digestibility assays are typically used to determine the ratio of the amount of protein consumed to the amount of protein that is absorbed by cells and broken down by digestive enzymes. The release of proteins from the food matrix, their aggregation state, their structure, and the existence of antinutritional substances all affect how digestible proteins are. Therefore, by controlling these different factors during processing, plant proteins' digestibility can be improved. The potential for many green processing techniques, such as enzymatic, sonication, microwave, HPP, PEF, and OH technologies, to be employed as pre-manufacturing treatments to improve protein digestibility and bioavailability.

4.5.2. Enzymatic Treatment

Protein digestibility and bioavailability can be improved by enzymatic treatments through the use of controlled hydrolysis or fermentation techniques.

4.5.2.1. Fermentation Process

Many microbial species, particularly yeasts and bacteria, are used in fermentation processes to improve the digestibility and bioavailability of plant proteins. During fermentation, microorganisms produce enzymes that hydrolyze proteins in a specific way. During fermentation, proteins are hydrolyzed, which greatly enhances protein bioavailability and digestibility. For example, fermenting proteins can increase their bioavailability and improve their low digestibility because they are normally insoluble in water.

4.5.2.2. Enzyme Hydrolysis

Proteins can be made more soluble, digestible, and bioavailable by partially hydrolyzing them with enzymes, which generally modifies their molecular structure and aggregation state. Enzymes that hydrolyze dietary fibers like cellulases and hemicellulases, which can impede protein digestion, can also improve the digestibility and bioavailability of proteins extracted from plant sources. To increase the digestibility and bioavailability of proteins, a variety of commercial proteases and peptidases, including Flavourzyme and Alcalase, have been used. Plant proteins function better when hydrolyzed under controlled conditions, according to several studies. The nutritional (digestibility and bioavailability) and functional (solubility, foaming, emulsifying, and gelling) qualities of plant proteins used in meat substitutes can therefore be improved by manufacturers through the use of these enzyme treatments. However, using enzymatically hydrolyzed proteins in meat analogs presents certain difficulties as the process may lead to the formation of bitter hydrophobic residues. Additionally, using enzymes can be rather costly.

4.5.3. Ultrasound and Microwave Treatments

The mechanism of action of microwave and ultrasonic treatments is first briefly discussed, followed by their possible effects on the digestibility and bioavailability of plant proteins. Though a lot of research and development has gone into it, commercial implementation of high-intensity ultrasound treatments for enhancing protein functionality is still required. Proteins have been demonstrated to undergo structural changes as a result of this technique, increasing their bioavailability and digestibility. This method primarily works on PBPs by cavitating microbubbles in the presence of high-intensity oscillating pressure waves. These pressure waves cause acoustic cavitation, which causes bubbles in a fluid to form and burst. The fast expansion and collapse of these bubbles, together with the extraordinarily high local pressures and temperatures, have thermomechanical effects on the nearby fluids.

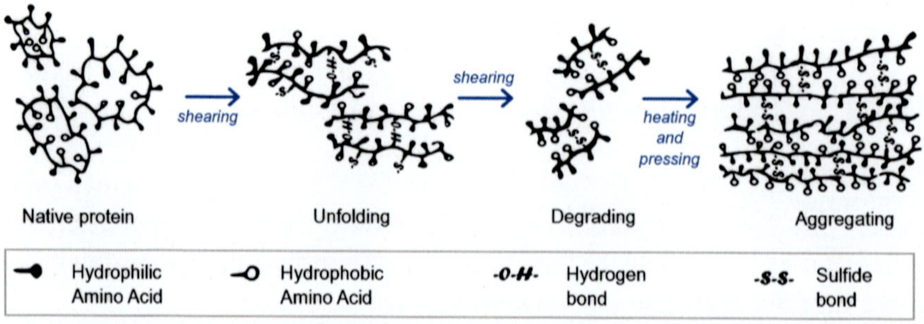

Figure 4.4. *Changes in protein conformation during extrusion of HMMA.*

Source: Zahari I, Östbring K, Purhagen JK, Rayner M. Plant-Based Meat Analogues from Alternative Protein: A Systematic Literature Review. Foods. *2022; 11(18):2870. https://doi.org/10.3390/foods11182870.*

The impact of ultrasound therapy on the composition and *in vitro* digestibility of protein isolates made from buckwheat. Buckwheat is a rich source of lysine, in contrast to most other cereals, such as wheat and corn, which are lacking in amino acids. It has been demonstrated that applying high-intensity ultrasonic waves (20 kHz) changed the proteins' secondary structures and broke up any protein aggregates, increasing the digestibility of the proteins. According to research by other scientists, sonication can improve the digestibility of buckwheat by breaking down intramolecular disulfide bonds, which encourages partial protein unfolding. It has also been demonstrated that sonication modifies the secondary structure of peanut proteins, improving protein digestibility.

It has also been investigated whether microwave treatments can increase the plant proteins' bioavailability and digestibility. Because of the heat produced by the motion of polar molecules in the presence of electromagnetic waves in this frequency range (300 MHz–300 GHz), microwave treatments encourage changes in protein structure (Divekar et al., 2017). Microwaves not only interact directly with proteins and transfer energy more efficiently than other heating methods, but they are also a more environmentally friendly technology

(Brittany and Hayes, 2002). Therefore, there is interest in using microwave processing to improve the plant proteins that will be used in meat substitutes' bioavailability and digestibility. Protease inhibitors and microbes may also be inactivated by microwave treatments, which could enhance the safety and nutritional value of plant-based diets.

One of the most significant antinutritional components in soybeans is trypsin inhibitors. These are compounds that have a potent interaction with chymotrypsin and trypsin, which lowers the digestibility of proteins. Vanga et al. (2020) looked into how microwave and ultrasonic processing affected the protein digestibility of soymilk. These researchers demonstrated that soymilk protein digestibility was increased to 84% by a 16-minute ultrasonic treatment and to 93% by a 10-minute microwave treatment at 85□. While little heat was produced during ultrasound processing, the inactivation of trypsin inhibitors and the increase in protein unfolding brought on by the high temperatures involved were blamed for the microwave treatment's superior performance. However, it's critical to maximize the processing parameters. It has been demonstrated that overheating soybeans in the microwave increases protein denaturation and aggregation, which lowers protein digestibility. A major factor influencing the inactivation of trypsin inhibitors during processing is the water content of soybeans. When dry soybeans (7% moisture) were microwave-heated, the result was a 70% inactivation of the enzyme, while wet soybeans (25% moisture) produced a 96% inactivation of the enzyme. Microwave processing of legumes results in the disruption of hydrogen bonds, an increase in dissolved ion mobility, and modifications to the secondary structure of proteins, all of which improve the functional characteristics of the legumes—such as their ability to retain water, foam, emulsify, and be digested. The length of time the product is processed determines how much of an improvement microwave treatments offer. Protein digestibility usually rises as processing times increase, but they can also fall over longer periods of time. Furthermore, prolonged treatment times may result in nonenzymatic browning, which is bad for the quality of the protein. After being microwave-treated, the digestibility of the proteins in several varieties of broad beans (Bacchus, Basta, and Windsor White) was enhanced; however, the optimal outcomes were again obtained with an intermediate processing time. It's crucial that the protein molecules partially unfold without clumping together.

4.5.4. High-Pressure Process (HPP)

As a green processing technique, HPP is thought to be able to alter the nutritional value and functional characteristics of proteins. Usually, the food matrix is subjected to high isostatic pressure (100–1000 MPa), which is transmitted to the food by water at approximately room temperature. When the pressure surpasses a certain threshold, HPP encourages the denaturation of proteins. It does not, however, encourage alterations to the primary sequence of proteins. Therefore, without compromising the profile of essential amino acids, it can be utilized to increase the protein bioavailability and digestibility. The impact of HPP on PBPs' functional characteristics has been the subject of numerous studies. However, research on its effects on protein bioavailability and digestibility is scarce. Bean protein isolate's *in vitro* digestibility is impacted by HPP. They discovered that the amount of pressure applied had an impact on how digestible proteins were. Because they encouraged excessive protein aggregation, relatively low-pressure treatments (>200 MPa) decreased the digestibility of proteins. As a result, it's critical to maximize the processing parameters employed. Although there was no change in the overall digestibility of proteins, HPP of lentil or fava bean proteins resulted in a similar or higher gastric digestibility than untreated controls.

4.5.5. Electric Field

4.5.5.1. Ohmic Heating (OH)

A new food technology called ohmic heating (OH) is effective in terms of processing time, energy, water, resources, and operating costs. Proteins undergo structural and sequential changes as a result of the uniform heat distribution produced by the moderate electric field (less than 1000 V/cm) applied to the food matrix. Through disulfide bond rearrangement, molecular interaction formation, and degradation, the OH can enhance the functional and nutritional qualities of plant proteins. The allergenic soybean trypsin inhibitor protein is remarkably acid- and thermo-stable, but it can be considerably reduced in immunoreactivity by treating it with OH at 50 Hz (95°C for 10 min). It is unclear, nevertheless, if the improvement results from electrochemical interactions or from protein aggregation. The meat substitutes made of soy protein were processed using an electric field of 30 V/cm by OH. It was determined that 3 minutes of cooking at 100°C produced a texture similar to that of restructured beef steaks and commercial chicken nuggets.

Through modifications to denaturation mechanisms, aggregation, and interaction patterns, electric field (EF) processing modifies the functional and nutritional characteristics of proteins. It has been determined how EF affects the allergenicity and digestibility of specific proteins. Due to its hydrophobic regions, β-conglycinin in soybeans exhibits pepsin resistance, which contributes to allergenicity by reducing the release of bioactive peptides during protein digestion. By altering the structure of proteins and increasing their susceptibility to digestive enzymes, OH can increase the digestibility of proteins. It is necessary to look into the likelihood of certain aggregated proteins having poorer digestibility or the production of cytotoxic and immunoreactive substances.

4.5.5.2. Pulsed Electric Field (PEF)

PEF is a non-thermal process that involves exposing a material to brief, high-voltage electrical pulses. Studies examining the effects of PEF treatments on food proteins—especially plant-based ones—are scarce. Strong PEF devices may ionize some of the chemical groups in proteins, changing their aggregation state and structure, which in turn affects digestibility and bioavailability. Modest PEF treatments primarily modify PBPs' secondary structure, which changes how digestible proteins are. The secondary structure of peanut protein was altered during PEF processing, increasing the protein's *in vitro* digestibility. A structural characteristic called protein surface hydrophobicity is used to evaluate conformational changes in proteins, which are linked to their digestibility and functional characteristics. The soy protein isolate's surface hydrophobicity may be impacted by the PEF treatment, changing the protein's conformation. Improved digestibility may result from these surface hydrophobicity-related alterations in addition to other conformational changes like molecular size, surface charge, and polydispersity (Amagliani et al., 2017).

REFERENCES

1. Abbas, Y., & Ahmad, A. (2018). Impact of processing on nutritional and antinutritional factors of legumes: A review. *Annals of Food Science and Technology, 19*, 199–215.
2. Ahmad, M., Qureshi, S., Akbar, M. H., Siddiqui, S. A., Gani, A., Mushtaq, M., et al. (2022). Plant-based meat alternatives: Compositional analysis, current development, and challenges. *Applied Food Research, 100154.*
3. Akharume, F. U., Aluko, R. E., & Adedeji, A. A. (2021). Modification of plant proteins for improved functionality: A review. *Comprehensive Reviews in Food Science and Food Safety, 20*, 198–224.
4. Alrosan, M., Tan, T. C., Easa, A. M., Gammoh, S., & Alu'datt, M. H. (2022). Recent updates on lentil and quinoa protein-based dairy protein alternatives: Nutrition, technologies, and challenges. *Food Chemistry, 132386.*
5. Amagliani, L., O'Regan, J., Kelly, A. L., & O'Mahony, J. A. (2017). The composition, extraction, functionality, and applications of rice proteins: A review. *Trends in Food Science & Technology, 64*, 1–12.
6. Arise, A. K., Aliyu, B. N., & Ajidagba, S. D. (2020). Effect of thermal processing and fermentation on the chemical composition, protein digestibility, and functional properties of bambara protein isolate. *Carpathian Journal of Food Science & Technology, 12.*
7. Arya, S. S., Salve, A. R., & Chauhan, S. (2016). Peanuts as functional food: A review. *Journal of Food Science & Technology, 53*, 31–41.
8. Avelar, Z., Vicente, A. A., Saraiva, J. A., & Rodrigues, R. M. (2021). The role of emergent processing technologies in tailoring plant protein functionality: New insights. *Trends in Food Science & Technology, 113*, 219–231.
9. Bartkiene, E., Juodeikiene, G., & Vidmantiene, D. (2012). Nutritional quality of fermented defatted soya and flaxseed flours and their effect on texture and sensory characteristics of wheat sourdough bread. *International Journal of Food Sciences & Nutrition, 63*, 722–729.
10. Bennetau-Pelissero, C. (2019). Plant proteins from legumes. In *Bioactive Molecules in Food* (pp. 223–265).
11. Berrazaga, I., Micard, V., Gueugneau, M., & Walrand, S. (2019). The role of the anabolic properties of plant-versus animal-based protein sources in supporting muscle mass maintenance: A critical review. *Nutrients, 11*, 1825.
12. Boye, J., Wijesinha-Bettoni, R., & Burlingame, B. (2012). Protein quality

evaluation twenty years after the introduction of the protein digestibility corrected amino acid score method. *British Journal of Nutrition, 108*, S183–S211.

13. Brittany, L., & Hayes, D. (2002). *Microwave Synthesis: Chemistry at the Speed of Light.* USA: Cem. Publ.
14. Cornet, S. H., Snel, S. J., Schreuders, F. K., van der Sman, R. G., Beyrer, M., & van der Goot, A. J. (2022). Thermo-mechanical processing of plant proteins using shear cell and high-moisture extrusion cooking. *Critical Reviews in Food Science and Nutrition, 62*, 3264–3280.
15. Day, L., Cakebread, J. A., & Loveday, S. M. (2021). Food proteins from animals and plants: Differences in the nutritional and functional properties. *Trends in Food Science & Technology, 119*, 428–442.
16. Dekkers, B. L., Boom, R. M., & van der Goot, A. J. (2018). Structuring processes for meat analogues. *Trends in Food Science & Technology, 81*, 25–36.
17. Dias, D. R., Abreu, C. M. P. D., Silvestre, M. P. C., & Schwan, R. F. (2010). *In vitro* protein digestibility of enzymatically pre-treated bean (*Phaseolus vulgaris* L.) flour using commercial protease and *Bacillus sp.* protease. *Food Science and Technology, 30*, 94–99.
18. Ditta, Y., & King, A. (2017). Recent advances in sunflower seed meal as an alternate source of protein in broilers. *World's Poultry Science Journal, 73*, 527–542.
19. Divekar, M. T., Karunakaran, C., Lahlali, R., Kumar, S., Chelladurai, V., Liu, X., et al. (2017). Effect of microwave treatment on the cooking and macronutrient qualities of pulses. *International Journal of Food Properties, 20*, 409–422.
20. Drulyte, D., & Orlien, V. (2019). The effect of processing on digestion of legume proteins. *Foods, 8*, 224.
21. Espinosa-Paéz, E., Alanis-Guzmán, M. G., Hernández-Luna, C. E., Báez-González, J. G., Amaya-Guerra, C. A., & Andrés-Grau, A. M. (2017). Increasing antioxidant activity and protein digestibility in *Phaseolus vulgaris* and *Avena sativa* by fermentation with the *Pleurotus ostreatus* fungus. *Molecules, 22*, 2275.
22. Farchi, S., De Sario, M., Lapucci, E., Davoli, M., & Michelozzi, P. (2017). Meat consumption reduction in Italian regions: Health co-benefits and decreases in GHG emissions. *PLOS ONE, 12*(8), 1–19.
23. Fawale, O. S., Gbadamosi, S. O., Ige, M. M., & Kadiri, O. (2017). Effects of cooking and fermentation on the chemical composition, functional, and antinutritional properties of kariya (*Hildergardia barteri*) seeds. *Food*

Sciences and Nutrition, 5, 1106–1115.

24. Fernández-López, J., Viuda-Martos, M., & Pérez-Álvarez, J. A. (2021). Quinoa and chia products as ingredients for healthier processed meat products: Technological strategies for their application and effects on the final product. *Current Opinion in Food Science, 40*, 26–32.
25. Freitas, J. B., Fernandes, D. C., Czeder, L. P., Lima, J. C. R., Sousa, A. G., & Naves, M. M. V. (2012). Edible seeds and nuts grown in Brazil as sources of protein for human nutrition. *Food and Nutrition Sciences, 3*.
26. Fu, Y., Chen, T., Chen, S. H. Y., Liu, B., Sun, P., Sun, H., et al. (2021). The potentials and challenges of using microalgae as an ingredient to produce meat analogues. *Trends in Food Science & Technology, 112*, 188–200.
27. Giami, S. (2004). Effect of fermentation on the seed proteins, nitrogenous constituents, antinutrients, and nutritional quality of fluted pumpkin (*Telfairia occidentalis* Hook). *Food Chemistry, 88*, 397–404.
28. Gu, J., Bk, A., Wu, H., Lu, P., Nawaz, M. A., Barrow, C. J., et al. (2022). Impact of processing and storage on protein digestibility and bioavailability of legumes. *Food Reviews International*, 1–28.
29. Habiba, R. (2002). Changes in anti-nutrients, protein solubility, digestibility, and HClextractability of ash and phosphorus in vegetable peas as affected by cooking methods. *Food Chemistry, 77*, 187–192.
30. Hall, A. E., & Moraru, C. I. (2021). Effect of high pressure processing and heat treatment on *in vitro* digestibility and trypsin inhibitor activity in lentil and faba bean protein concentrates. *Lebensmittel-Wissenschaft & Technologie, 152*, 112342.
31. Han, F., Wang, Y., Fan, L., Song, G., Chen, X., et al. (2019). Digestible indispensable amino acid scores of nine cooked cereal grains. *British Journal of Nutrition, 121*, 30–41.
32. Hernandez-Infante, M., Sousa, V., Montalvo, I., & Tena, E. (1998). Impact of microwave heating on hemagglutinins, trypsin inhibitors, and protein quality of selected legume seeds. *Plant Foods for Human Nutrition, 52*, 199–208.
33. House, J. D., Neufeld, J., & Leson, G. (2010). Evaluating the quality of protein from hemp seed (*Cannabis sativa* L.) products through the use of the protein digestibility-corrected amino acid score method. *Journal of Agricultural and Food Chemistry, 58*, 11801–11807.
34. Indurkar, A., Pandit, A., Jain, R., & Dandekar, P. (2021). Plant-based biomaterials in tissue engineering. *Bioprinting, 21*, e00127.
35. Ishaq, A., Irfan, S., Sameen, A., & Khalid, N. (2022). Plant-based meat

analogs: A review with reference to formulation and gastrointestinal fate. *Current Research in Food Science*.

36. Jin, J., Okagu, O. D., Yagoub, A. E. A., & Udenigwe, C. C. (2021). Effects of sonication on the *in vitro* digestibility and structural properties of buckwheat protein isolates. *Ultrasonics Sonochemistry, 70*, 105348.
37. Joye, I. (2019). Protein digestibility of cereal products. *Foods, 8*, 199.
38. Jung, A. H., Hwang, J. H., Jun, S., & Park, S. H. (2022). Application of ohmic cooking to produce a soy protein-based meat analogue. *Lebensmittel-Wissenschaft & Technologie, 160*, 113271.
39. Kang, S., Zhang, J., Guo, X., Lei, Y., & Yang, M. (2022). Effects of ultrasonic treatment on the structure and functional properties of chickpea protein isolate and its digestibility *in vitro*. *Foods, 11*, 880.
40. Kapravelou, G., Martínez, R., Andrade, A. M., Sánchez, C., Chaves, C. L., López-Jurado, M., et al. (2013). Health promoting effects of lupin (*Lupinus albus* var. *multolupa*) protein hydrolyzate and insoluble fiber in a diet-induced animal experimental model of hypercholesterolemia. *Food Research International, 54*, 1471–1481.
41. Kato, A., & Nakai, S. (1980). Hydrophobicity determined by a fluorescence probe method and its correlation with surface properties of proteins. *Biochimica et Biophysica Acta (BBA)–Protein Structure, 624*, 13–20.
42. Ketnawa, S., & Ogawa, Y. (2021). *In vitro* protein digestibility and biochemical characteristics of soaked, boiled, and fermented soybeans. *Scientific Reports, 11*, 1–11.
43. Khatkar, A. B., Kaur, A., & Khatkar, S. K. (2020). Restructuring of soy protein employing ultrasound: Effect on hydration, gelation, thermal, *in-vitro* protein digestibility, and structural attributes. *Lebensmittel-Wissenschaft & Technologie, 132*, 109781.
44. Kyriakopoulou, K., Dekkers, B., & van der Goot, A. J. (2019). Plant-based meat analogues. In *Sustainable Meat Production and Processing* (pp. 103–126). Elsevier.
45. Kyriakopoulou, K., Keppler, J. K., & van der Goot, A. J. (2021). Functionality of ingredients and additives in plant-based meat analogues. *Foods, 10*, 600.
46. Lafarga, T., Acién-Fernández, F. G., & Garcia-Vaquero, M. (2020). Bioactive peptides and carbohydrates from seaweed for food applications: Natural occurrence, isolation, purification, and identification. *Algal Research, 48*, 101909.
47. Li, Y., Chen, Z., & Mo, H. (2007). Effects of pulsed electric fields on

physicochemical properties of soybean protein isolates. *LWT–Food Science and Technology, 40*, 1167–1175.

48. Liceaga, A. M., & Hall, F. (2019). Nutritional, functional, and bioactive protein hydrolysates.
49. Marinova, D., & Bogueva, D. (2019). Planetary health and reduction in meat consumption. *Sustainability in Earth, 2*(3).
50. Sanchez-Sabate, R., & Sabate, J. (2019). Consumer attitudes towards environmental concerns of meat consumption: A systematic review. *International Journal of Environmental Research and Public Health, 16*, 1220.
51. Xie, Y., Cai, L., Zhao, D., Liu, H., Xu, X., Zhou, G., et al. (2022). Real meat and plant-based meat analogues have different *in vitro* protein digestibility properties. *Food Chemistry, 387*, 132917.
52. Zhang, J., Liu, L., Liu, H., Yoon, A., Rizvi, S. S., & Wang, Q. (2019). Changes in conformation and quality of vegetable protein during the texturization process by extrusion. *Critical Reviews in Food Science and Nutrition, 59*, 3267–3280.

Chapter 5

Fat Content and Composition

Contents

Introduction 166
5.1. Dietary Fats 167
5.2. Cis/trans-Fatty Acid Content of Red Meats and the Related Effects on Meat Quality and Human Health 170
5.3. Plant-Based Fat Sources 186
5.4. The Development of Plant-Based Fat Substitutes 189
References 194

INTRODUCTION

Together with carbohydrates and proteins, fats make up the majority of common food items like milk, butter, tallow, lard, salt pork, and cooking oils. Fats are one of the three main macronutrient groups in the human diet. They serve crucial structural and metabolic roles in the majority of living things, such as energy storage, waterproofing, and thermal insulation. They are a significant and dense source of food energy for many animals. Except a few essential fatty acids that must be consumed through the diet, the human body can produce the necessary amount of fat from other food ingredients. Certain vitamins and flavoring and aroma ingredients that are not water soluble are also transported by dietary fats.

In many parts of the world, meat is a major source of high-value animal protein. Globally, comparatively more urbanized populations' diets tend to be more heavily concentrated in meat, poultry, and other animal products than the less varied diets of rural areas. The amount of fat in meat and meat products varies significantly depending on the species, age, and portion of the carcass utilized. Animal feeding also affects the fat content and composition; this fact is used to modify the fatty acid composition of meat, with single-stomached pigs and poultry showing the best results in this regard.

People who are vegetarians are more accepting of the unique taste of plant meat. But omnivorous consumers, who merely wish to consume less meat, have higher standards for the flavor of meat substitutes. Despite the growing availability of meat substitutes in eateries and supermarkets, the market share of plant-based meat substitutes is still small because plant-based ingredients lack sufficient texture and flavor. To create the fibrous structure of meat analogs, plant-based components must typically be structured using mechanical tools like extrusion, shearing, spinning, etc. Another structuring technique is the processing of single or complex plant proteins and plant polysaccharides, which results in the formation of polymers with cross-linked structures.

Emulsified meat, or minced meat, is frequently processed into sausages, patties, chicken nuggets, and other meat products using polymer colloidal systems.

Meat's basic constituents are primarily animal fat and protein, with trace amounts of carbohydrates, dietary fiber, and other elements. Furthermore, consumers who choose plant-based fat substitutes will continue to consume meat, meaning that their protein intake won't change much. For this reason, plant-based fat substitutes are not taken into consideration. Protein consumption and digestion differ markedly between omnivorous and vegan consumers.

5.1. DIETARY FATS

Though they should be a significant portion of your diet, not all fats are created equal. You can reduce your risk of heart attack, stroke, and other serious health issues by selecting healthy fats from vegetable sources more frequently than less healthy types from animal products.

One kind of nutrient that comes from your diet is fat. While eating some fats is necessary, consuming too much can also be harmful. Your body needs fats to provide the energy it needs to function properly. Your body uses the calories from the carbohydrates you eat during physical activity. But after 20 minutes, your body starts to rely on fat calories to sustain your exercise.

Your body needs the essential fatty acids linoleic and linolenic acid, which are found in fats that you eat. Because your body cannot produce them on its own or function without them, they are referred to as essentials. They are necessary for blood coagulation, inflammation management, and brain development.

Protein and carbs have 4 calories per gram, while fat has 9 calories, more than twice as many as these two macronutrients. Saturated and unsaturated fatty acids make up all fats. The term "saturated" or "unsaturated" refers to the relative amounts of each type of fatty acid in fat.

5.1.1. Unsaturated Fats

Liquid at room temperature, unsaturated fats are thought to have many positive effects, including lowering blood cholesterol, reducing inflammation, regulating heart rhythms, and many more. Plant-based foods like vegetable oils, nuts, and seeds are the main sources of unsaturated fats.

There are two types of "good" unsaturated fats:

1. Monounsaturated fats are found in high concentrations in:
   - Olive, peanut, and canola oils.
   - Avocados.
   - Nuts such as almonds, hazelnuts, and pecans.
   - Seeds such as pumpkin and sesame seeds.
2. Polyunsaturated fats are found in high concentrations in:
   - Sunflower, corn, soybean, and flaxseed oils.
   - Walnuts.
   - Flax seeds.
   - Fish Canola oil – though higher in monounsaturated fat, it's also a good source of polyunsaturated fat.

One significant kind of polyunsaturated fat is omega-3 fat. They must come from food because the body is unable to produce them.

- Eating fish two to three times a week is a great way to obtain omega-3 fats.
- Walnuts, canola or soybean oil, flax seeds, and walnuts are good plant sources of omega-3 fats.
- A study conducted by faculty at HSPH found that among older adults, higher blood levels of omega-3 fats are linked to a lower risk of premature death.

The majority of people don't consume enough good unsaturated fats. There is evidence that eating more polyunsaturated fat—up to 15% of daily calories—instead of saturated fat can reduce the risk of heart disease. The American Heart Association recommends that 8–10% of calories come from polyunsaturated fats.

- An analysis of 60 trials examining the effects of different fats and carbohydrates on blood lipid levels was carried out by Dutch researchers. In studies where monounsaturated and polyunsaturated fats were consumed in place of carbohydrates, these healthy fats raised levels of protective HDL and lowered levels of dangerous LDL. (Mensink et al., 2003)
- More recently, the Optimal Macronutrient Intake Trial for Heart Health, a randomized trial, demonstrated that switching from a diet high in carbohydrates to one high in unsaturated fats—mostly monounsaturated fats—lowers blood pressure, enhances lipid profiles, and lowers the estimated risk of cardiovascular disease.

5.1.2. Saturated Fats

Every food that contains fat has a combination of different kinds of fat. Saturated fat is present in small amounts in even healthful foods like nuts and chicken, but much less than in beef, cheese, and ice cream. Animal foods contain the majority of saturated fat, but certain plant foods—like coconut, coconut oil, palm oil, and palm kernel oil—also contain significant amounts of saturated fat.

- It is advised by the Dietary Guidelines for Americans to consume less than 10% of daily calories from saturated fat.
- Going one step further, the American Heart Association suggests keeping saturated fat to no more than 7% of total calories.
- Reducing saturated fat, however, is probably useless if people substitute refined carbohydrates for saturated fat. While replacing

saturated fat with refined carbohydrates lowers "bad" LDL cholesterol, it also raises triglycerides and lowers "good" HDL cholesterol. The net effect is as bad for the heart as eating too much saturated fat.

Pizza and cheese, whole, and reduced-fat milk, butter, and dairy desserts, meat products (sausage, bacon, beef, and hamburgers), cookies, and other grain-based desserts, and a variety of mixed fast food dishes are the main sources of saturated fat in the American diet.

Even though saturated fat was advised to be avoided for decades, there has been a shift in this belief in recent years. A number of studies, including one that examined the results of 21 studies that tracked 350,000 participants for up to 23 years, indicate that diets heavy in saturated fat do not increase the risk of heart disease.

- Researchers examined the connection between consumption of saturated fat and stroke, CVD, and coronary heart disease (CHD). They came to the contentious conclusion that there was not enough data from prospective epidemiologic studies to conclude that dietary saturated fat raised the risk of CVD, CHD, or stroke.
- A well-known 2014 study cast doubt on the connection between heart disease and saturated fat, but HSPH nutrition experts found the paper to be gravely misleading.

The main takeaway is that reducing saturated fat intake can benefit health if it is replaced with healthy fats, particularly polyunsaturated fats. Consuming healthy fats instead of saturated fat reduces the risk of heart disease by lowering "bad" LDL cholesterol and improving the ratio of total cholesterol to "good" HDL cholesterol.

Consuming healthy fats rather than saturated fats can also aid in preventing insulin resistance, which is a risk factor for diabetes. Therefore, evidence clearly indicates that unsaturated fat is still the healthiest type of fat, even though saturated fat may not be as harmful as previously believed.

5.1.3. Trans Fats

Trans fatty acids, or trans fats as they are more widely known, are produced through a process known as hydrogenation that involves heating liquid vegetable oils while hydrogen gas and a catalyst are present.

- Vegetable oils become more stable and less prone to going rancid when they are partially hydrogenated. Additionally, this process solidifies the oil, enabling it to be used as shortening or margarine.

- Partially hydrogenated oils are perfect for frying fast food because they can tolerate repeated heating without degrading.
- Due to these factors, partially hydrogenated oils have become standard ingredients in bakeries, restaurants, processed snack foods, and margarine.

Trans fats in our diets come from a variety of sources besides partially hydrogenated oil. Small amounts of dairy and beef fat are also naturally occurring sources of trans fats.

The worst kind of fat for the heart, blood vessels, and body as a whole are trans fats because they:

- Raise bad LDL and lower good HDL.
- Create inflammation, – a reaction related to immunity – which has been implicated in heart disease, stroke, diabetes, and other chronic conditions.
- Contribute to insulin resistance.
- Can have harmful health effects even in small amounts – for each additional 2% of calories from trans-fat consumed daily, the risk of coronary heart disease increases by 23%.

5.2. CIS/TRANS-FATTY ACID CONTENT OF RED MEATS AND THE RELATED EFFECTS ON MEAT QUALITY AND HUMAN HEALTH

Because of their perceived high fat content and composition of saturated fatty acids, red meats are stigmatized as being unhealthy. Consumers may be discouraged from eating red meat if they have doubts about the trans-fatty acid content of the meat, particularly if they have non-communicable diseases like obesity, hypertension, or cardiovascular disease. A large portion of the criticism leveled at red meat's fat content is baseless and typically misinformed.

All unsaturated fatty acids with one or more isolated, non-conjugated double bonds in the trans geometric configuration are combined to form trans-fatty acids (TFA); conjugated fatty acids (CLAs) are not included in this total. The European Food Safety Authority (EFSA) claims that because TFAs are neither synthesized nor needed by the body, they do not perform any essential roles in human health. Health experts advise consuming more omega-3 polyunsaturated fatty acids (PUFAs) and reducing total intake of saturated fatty acids (SFA), trans fats, and cholesterol. (Griel and Kris-Etherton, 2006) TFA consumption specifically has negative health effects linked to heart disease (CHD), cardiovascular disease (CVD), and other related conditions. Consumption of

TFA exceeding 5 grams per serving (100 g) is linked to a higher risk of coronary heart disease. The total fatty acids (TFAs) found in the human diet typically come from two sources: either the partial hydrogenation of vegetable oils and, to a lesser extent, fish oils, which occurs through deep-fat frying and other heat treatments used in the fast-food industry, or a select few naturally occurring TFA types that are produced from polyunsaturated fatty acids (PUFA) by anaerobic bacteria in the rumen of animals with cloven hooves.

It has been reported that the total fatty acid content (TFA) in certain industrial foods can reach up to 60% and occasionally even higher levels. Comparatively, the TFAs derived from ruminants that are present in ruminant fat rarely surpass 6%; for example, the TFA content of milk fat ranges from 4% to 6%. Up to 20% of the TFA content in ruminant fat may be made up of the C16:1 trans-isomer range, which is absent from industrial TFA profiles. Elaidic acid (C18:1), vaccenic acid (C18:1 t-11), and rumenic acid (C18:2) are the most prevalent TFA-isomers.

The trans isomers that are unique to fats derived from ruminants are vaccenic acid and rumenic acid. Because ruminant-derived trans vaccenic acid is easily converted to CLA isomers, which have many health benefits, research suggests that it is not linked to an increased risk of CHD or CVD. Excessive consumption of trans unsaturated fatty acids is linked to increased plasma LDL cholesterol and decreased HDL cholesterol, as well as CVD, CHD, and Type II diabetes mellitus.

5.2.1. Lipid Composition of Animal Fats

Triacylglycerols make up the majority of the lipid subgroup which includes the fats found in animal tissues. Animal fats are stored at various physiological stages of growth and development in distinct adipose tissue (fat) depots. Animals' subcutaneous (fat beneath the skin), intramuscular, and various internal fat depots, such as the perirenal, cardiac, omental, and pelvic fat depots, are the main adipose tissue depots.

An allometric accumulation of bone, muscle, and fat, in that order, characterizes an animal's growth; fat accumulates and matures more slowly and later than the other tissues. The rates of adipose depot maturation vary, with the subcutaneous and intramuscular fat depots maturing later and more slowly than the internal tissue depots.

It is a common misconception that red meat is high in fat, particularly saturated fatty acids (SFA), which have been linked to several non-communicable human diseases, including the metabolic syndrome. However, lean meat actually contains very little intramuscular fat—only 2–3% in lean beef, 5–7% in conventional beef, and up to 15% in marbled beef. Lamb has an intramuscular fat content that ranges from 8 to 14%. The main saturated fatty acids (SFA) found in red meats are myristic acid (C14:0), palmitic acid (C16:0), and stearic acid (C18:0). When compared to the average fat content of processed foods in the United States, the intramuscular fat contents of red meats are, respectively, 80% and 40% lower in conventional and marbled beef. The value of animal fats is greatly undervalued, particularly when considering that they are a source of essential fatty acids, act as a transporter of fat-soluble vitamins A, D, E, and K, aid in the synthesis of steroidal hormones, provide energy for metabolism, and enhance the flavor, aroma, and texture of meat.

5.2.1.1. The Nutritional Value of Fat in Meat

Red meat's lean portion is an extremely biologically valuable and highly digestible source of protein for humans. For example, red meat has a digestibility value of 94%, which is significantly higher than that of whole wheat (86%) or beans (78%). While red meat is low in fat and sodium, it is high in antioxidants and other nutritionally significant bioactive substances like taurine, carnitine, carnosine, ubiquinone, glutathione, and creatine. These substances are vital to human health because they have anti-inflammatory and antioxidant properties that support immune, neurological, muscular, and retinal functions. Beef contains high levels of taurine, carnitine, carnosine, and creatine, which are absent from foods derived from plants.

The proportion of saturated (SFA), monounsaturated (MUFA), and polyunsaturated (PUFA) fatty acids, as well as the balance between n-6 and n-3 fatty acids, determine the nutritional value of meat fats. The conventional view is that red meats are entirely composed of saturated fat. Numerous research teams have conducted in-depth studies on animal fats in Australia, Europe, the UK, and South Africa. Everyone agrees that not all animal fats found in red meat species are saturated. This is arguably one of the most false beliefs about red meats.

Table 5.1 summarizes the SFA, MUFA, and PUFA content of meat from sheep, goats, and cattle that have previously been reported. These findings show that meat from goats, lambs, or cattle is not all saturated. These red meats typically have normal saturation levels of 45 to 54%, with notable amounts of the mono-unsaturated fatty acid oleic acid ranging from ca. 38–44%. In general, there are less than 3 grams of SFA in every 100 grams of edible raw red meat.

Table 5.1. *Summary of Saturated and Unsaturated Fatty Acids Previously Reported for Meat From Cattle, Sheep, and Goats*

Meat Species	Fatty Acid Fraction (W/W%)			Reference
	SFA	MUFA	PUFA	
Beef				
Feedlot steers#	49.6 ± 4.4	38.5 ± 3.4	8.7 ± 1.5	(Webb et al., 1995)
*Belgian white blue**	46.5 ± 4.2	38.4 ± 4.1	15.0 ± 3.9	(De Smet et al., 2000)
Sheep				
Dorper sheep	52.8 ± 1.9	43.9 ± 1.3	3.3 ± 0.5	(Tshabalala et al., 2003)
Damara sheep	51.8 ± 1.8	44.3 ± 1.3	3.9 ± 0.2	(Tshabalala et al., 2003)
Dorper sheep	50.8 ± 2.9	42.7 ± 3.7	4.7 ± 0.9	(Webb et al., 1995)
Merino sheep	52.2 ± 3.9	40.0 ± 3.2	5.21 ± 1.1	(Webb et al., 1994)
Dorper (Karoo grass)	52.9 ± 4.5	43.1 ± 3.6	3.6 ± 0.9	(Meyer et al., 2003)
Dorper (Karoo browse)	54.7 ± 2.9	41.3 ± 1.9	3.7 ± 1.5	(Meyer et al., 2003)
Goats				
Boer goats	54.7 ± 2.2	41.9 ± 0.9	3.4 ± 0.4	(Tshabalala et al., 2003)
Indigenous goats	53.6 ± 2.8	42.5 ± 1.1	3.9 ± 0.4	(Tshabalala et al., 2003)

#Conventionally raised beef cattle.

*High lean beef breed.

Source: C. Webb, E. (2021). Cis/Trans-Fatty Acid Content of Red Meats and the Related Effects on Meat Quality and Human Health. IntechOpen. doi: 10.5772/intechopen.96726.

In the past, doctors have advised lowering total fat intake to reduce obesity and lower the risk of coronary heart disease (CHD). Nevertheless, rather than emphasizing quantity, the recommendations now focus on the makeup of fatty acids. For instance, it is emphasized that, in middle-aged men, MUFAs, and PUFAs are more significant in lowering the risk of cardiovascular heart disease than total fat consumption. Subcutaneous and intramuscular fat depots are still present in the meat that is sold to consumers, even though the majority of the internal carcass fat is removed during the slaughter process. Although it is thought to be a wasteful procedure to remove carcass fat, finishing livestock to the necessary body condition (grade) to produce the most palatable and flavorful meat inevitably results in the fat building up in the carcass (Laaksonen et al., 2005).

Various techniques and biotechnologies have been utilized to maximize livestock fattening, reduce the build-up of surplus internal carcass fat, and enhance production efficiency. In addition to using biotechnologies like better animal feeding, ration formulation, steroidal hormone implants, and feed additives like probiotics, prebiotics, and repartitioning agents (beta-adrenergic agonists), the techniques involve manipulating age, sex, and production systems. The latter molecules promote muscle anabolism and inhibit muscle catabolism, which redistributes energy from the accumulation of fat to the accumulation of protein (muscle). Due to their significant reductions in carcass fat content and efficient energy repartitioning towards lean gain (muscle growth), these molecules are becoming more and more popular. Furthermore, these molecules lessen the environmental impact of animal production systems by enhancing the efficient growth and utilization of natural resources.

5.2.1.2. Fatty Acid Composition of Red Meats

Triacylglycerols, another name for neutral lipids, are the predominant lipid class found in animal fats, whereas phospholipids are primarily found in muscle tissue and cell membranes. The body uses triacylglycerols as a concentrated source of energy. Because phospholipids are a component of cellular membranes, they have a lower SFA content and much higher levels of PUFA. Stearoyl CoA desaturase converts stearic acid (C18:0) into oleic acid, which is a primary constituent of neutral lipids. Lipids accumulate primarily as triacylglycerols in adipocytes (fat cells) found in abdominal fat depots like omental and perirenal adipose tissue, as well as subcutaneous, intermuscular, and intramuscular adipose tissue depots in ruminant animals. Notwithstanding rumen biohydrogenation, dietary n-6 and n-3 PUFA are integrated into adipose and muscle tissues. Muscle tissues are the preferred location for ruminant animals to absorb essential fatty acids as opposed to adipose tissue.

Because of rumen biohydrogenation, the diet of ruminant animals affects the fatty acids in triacylglycerols in a minor but significant way. On the other hand, PUFAs are found in phospholipids in higher proportions than triacylglycerols, and they are less influenced by diet. Essential fatty acids, such as C18:2n-6 and C18:3n-3, and their long-chain derivatives, eicosapentaenoic acid (C20:5n-3; EPA) and decosahexaenoic acid (C22:6n-3; DHA), make up the PUFAs in phospholipids. The firmness and oiliness of adipose tissue as well as the oxidative stability of muscle tissues are determined by the fatty acid composition, which also affects the flavor and color of the meat product.

The PUFA/SFA and n-6/n-3 ratios have an impact on the lipids' nutritional value in red meats. It is thought that fats with higher PUFA and CLA content and lower SFA content are optimal. The main metrics used to assess the nutritional value of foods for human nutrition are the n-6/n-3 and PUFA/SFA (also known as P:S) ratios. It is advised to use a benchmark value of >0.7 for the PUFA:SFA (or P:S) ratio and <5 for the n-6/n-3 ratio.

A great source of long-chain omega-3 polyunsaturated fatty acids is lean red meat. Beef contains between 14 and 24% polyunsaturated fats (PUFA) and roughly 3% trans-fatty acid (C18:1) in its overall fatty acid profile. Naturally found in meat and dairy products, conjugated linoleic acid (CLA) has been linked to improvements in type II diabetes, weight loss, and muscle growth. Omega 6-fatty acids, or CLAs, come in both cis and trans configurations and have a total of roughly 28 isomers. Conjugated fatty acids, or CLAs, are not included in TFAs because the two "conjugated" double bonds are not divided by a methylene group.

5.2.1.3. Overview of Trans-Fatty Acids in Food

The total amount of unsaturated fatty acids with one or more isolated, non-conjugated double bonds in the trans-geometric configuration is known as trans-fatty acids, or TFAs. The European Food Safety Authority claims that because the body cannot produce or use TFAs, they are not essential to human health. Globally, medical professionals advise consuming fewer SFAs, TFAs, and cholesterol overall and consuming more n-3 polyunsaturated fatty acids (PUFAs). Regretfully, TFAs do not yet have an established "Population Reference Intake (PRI)," "Average Requirement (AR)," or "Adequate Intake (AI) value."

As of right now, the negative health effects of TFA consumption are primarily linked to non-communicable illnesses like cardiovascular disease (CVD and CHD). According to estimates, heart failure (CHD) is the leading cause of death worldwide. Consuming more than five grams of TFA per serving is linked to a higher risk of developing CHD. The consumption of industrial TFA was found to be positively correlated with an elevated risk of CHD. Industrial TFAs are frequently found in products like margarines, spreads, and shortenings, as well as baked goods and deep-fried foods like those found in fast food restaurants. They are created during the partial hydrogenation of vegetable fats and fish oils.

Relatively little is known about TFAs in livestock or the particular variables affecting TFA concentrations in red meat.

5.2.1.4. Effects of Production Systems on Fat and Fatty Acid Composition of Red Meats

Either concentrate diets or rangelands are commonly used to fatten cattle. Livestock are raised using various combinations of intensive and pasture-based systems; the extent to which these are used depends on the resources available and the climate. The benefits of pasture feeding on the fatty acid profile of meat have led to a rise in popularity; however, these benefits are marginal and come at the expense of higher production efficiency.

Alpha-linolenic acid (C18:3n-3) is an essential fatty acid that is abundant in grasses and, in spite of rumen biohydrogenation, is consistently concentrated in muscle tissue. Cattle fed concentrate tend to accumulate slightly less C18:3n-3, which could lead to a slight decrease in the n-6/n-3 ratio. When feeding with grass and pasture, the n-6/n-3 ratio decreases. Because grass-fed animals have higher concentrations of natural antioxidants like vitamin E, which stabilizes the PUFAs, their meat products have better oxidative stability. One of the main fatty acids in concentrate feeds (grains and oilseeds) is linoleic acid (C18:2n-6). It is incorporated into muscle and adipose tissue in relatively high concentrations after being broken down into MUFAs and SFAs in the rumen.

The primary advantage of concentrate feeding animals is that it results in higher growth rates and feed efficiency, which produce high-value carcasses at a younger age than extensive (pasture-based) production systems. This usually depends on more efficient use of feed resources as a result of careful mixing and formulation of feed. Cattle that are concentrate fed typically exhibit superior carcass features, such as i. e. superior to steers finished on pasture in terms of carcass weight, dressing percentage, conformation scores, and carcass grading or classification. Cattle raised on pasture produce meat that is darker and more yellow in color. Because of rumen biohydrogenation, there aren't many changes in the fatty acid content between cattle raised on pastures and those fed concentrate.

In South Africa and France, comparable outcomes for lamb and mutton fed on pasture as opposed to concentrate feeding were documented. Compared to lambs fed grass, concentrate-fed lambs yielded heavier carcasses with superior muscular conformation scores and greater body fat. When compared to lambs from grazing natural pastures, the meat from concentrate-fed lambs is more succulent and juicy. In comparison to lambs fed grain, the meat of pasture-fed lambs is darker and the subcutaneous fat is harder and yellower. Lambs raised on grass had meat with slightly lower levels of triacylglycerols and higher levels of phospholipids. The proportions of C18:0, C18:3n-3, and CLA are slightly higher in the triglyceride fraction, while the proportions of MUFAs, C18:2n-6, and other n-6 PUFAs are lower. Lower MUFAs, C18:2n-6 and other n-6 PUFAs, and higher levels of C18:3n-3 are found in the phospholipids. It's comparatively new to background feed before concentrate feeding starts. Studies show that backgrounding techniques improve the lipid profiles and CLA content of meat from Angus heifers in the long run.

5.2.1.5. Effects of Exogenous Growth-regulating Molecules on Fat and Fatty Acid Content of Red Meats

For more than 50 years, companies that produce red meat have used growth-promoting agents to increase average daily weight gain, improve animal feed efficiency, and improve muscle leanness. In the beef industry (apart from the European Union), hormonal growth implants and feed additives like anabolic steroids and beta-adrenergic agonists are used to increase growth rates and feed efficiency during the fattening phase, which leads to better carcass composition and quality. By producing meat that more closely satisfies consumer demands for optimizing carcass fat accumulation and resulting in a leaner product, the use of these exogenous growth modifiers has increased the effectiveness of red meat production. Since beef is now more reasonably priced than other sources of dietary protein, this has greatly improved the efficiency of animal production and benefited beef producers, retailers, and consumers.

Similar to endogenous catecholamines found in the body (norepinephrine and epinephrine), beta-adrenergic agonists are added to feed to increase feed efficiency by prioritizing protein synthesis over fat accretion towards the end of the fattening phase. Similar production benefits are offered by these synthetic products, such as ractopamine hydrochloric acid and zilpaterol hydrochloric acid, but they differ in application and mode of action. These treatments produce leaner carcasses with better conformation in the last twenty to 34 days of the fattening period. L-644,969, a beta-adrenergic agonist, was used to increase the amount and distribution of lean meat. It has been demonstrated that zelpaterol hydrochloride reduces the amount of fat in the carcass and improves the fatty acid composition of beef. Steer tissues supplemented with zilpaterol hydrochloride had higher amounts of oleic acid (C18:1) and lower amounts of C14:0 and C16:0, reducing the amount of saturated fatty acids in the meat.

5.2.2. Analysis of Cis/trans-Fatty Acid Composition of Red Meats

5.2.2.1. Cis/trans Fatty Acid Content of Beef and Lamb

This study set out to ascertain the fat and fatty acid composition of beef and mutton, paying particular attention to the ratio of cis to trans fatty acids.

Analysis was done on the cis/trans fatty acid content of lamb and beef from South Africa raised in conventional and range systems. For proximate and fatty acid analyses, representative samples of lamb loin chops (n = 80) and beef sirloin steaks (n = 60)—the lumbar region of the longissimus dorsi muscle—were taken from carcasses from multiple randomly chosen abattoirs in the Gauteng region of South Africa. The samples, which came from feedlot systems for the beef and extensively reared lamb production systems for the lamb, were fairly representative of the beef and lamb that could be found at meat shops in South Africa.

5.2.2.2. Materials and Methods

As previously mentioned, proximate analyses of medium- and long-chain fatty acids, as well as analyses of beef sirloins and lamb loin chops, were carried out. Using gas chromatography, the fatty acid methyl esters from samples of lamb and beef were examined. Using a scalpel, all visible fat was removed from the meat samples. Each sample was then minced and blended separately into a homogenous mixture using a BUCHI Mixer B-400.

An analysis was conducted to ascertain the nutritional contribution of intramuscular fat to the human diet. Warm, soapy water was used to thoroughly clean the glass cutting board, scalpel, mixer beaker, and blades that were used to dissect the meat samples. All mixed meat samples were kept in the refrigerator overnight in airtight receptacles pending lipid extraction and proximate analysis. The remaining samples were freeze-dried and then ground in a Hamilton Beach Commercial blender for about 10 seconds to a fine powder. The milled samples were labeled and put into sizable plastic containers with screw tops.

Using the Soxtec method and the Tecator Soxtec System 1040 Extraction unit, the fat content was determined by ether extract at an 80°C temperature. A duplicate ether extraction procedure was carried out utilizing subsamples of the minced, freeze-dried meat samples. The lipids were extracted from the freeze-dried meat samples using the previously described modified chloroform: methanol (2:1, v/v) method to perform fatty acid analyses. Using 14% BF_3/CH_3OH, the fatty acids were saponified and methylated. A Varian 3300 FID gas chromatograph equipped with a 100-meter WCOT fused silica capillary column (CP-Sil 88, 100 m x 0.25 mm DF 0.2 μm) was then used for analysis.

Fatty acids were identified based on the retention times of known fatty acid methyl ester standards obtained from Nu Chek Prep.Inc., Elysim, Minnesota (USA). The gas chromatograph program utilized was the same as the one previously mentioned, and helium was utilized as the carrier gas at a flow rate of 50 milliliters per minute. Fatty acids were reported as gravimetric concentrations (mg/g of tissue sample) and as the percentage of total medium- and long-chain fatty acids (w/w, percent).

An Excel spreadsheet was used to record the raw data, and IBM SPSS Statistics Windows, version 26 (SPSS Inc., Chicago, IL, USA) was used for all statistical operations. The molar proportions (w/w percent) and gravimetric content (mg/g meat) of the identified fatty acids, as well as the means and standard deviations of each lipid fraction, were calculated through quantitative analysis of the lamb and beef samples.

5.2.2.3. Composition of cis/trans-Fatty Acids in Beef and Lamb

Tables 5.2–5.4 present the findings of the cis/trans-fatty acid analysis of lamb and beef in South Africa, respectively. The findings shown in Table 5.2 show that the recommended threshold of 5 g per 100 g portion, which is the threshold for increased risk of CHD, is about 70% higher than the fatty acid composition of conventional beef (i.e., 1.4 g/100 g beef; Table 5.2). It should be noted that the beef used to obtain these TFA values came from cattle raised in South Africa's traditional production systems, i.e., carcass composition to satisfy market requirements is achieved by concentrate feeding for approximately 120 days after extensive grazing, resulting in carcasses with roughly 15–18% carcass fat and 5–7% intramuscular fat (measured in the longissimus dorsi (loin) muscle).

Table 5.2. *The Lipid Composition of Beef Sirloin (Longissimus Dorsi Samples) from Conventional Production Systems in South Africa (N = 60)*

Lipid Fraction	**Content (%)**
IMF SFA	6.5% 49.61%
MUFA	46.96%
PUFA	3.41%
TFA	4.19%
TFA/100 g meat	1,4 g

Source: C. Webb, E. (2021). Cis/Trans-Fatty Acid Content of Red Meats and the Related Effects on Meat Quality and Human Health. IntechOpen. doi: 10.5772/ intechopen.96726.

Table 5.3. *The Molar (W/W%) and Gravimetric (Mg/G) Content of Cis-and Trans-Fatty Acids in Beef Sirloin (Longissimus Dorsi Samples) from Conventional Production Systems in South Africa (w/w%) (n = 60)*

Fatty Acid	**Molar% (w/w%; n = 60)**	**Gravimetric Content (mg /g meat; n = 60)**
C14:0	3.2 ± 0.569	6.36 ± 2.69
C16:0	25.97 ± 3.205	51.67 ± 21.900
C18:0	20.44 ± 5.589	40.01 ± 17.285
C18:1(n-9)t	4.82 ± 1.930	9.32 ± 4.8
C18:1(*n-9*)c	40.71 ± 8.793	83.06 ± 44.641
C18:2 (*n-6*)	3.02 ± 0.955	5.98 ± 3.057
C18:2 (t-9,t-12)	0.14 ± 0.134	0.29 ± 0.206
C18:2 (c-9,t-12)	0.12 ± 0.097	0.27 ± 0.270
C18:2 (t-9,c-12)	0.01 ± 0.029	0.03 ± 0.109
C18:2 (c-9,c-12)	3.75 ± 0.923	7.35 ± 2.724
C18:3 (*n-3*)	0.39 ± 0.183	0.86 ± 0.594
C18:3 (t-9,t-12,t-15)	0.09 ± 0.098	0.20 ± 0.159
C18:3 (c-9,t-12,c-15)	0.07 ± 0.059	0.17 ± 0.179
C18:3 (t-9,c-12,c-15)	0.12 ± 0.094	0.27 ± 0.241
C18:3 (c-9,c-12,c-15)	0.09 ± 0.064	0.22 ± 0.153
CLA (*n-6*)	1.43 ± 0.234	0.90 ± 0.738

Source: C. Webb, E. (2021). Cis/Trans-Fatty Acid Content of Red Meats and the Related Effects on Meat Quality and Human Health. IntechOpen. doi: 10.5772/intechopen.96726.

Table 5.4. *The Lipid Composition of Lamb Loin Chops (Longissimus dorsi samples) from Conventional Production Systems in South Africa (n = 80)*

Lipid Fraction	**Content (%)**
Intramuscular fat%	13.95 ± 4.18%
SFA	52.13 ± 1.22%
MUFA	38.98 ± 2.65%

PUFA	3.24 ± 0.857%
CLA	0.70 ± 0.148%
TFAs	4.65 ± 1.284%
TFA's/100 g lamb	0.22 g

Source: C. Webb, E. (2021). Cis/Trans-Fatty Acid Content of Red Meats and the Related Effects on Meat Quality and Human Health. IntechOpen. doi: 10.5772/intechopen.96726.

Table 5.5. *The Molar (w/w%) and Gravimetric (mg/g) Content of Cis-and Trans-Fatty Acids in Lamb Shops (Longissimus dorsi samples) from Conventional Production Systems in South Africa (w/w%) (n = 80)*

Fatty Acid Composition	**Molar% (w/w%)**	**Gravimitric Content (mg/g meat)**
C12:0	0.146 ± 0.057	0.081 ± 0.040
C14:0	3.368 ± 0.673	1.742 ± 0.710
C14:1	0.084 ± 0.024	0.050 ± 0.024
C16:0	24.23 ± 1.788	16.612 ± 4.562
C16:1	1.023 ± 0.192	0.908 ± 0.279
C17:0	1.774 ± 0.525	0.545 ± 0.152
C18:0	22.30 ± 3.321	10.740 ± 2.373
C18:1 (n-11 t)	3.910 ± 1.256	1.735 ± 0.584
C18:1 (n-9c)	33.27 ± 2.809	24.348 ± 5.668
C18:l(n-11c)	0.631 ± 0.114	0.608 ± 0.124
C18:2 (n-6 t)	0.046 ± 0.014	0.019 ± 0.007
C18:2(n-6c)	2.029 ± 0.772	1.918 ± 0.385
C18:2 (n-10 t, n-12c)	0.592 ± 0.133	0.314 ± 0.147
C18:2 (n-10c, n-12c)	0.006 ± 0.001	0.003 ± 0.001
C18:2 (n-9c, n-11 t)	0.013 ± 0.008	0.008 ± 0.006
C18:3(n-6)	0.026 ± 0.005	0.022 ± 0.004
C18:3(n-3)	0.331 ± 0.069	0.176 ± 0.056
C20:0	0.131 ± 0.026	0.052 ± 0.011

C20:1(n-9)	0.060 ± 0.016	0.034 ± 0.006
C20:2	0.029 ± 0.007	0.030 ± 0.006
C22:0	0.011 ± 0.004	0.018 ± 0.004
C20:4(n-6)	0.081 ± 0.020	0.520 ± 0.147

Source: C. Webb, E. (2021). Cis/Trans-Fatty Acid Content of Red Meats and the Related Effects on Meat Quality and Human Health. IntechOpen. doi: 10.5772/intechopen.96726.

Figure 5.1 displays the regression of TFA accumulation with increasing carcass fat content in beef carcasses ($r^2 = 0.64$, $P < 0.05$). This shows that TFA concentrations rise gradually as the amount of carcass fat increases, but they reach a tipping point at around 25% of the carcass is fat. It appears that the concentration of TFAs reaches a turning point, possibly because ruminants preferentially incorporate other dominant fatty acids like MUFA and CLA metabolism and absorption, even though the proportions of fatty acids in ruminant fats generally remain relatively unchanged due to rumen biohydrogenation.

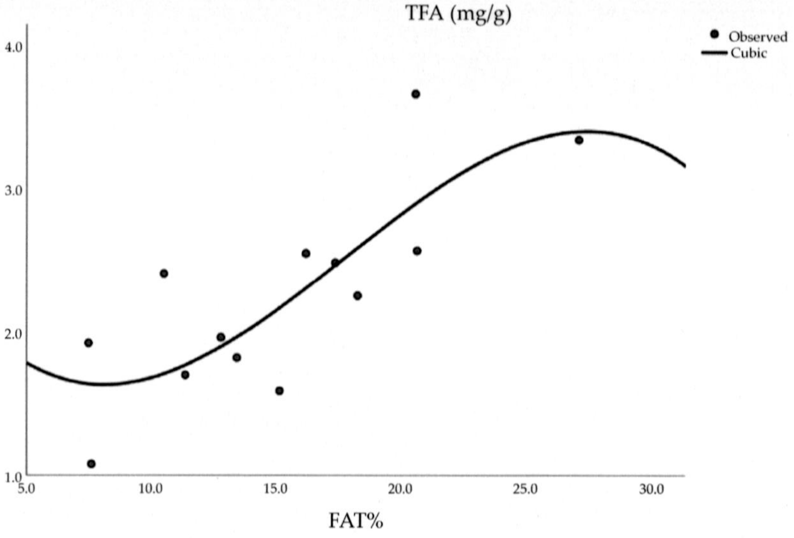

Figure 5.1. *Regression of trans-fatty acid content of beef loin samples (mg/g) with increasing carcass fat content in conventionally fed cattle.*

Source: C. Webb, E. (2021). Cis/Trans-Fatty Acid Content of Red Meats and the Related Effects on Meat Quality and Human Health. IntechOpen. doi: 10.5772/intechopen.96726.

Less than 50% of the fatty acids in beef produced using conventional production methods are saturated fatty acids; the remaining fatty acids are made up of heart-healthy PUFAs and MUFAs. There are about in trans-fatty acids in beef 4:19%, of which the necessary conjugated linoleic acid isomers make up a sizable portion.

Evaluation of the medium- and long-chain fatty acids in beef shows that the SFAs are made up of roughly 25% palmitic acid (C16:0), 20% stearic acid (C18:0), and less than 3.5% myristic acid (C14:0). Consequently, less than one-third of the fatty acid content of beef is made up of the medium-chain fatty acids (MCFA; C14:0 and C16:0). This is good news for heart health because MCFAs have been linked to a slight increase in low-density lipoprotein (LDL) cholesterol levels. Since stearic acid is a fatty acid that is neutral for cholesterol, it is advantageous to consume essential fats. Stearic acid appears to lower the risk of cancer and heart attacks in people, according to recent studies. It is also believed that this advantageous fatty acid signals cells' mitochondria and promotes the beta-oxidation of fatty acids. Because oleic acid (C18:1) lowers LDL cholesterol and makes up 40% of the MUFAs in beef, it has no negative effects on CVD or CHD. CLA isomers, which are mostly found in milk and meat from ruminant animals like cattle and sheep, also make up a sizable portion of PUFAs.

It is commonly known that oleic acid (C18:1) reduces the risk of cardiovascular disease. It is highlighted in research that substituting oleic acid for 5% of SFAs may lower the risk of coronary heart disease by 20–40%. Beef has a nearly 45% C18:1 component. It's crucial to understand that adding zilpaterol hydrochloride to cattle diets through concentrate feeding raises the oleic acid content of beef by about 2%.

The accumulation of MUFAs in beef loin muscle is highly correlated with an increase in carcass fat percentage ($r^2 = 0.94$, $P < 0.001$), as shown in Figure 5.2(a). Similarly, as shown in Figure 5.2(b), there is a significant correlation ($r^2 = 0.709$, $P < 0.001$) between the CLA content of beef loin muscle (mg/g of loin muscle) and rising carcass fat content in beef carcasses.

This suggests that, in addition to the established dietary effects on beef CLAs, increasing maturity and fat accumulation in cattle have a major impact on the accumulation of both MUFAs and CLAs in beef muscle. In most cases, cattle carcasses are fed 18–25% carcass fat in Europe, Australia, and South Africa, and 25–35% carcass fat in the United States. The CLA content of beef benefits from backgrounding on pastures in conjunction with high concentrate feeding.

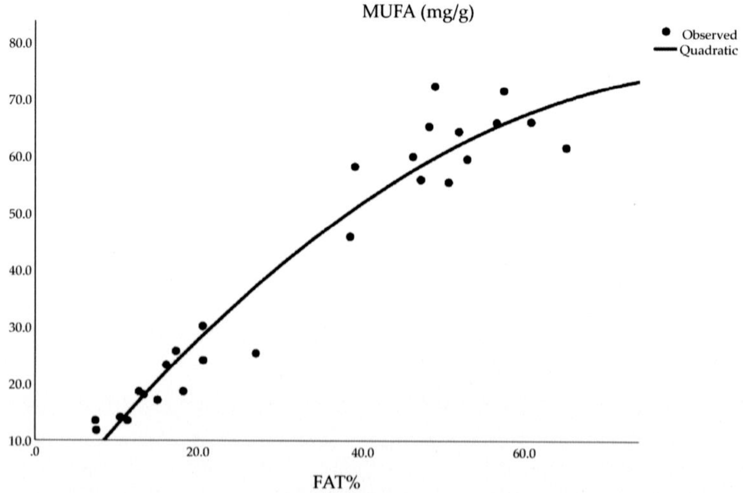

(a)

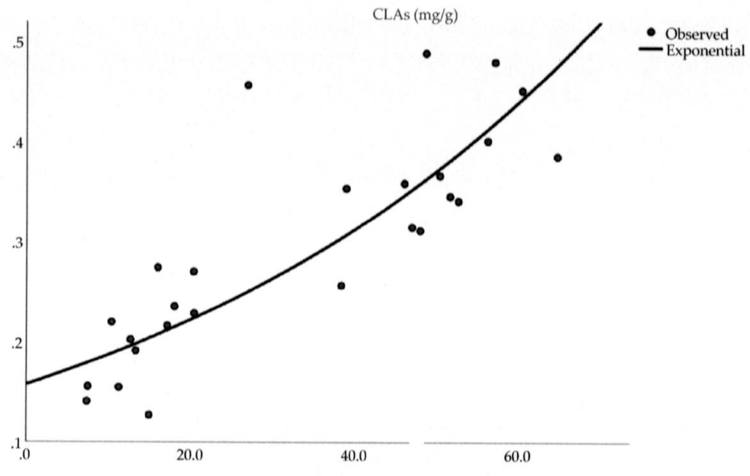

(b)

Figure 5.2. *Regression equations of (a) mono-unsaturated fatty acid content (MUFA, w/w%) of beef loin muscle with increasing carcass fat content, and (b) conjugated linoleic acid content (CLA, w/w%) of beef loin muscle with increasing carcass fat content.*

Source: C. Webb, E. (2021). Cis/Trans-Fatty Acid Content of Red Meats and the Related Effects on Meat Quality and Human Health. IntechOpen. doi: 10.5772/intechopen.96726.

Compared to beef, lamb meat has a higher intramuscular fat content. Contrary to popular belief, not all of the fat is saturated; there is a low trans-fatty content, roughly 40% MUFAs, 3.2% PUFAs, and SFAs that rarely exceed 53%. Data from common breeds of mutton sheep raised primarily on large pasture systems in South Africa show a low TFA content of 0.22 g/100 g meat, which is significantly less than the 5 g/100 g portion threshold.

Trans vaccenic acid derived from ruminants has been suggested not to be linked to an elevated risk of CVD or CHD because it is transformed into CLA isomers, which have several health advantages. In adipose tissues, vaccenic acid is transformed into CLA by the enzyme stearoyl-CoA desaturase. Additionally, oleic acid (C18:1c-9) is changed into stearic acid (C18:0) by this enzyme.

The TFA content of lamb loin muscle consistently stays low during the typical target weight ranges for slaughter, which is 40 to 55 kg live weight, as shown in Figure 5.3.

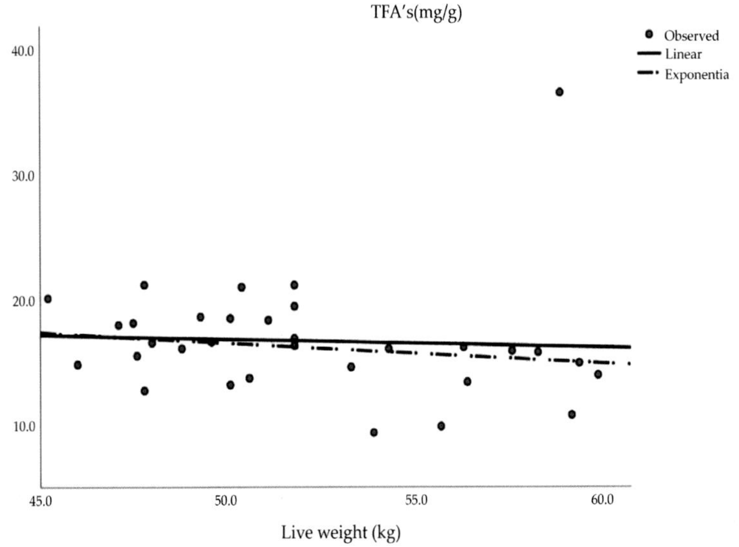

Figure 5.3. *Trans-fatty acid content (TFA, mg/g) of lamb loin muscle with increasing live weight.*

Source: C. Webb, E. (2021). Cis/Trans-Fatty Acid Content of Red Meats and the Related Effects on Meat Quality and Human Health. IntechOpen. doi: 10.5772/intechopen.96726.

5.3. PLANT-BASED FAT SOURCES

Research on plant lipids as well as plant protein components is required in relation to plant-based meat substitutes. The oil found in plant fruits or seeds, such as cocoa butter, sorghum fat, and shea butter, is referred to as vegetable fat. It is well-recognized that unsaturated fatty acid-rich vegetable fats can reduce blood cholesterol. Research is being done on this topic as more and more vegetable oils with low trans fats are being used to replace animal fats. It is well known that animal fat influences the flavor, aroma, and texture of meat products. It is ideal to use vegetable fats that can preserve the distinct texture, flavor, and appearance of animal fats when creating plant-based foods. There is comparatively little research on vegetable fats in comparison to that on vegetable proteins. Therefore, the purpose of research is to investigate traditional fat replacers and look at the use of vegetable oil in place of animal fat in meat alternatives (Lima et al., 2022).

5.3.1. Vegetable Fat Sources

Vegetable fats are often used in the form of medium-chain triglycerides (MCT) oil and canola oil. These high-content, refined vegetable oils are used because of their ability to hold water, induce gelation, and maintain emulsion stability. This has the benefit of giving meat substitutes a texture that is comparable to that of real meat. There are two primary types of fat substitutes, aside from vegetable oil: those based on fats and those based on proteins and carbs. The general term "fat replacer" is frequently used to refer to both kinds. Substances that mimic the physical or flavor characteristics of fat, especially in products with a high moisture content, fall under the category of fat mimetics. Proteins and carbohydrates are two examples of fat mimics. Another group consists of triglycerides that have undergone chemical and physical reconstitution, or fat substitutes. These claim to be low in calories but still provide all the functional qualities of fat. Directly derived fat substitutes fall under this category as well.

5.3.2. Vegetable Oils

MCT oil is used in medicine and nutrition. It is absorbed by the body and is primarily transformed into an energy source. It also aids in keeping the flavoring ingredients in their emulsion state. In comparison to unsaturated fatty acids, flaxseed, and chia seed oils have more than 0.85% saturated fatty acids, which helps to prevent arteriosclerosis and blood clot formation. When it comes to overall palatability, juiciness, and tenderness, canola oil is a great fat substitute. Sunflower oil also improves nutritional qualities and encourages the synthesis of phenolic compounds. Olive oil lowers the risk of cardiovascular illnesses and is high in unsaturated fatty acids, such as oleic acid. Meat substitutes have also been flavored with other vegetable oils, like margarine, shortening, orange oil, palm oil, and castor oil. Regardless of the quantity of oil added, meat analogs with orange oil had high moisture content, a comparatively high liquid retention capacity, and high DPPH radical scavenging activity.

Additionally, sensory analysis revealed that the juiciness of the soybeans increased and their smell decreased. Orange oil increased the meat analog juiciness and lessened their soybean odor, which enhanced the product's quality and appeal. Additionally, low-fat chicken sausages were made using garlic inulin as a fat substitute. Cold-pressed pumpkin seed oil is widely used to replace beef fat in the production of sausages resembling Bologna, canola oil is used as a solid fat substitute in plant-based cheese, and carrots and lemon are used as fat substitutes to improve the physicochemical, textural, and sensory aspects of low-fat beef burgers. These applications also make use of fibers (Jayarathna et al., 2022)

5.3.3. Fat Replacer: Fat Mimetics

Carbohydrate-based fat mimetics—which include corn syrup, gums, starch, pectin, cellulose, and polyol—have been used as thickeners and stabilizers for millennia and are primarily classified as Generally Recognized as Safe (GRAS) materials. More than 40 different types of starch-based alternative fats exist, using native and modified starch in instant or pregelatinized forms. Most notably, they have the benefit of not causing the hardening effect that is frequently seen when using products like bread and cookies in a microwave. Salad dressings and sauces use maltodextrin, which is made from hydrolyzed corn starch with acid or enzymes. Pfizer's "Litesse" is a polymer of glucose, sorbitol, and citric acid known as polydextrose. Oatrim, which is sold under the brand name Staley's "TrimChoice," is made by partially hydrolyzing the starch-containing portion of oat bran and outer skin with enzymes. Z-trim, created by the USDA G.C. Inglett, is an insoluble fiber that is not digested and is obtained from high-fiber soybeans, rice, and corn and wheat bran. Because it is an indigestible carbohydrate, inulin can act as a fat substitute and prebiotic. As a result, it is used in the food industry either alone or in combination with other pertinent ingredients, fulfilling a variety of functions as dietary fiber, low-calorie sweetener, fat replacement, gelling agent, viscosity modifier, and texture-modifying agent in a variety of foods.

Microparticulated protein products (MPP) derived from egg, milk, whey, soybean, and wheat proteins are formed into small, spherical particles (diameter 1~15 μm; 50 billion/1 tsp) and are referred to as protein-based fat mimetics. These particles have a creamy taste and texture in the mouth due to their small size, and they have freeze-thaw stability. In this category, NutraSweet's Simplesse is acknowledged as the leader in alternative fats.

5.3.4. Fat Replacer: Fat Substitutes

The production of sucrose fatty acid polyester (SPE) depends on the strong bond that forms between the hydroxyl and carboxyl groups of fatty acids. It is created when a single sugar molecule forms ester bonds with six to eight fatty acids. The type and form of the combined fatty acid ultimately determines the physical characteristics, functionality, and usability of SPE. Procter and Gamble's Olestra, which is made from fatty acids present in vegetable oils like soybean, corn, and cottonseed oil, is one well-known example. Sucrose fatty acid ester (SFE) is a useful stabilizer and emulsifier due to its exceptional emulsifying properties.

Alternative fats called carbohydrate fatty acid ester and polyol fatty acid ester are created when fatty acids and polyols with four or more hydroxyl groups—such as sorbitol, trehalose, raffinose, and stachyose—ester bond with each other. Triglycerides with high, intermediate, and low fatty acid compositions that are produced by random transesterification, chemical processes, or enzymes are referred to as structured lipids. Structured lipids have a wide melting point range and a broad range of applicable fatty acids, making them suitable for use in a variety of food products. Several fat substitutes are used in dairy products, such as protein-based microparticulated protein, modified whey protein concentrate, gums, inulin, maltodextrins, maltose, oatrim, polydextrose, and fat-based emulsifiers like olestra. For meat and poultry, carbohydrate-based gums, starches, oatrim, maltodextrins, and polydextrose are among the fat substitutes.

5.3.5. Features of Development Technology for Vegetable Fats

Vegetable oil can be used to replace animal fat in hamburger patties or sausages to soften them up. Hardness usually causes elasticity, gumminess, and chewiness to decrease. Furthermore, there is not much of a change in patty thickness in the vegetable oil-treated group. According to earlier studies, the rate at which regular patties lose thickness is significantly higher than that of low-fat patties that contain a partial amount of different vegetable oils. The kind and amount of fat used also affect the color of the patties. Later studies confirmed that the brightness values of sausages made with vegetable oil were higher than those of sausages made with pork fat. Little droplets of vegetable oil reflect more light than large droplets of animal fat, which accounts for the difference in color parameters. Furthermore, a study on the sensory evaluation and quality assessment of madeleines made with vegetable oil supported the finding that madeleines made with soy milk and vegetable oil instead of animal ingredients had a brighter color. Tests of physical properties revealed that products containing lecithin had much lower values for hardness, cohesiveness, adhesiveness, and chewiness. Principal component analysis was used to test the sensory experience and found that as the sample's juiciness and softness increased, so did overall preference. It was established that, in comparison to the other samples, the one with the additional emulsion had higher values for juiciness, softness, and overall preference. Samples for sensory tests were compared to their physicochemical properties using partial least squares regression analysis. As heat loss increased, so did the samples' juiciness and softness. It was established that the characteristics that showed the opposite trend to the samples' juiciness and softness—hardness, cohesiveness, elasticity, adhesiveness, and chewiness—had a positive correlation with liquid holding capacity, except adhesion.

Short- or medium-chain fatty acids can be combined to create fat-derived substitutes while maintaining the triglyceride structure. Or, they could be recently synthesized to have no calories and be able to express their function through the efficient use of emulsifiers. Foods that contain fat substitutes have a variety of functions, including preserving moisture, postponing aging, improving texture and physical characteristics, maintaining the stability of emulsions, and lowering the absorption of oil.

5.4. THE DEVELOPMENT OF PLANT-BASED FAT SUBSTITUTES

Partially hydrogenated oils (PHOs) were widely used in food processing in the past. Nonetheless, there has been a movement in favor of removing these oils from food products in recent years. Since PHOs are currently the main dietary source of artificial trans fatty acids, this removal of these fats is an attempt to remove industrially produced trans fats from the food supply. The task for food companies has been to find PHO replacements that can mimic their functionality in food processing without compromising organoleptic properties or consumer acceptance. This poses a serious technological challenge, especially for food systems that rely on the special functional qualities of these fats (such as the melting profiles of confectionary fats and the plasticity of laminating shortenings used in puff pastries).

5.4.1. Food Polymer Oleogels

Oleogels, also known as structured oils or organogels, are lipid-based materials that resemble solids and are made up of a lot of oil that has been organized into a three-dimensional thermoreversible gel network by oleogelators. The majority of structurants that are capable of forming networks are classified as polymers or as low-molecular-weight organogelators (LMOG). Since many polymers are affordable and of food grade compared to highly purified LMOGs, they have found extensive application in the food industry. Edible oils, which are mostly structured by oleogelators and have physicochemical characteristics akin to solid fats, make up the continuous liquid phase of edible oleogels. Edible oleogels, which have unique structures and physical characteristics, have the potential to replace solid fats, which are comparatively high in saturated and trans fatty acids. As mentioned earlier, trans, and saturated fats have been linked to an increased risk of several diseases. These include metabolic disorders like diabetes and obesity, as well as cancers of the breast, colon, and other organs. In addition to impairing newborn vision, trans, and saturated fats have also been linked to allergies, shortened gestational ages, preeclampsia risks, and nervous system disorders. Many commercial products (such as bakery, confectionery, and meat products) contain the majority of saturated and trans fatty acids. Nonetheless, it is now crucial for the food industry to look for alternatives and/or substitutes for these fats in food product formulations as more people become aware of the negative health effects of these fats. According to some authors, substituting polyunsaturated fats, like vegetable oils, for 5% of fat may lower human CVD risk by as much as 22–37%. It is important to remember, though, that trans and saturated fats are primarily added to food because they offer functional and technological benefits that are difficult to replicate, like crispness, flavor, extended shelf life, and satiety. To replace saturated and trans fats in food, many lipid modification techniques—including interesterification, fractionation, and blending—have been investigated. However, these techniques have not been able to effectively lower or eliminate the amounts of trans and saturated fat without changing the product's qualities. This may have something to do with the intrinsic distinctions between fats and oils, some of which are brought about by traits present in both their structural and functional attributes.

Foods containing palm and palm kernel have been used widely and effectively to partially replace hydrogenated fats. The use of these fats has drawn criticism and scrutiny due to sustainability concerns, and it has been demonstrated that they contain a high level of saturated fatty acids. As a result, there is growing concern about their use. It has also been investigated whether starch and other polysaccharides, or natural gums, can completely or partially replace shortenings in confectionery products. Refined carbohydrates do not give the product the same qualities as saturated fats, and this can have an impact on the final product's quality attributes and, ultimately, the product's acceptance by consumers. Furthermore, considering that diets high in refined, low-fat mono- and disaccharides have been linked to several health issues, including dyslipidemia, a condition that is a part of the metabolic syndrome, the use of carbohydrates does not meet the current nutritional demands. Therefore, the food industry's ongoing and vital task is to find sources of fat that are both technically sound and able to satisfy the demands of modern nutrition (Silva et al., 2023).

Research has been done on a number of structuring agents for edible oil structuring. These comprise, among other things, phytosterols, lecithin, vegetable waxes, and monoglycerides. Numerous writers have also studied the combination of these gelators. These various molecules combine to form fundamental supramolecular assemblies that belong to one of the pre-established oleogel structuration groups listed below:

- (i) Low-molecular weight compounds that have the potential of forming self-assembled structures,
- (ii) Crystalline particles,
- (iii) Polymeric or polymer strands of self-assembled structures, and
- (iv) Miscellaneous structures that include emulsion droplets and colloidal particles.

The ability to replicate the high saturated behavior of fat and the high structuration facilitated by various oleogel routes have been tested on a variety of food products, including spreads and margarine, ice cream, bakery goods, confections, and meat products.

It is important to keep in mind, though, that although many studies have been conducted, very few have been able to successfully replace fat with another substance while meeting all the necessary criteria, which include being completely replacement, having good technological qualities, being acceptable in terms of sensory aspects, and being commercially feasible. Therefore, more investigation is still required to identify effective substitutes that are both reasonably priced and organoleptically acceptable to the end user.

Based on their textural characteristics and sensory qualities, various combinations of oleogelator and oil were tested on various food products; the outcomes demonstrated the potential of oleogels as a solid fat substitute. It has been reported that the final application, which is dependent on how the oleogelators interact with other ingredients in the food matrix like sugar and water, is greatly influenced by the concentrations and types of oleogelators used. Vegetable waxes and monoglycerides work well as oleogelators to replace fat in baked goods, margarines, spreads, and other applications where it is possible to successfully replace all of the physical, chemical, and sensory aspects of the original fat. Research on meat product-based oleogels has also advanced, with different polymers and proteins demonstrating some encouraging outcomes. Nonetheless, there have been some encouraging findings for dairy and confectionery products, including ethyl cellulose and monoglycerides in chocolates, waxes in ice cream and chocolate paste, and monoglycerides in fillings. If the objective is to achieve 100% replacement, more research is necessary to validate the results, particularly in the areas of shelf life, heat stability, and sensory acceptance. To prepare oleogels for use in the production of food products, further research on scaling up some of the technologies used in oleogel fabrication and techno-functional property testing needs to be done.

5.4.2. Protein–Polysaccharides Oleogels

Because of their excellent solubility in water across a wide pH range and nutritional value, proteins are primarily used in the formulation of various food products. Under specific physicochemical conditions that rely on both their amino acid sequence and secondary structure, proteins can also form well-structured aqueous gels. An imbalance between repulsive and attractive interactions, which frequently result in protein aggregation, causes gelation. This process is controlled by many variables, including pH variation, ionic concentration, and the rate of heating and cooling. Protein networks tend to form more finely when the protein is farther from the point of intersection than when it is closer to the pI. The two primary types of proteins required for the synthesis of oleogels are fibrillar and globular proteins. Intermolecular interaction energy plays a major role in heat denaturation, which is the primary mechanism responsible for the gelation of globular proteins like soy proteins. A three-dimensional network is formed when globular proteins are heated because this causes the peptide chains to move energetically and interact with one another. The amino acids then bind to other proteins. Protein-based oleogels have been made using animal proteins like whey and egg proteins. Some examples of plant-based proteins that have been utilized in the creation of oleogels are soy protein and canola protein isolate. Nevertheless, there are still difficulties in creating oleogels using plant proteins. As mentioned earlier, the solubility, emulsifying potential, gelling or foaming ability, and water-holding capacity of plant-based proteins are generally regarded as being lower than those of animal sources. These variations result from their unique natural surroundings. As a result, it is typically difficult to directly replace animal proteins with plant proteins. Their solubility is a common illustration of a problem that occurs when plant proteins are substituted for animal proteins. Plant proteins are primarily classified as storage proteins because of their larger, more compact structures, increased hydrophobicity, and resulting decreased solubility.

Feichtinger, Scholten, Feichtinger, and others tried to produce oleogels made of plant-based proteins. Oleogels made of plant-based proteins demonstrated similar qualities to animal-based proteins in similarly effective manners.

Oleogels have also been made via the complexation of proteins (like soy protein) and polysaccharides (like kappa-carrageenan). Protein and polysaccharide complexation is achieved by noncovalent bonding, which results in the formation of well-defined nanoparticles with good dispersity. After the protein and polysaccharides are incorporated into the gel matrix, complexes between them are formed. These complexes can trap oil, which facilitates the interaction between oil droplets and increases structural integrity. According to certain authors, the main factors influencing the ability of polymer-based emulsions to withstand coalescence or even the separation of oil during dehydration are interfacial rigidity and network strength. Therefore, polymer complexation offers the best means of enhancing emulsion stability. Furthermore, distinct hydration, structure, and surface properties can be enabled by the interaction of proteins and polysaccharides in solution and at the oil–water interface. These properties can then be applied in a variety of domains, including oleogels, hydrogels, and emulsion gels. While the structural integrity of oleogels has been clarified, given that they are thermoreversible, there is still a lack of knowledge regarding the digestibility of these gels and the stability of the oils against lipid oxidation when used in various processes that would require high heat (Webb, 2021).

REFERENCES

1. Adams, D. J. (2022). Personal perspective on understanding low molecular weight gels. *Journal of the American Chemical Society, 144*(25), 11047–11053.
2. Andreani, G., Sogari, G., Marti, A., Froldi, F., Dagevos, H., & Martini, D. (2023). Plant-based meat alternatives: Technological, nutritional, environmental, market, and social challenges and opportunities. *Nutrients, 15*(2), 452.
3. Bai, M., Qin, G., Sun, Z., & Long, G. (2015). Relationship between molecular structure characteristics of feed proteins and protein *in vitro* digestibility and solubility. *Asian-Australasian Journal of Animal Sciences, 29*(8), 1159–1165.
4. Barac, M. B., Pesic, M. B., Stanojevic, S. P., Kostic, A. Z., & Bivolarevic, V. (2015). Comparative study of the functional properties of three legume seed isolates: Adzuki, pea and soy bean. *Journal of Food Science and Technology, 52*, 2779–2787.
5. Barbut, S., Wood, J., & Marangoni, A. (2016). Potential use of organogels to replace animal fat in comminuted meat products. *Meat Science, 122*, 155–162.
6. Berghout, J., Boom, R. M., & Van der Goot, A. J. (2014). The potential of aqueous fractionation of lupin seeds for high-protein foods. *Food Chemistry, 159*, 64–70.
7. Berrazaga, I., Micard, V., Gueugneau, M., & Walrand, S. (2019). The role of the anabolic properties of plant-versus animal-based protein sources in supporting muscle mass maintenance: A critical review. *Nutrients, 11*(8), 1825.
8. Bessada, S. M., Barreira, J. C., & Oliveira, M. B. P. (2019). Pulses and food security: Dietary protein, digestibility, bioactive and functional properties. *Trends in Food Science & Technology, 93*, 53–68.
9. Boukid, F. (2021). Plant-based meat analogues: From niche to mainstream. *European Food Research and Technology, 247*(2), 297–308.
10. Byanju, B., Rahman, M. M., Hojilla-Evangelista, M. P., & Lamsal, B. P. (2020). Effect of high-power sonication pretreatment on extraction and some physicochemical properties of proteins from chickpea, kidney bean, and soybean. *International Journal of Biological Macromolecules, 145*, 712–721.
11. Chen, L., Chen, J., Ren, J., & Zhao, M. (2011). Effects of ultrasound pretreatment on the enzymatic hydrolysis of soy protein isolates and on the emulsifying properties of hydrolysates. *Journal of Agricultural and Food Chemistry, 59*(6), 2600–2609.

12. Chen, L., Chen, J., Yu, L., Wu, K., & Zhao, M. (2018). Emulsification performance and interfacial properties of enzymically hydrolyzed peanut protein isolate pretreated by extrusion cooking. *Food Hydrocolloids*, *77*, 607–616.
13. Da Pieve, S., Calligaris, S., Co, E., Nicoli, M. C., & Marangoni, A. G. (2010). Shear nanostructuring of monoglyceride organogels. *Food Biophysics*, *5*, 211–217.
14. Day, L., Cakebread, J. A., & Loveday, S. M. (2022). Food proteins from animals and plants: Differences in the nutritional and functional properties. *Trends in Food Science & Technology*, *119*, 428–442.
15. De Smet, S., Webb, E. C., Claeys, E., Uytterhaegen, L., & Demeyer, D. I. (2000). Effect of dietary energy and protein levels on fatty acid composition of intramuscular fat in double-muscled Belgian Blue bulls. *Meat Science*, *56*, 73–79.
16. Dhaka, V., Gulia, N., Ahlawat, K. S., & Khatkar, B. S. (2011). Trans fats–Sources, health risks and alternative approach–A review. *Journal of Food Science and Technology*, *48*, 534–541.
17. Feichtinger, A., & Scholten, E. (2020). Preparation of protein oleogels: Effect on structure and functionality. *Food*, *9*(12), 1745.
18. Geerts, M. E., Dekkers, B. L., van der Padt, A., & van der Goot, A. J. (2018). Aqueous fractionation processes of soy protein for fibrous structure formation. *Innovative Food Science & Emerging Technologies*, *45*, 313–319.
19. Giacomozzi, A. S., Carrín, M. E., & Palla, C. A. (2018). Muffins elaborated with optimized monoglycerides oleogels: From solid fat replacer obtention to product quality evaluation. *Journal of Food Science*, *83*(6), 1505–1515.
20. Godoi, K. R. R., Basso, R. C., Ming, C. C., Silva, V. M. d., Cunha, R. L. d., Barrera-Arellano, D., & Ribeiro, A. P. B. (2019). Physicochemical and rheological properties of soybean organogels: Interactions between different structuring agents. *Food Research International*, *124*, 108475.
21. Goldstein, B., Moses, R., Sammons, N., & Birkved, M. (2017). Potential to curb the environmental burdens of American beef consumption using a novel plant-based beef substitute. *PLOS One*, *12*(12), e0189029.
22. Griel, A. E., & Kris-Etherton, P. M. (2006). Beyond saturated fat: The importance of the dietary fatty acid profile on cardiovascular disease. *Nutrition Reviews*, *64*, 257–262.
23. Grossmann, L., & Weiss, J. (2021). Alternative protein sources as techno-functional food ingredients. *Annual Review of Food Science and Technology*, *12*, 93–117.

24. He, X., Liu, H., Liu, L., Zhao, G., Wang, Q., & Chen, Q. (2014). Effects of high pressure on the physicochemical and functional properties of peanut protein isolates. *Food Hydrocolloids*, *36*, 123–129.
25. Jayarathna, G. N., Jayasena, D. D., & Mudannayake, D. C. (2022). Garlic inulin as a fat replacer in vegetable fat incorporated low-fat chicken sausages. *Food Science and Animal Resources*, *42*, 295–312. https://doi.org/10.5851/kosfa.2022.e5.
26. Laaksonen, D. E., Nyyssonen, K., Niskanen, L., Rissanen, & Salonen, J. T. (2005). Prediction of cardiovascular mortality in middle-aged men by dietary and serum linoleic and polyunsaturated fatty acids. *Archives of Internal Medicine*, *165*, 193–199.
27. Lima, T., Costa, G. F., Alves, R., de Araújo, C., Silva, G., Ribeiro, N., et al. (2022). Vegetable oils in emulsified meat products: A new strategy to replace animal fat. *Food Science and Technology*, *42*, 103621. https://doi.org/10.1590/fst.103621.
28. Mensink, R. P., et al. (2003). Effects of dietary fatty acids and carbohydrates on the ratio of serum total to HDL cholesterol and on serum lipids and apolipoproteins: A meta-analysis of 60 controlled trials. *American Journal of Clinical Nutrition*, *77*(5), 1146–1155.
29. Meyer, B. J., Mann, N. J., Lewis, J. L., Milligan, G. C., Sinclair, A. J., & Howe, P. R. C. (2003). Dietary intakes and food sources of omega-6 and omega-3 polyunsaturated fatty acids. *Lipids*, *38*, 391–397.
30. Silva, R. C. D., Ferdaus, M. J., Foguel, A., & da Silva, T. L. T. (2023). Oleogels as a fat substitute in food: A current review. *Gels*, *9*(3), 180.
31. Tshabalala, P. A., Strydom, P. E., Webb, E. C., & De Kock, H. L. (2003). Meat quality of designated South African indigenous goat and sheep breeds. *Meat Science*, *65*, 563–570.
32. Webb, C. (2021). Cis/Trans-Fatty Acid Content of Red Meats and the Related Effects on Meat Quality and Human Health. *IntechOpen*.
33. Webb, E. C., & Casey, N. H. (1995). Fatty acids in carcass fat of steers treated with β-adrenergic agonist individually or in combination with trenbolone acetate + oestradiol-17b. *Meat Science*, *41*(1), 69–76.
34. Webb, E. C., & Casey, N. H. (1995). Genetic differences in fatty acid composition of subcutaneous adipose tissue in Dorper and SA Mutton Merino wethers at different live weights. *Small Ruminant Research*, *18*, 81–88.
35. Webb, E. C., Casey, N. H., & Van Niekerk, W. A. (1994). Fatty acids in the subcutaneous adipose tissue of intensively fed SA Mutton Merino and Dorper wethers. *Meat Science*, *38*, 123–131.

Chapter 6

Vitamins and Minerals

Contents

Introduction 198
6.1. Key Vitamins and Minerals in Traditional Meats 198
6.2. Vitamin and Mineral Content in Plant-Based Alternatives 205
6.3. Bioavailability and Absorption Differences 206
References 209

INTRODUCTION

Plant-based meat alternatives (PBMAs) are becoming more and more popular and can give certain populations the essential nutrients they need. As a result, the nutritional composition of several popular PBMA formulations, including the Beyond Meat Burger (BMB1 and BMB2), Impossible Foods Burger (IFB1 and IFB2), and Morning Star's Black Bean Burger (BBB), was evaluated in contrast to conventional Animal-Based Meats (ABM), like 80/20 Ground Pork (GP). All PBMAs had significantly higher sodium contents ($P<0.05$) than GP, and all PBMAs had numerically higher total saturated fat contents than GP. All PBMAs had a vitamin E content that was numerically higher than ABMs. While anabolically important EAA, like methionine and lysine, were significantly greater ($P<0.05$) in GP, the total content of Essential Amino Acids (EAA) was numerically higher in BMB2 than in ABMs. Although PBMAs and ABMs were similar in terms of nutrients, more research needs to be done on bioavailability.

6.1. KEY VITAMINS AND MINERALS IN TRADITIONAL MEATS

When people think of nourishing, nutrient-dense foods, meat is frequently disregarded due to a trend against it in popular dietary guidelines. However, the meat that is highest in nutrients on the planet is actually meat. The most important nutrients that are only present in meat, or only in significant amounts, are listed below.

- D3.
- B12.
- Vitamin A (Retinol).
- Creatine.
- Carnitine.
- Carnosine.
- Heme iron.
- Docosahexaenoic acid (DHA).
- Eicosapentaenoic Acid (EPA).
- Taurine.

6.1.1. D3

Vitamin D3 is a hormone as well as a vitamin. When you get enough sun exposure, your skin can produce vitamin D in addition to ingesting it through food. This means that if you live in regions with cold, cloudy winters and in the far north or south, vitamin D becomes even more crucial. Ergocalciferol (D2), which is found in plant foods, and cholecalciferol (D3), which is found in animal foods, are the two forms of dietary vitamin D. Research indicates that animal-based cholecalciferol (D3) has a significantly higher absorption rate than its plant-based equivalent. The best natural sources of D3 are egg yolks, beef organs, and fatty fish. Individuals with inadequate intake of D3 from food and sunlight may encounter a range of adverse symptoms, such as immune system dysfunction, multiple sclerosis, and depression. According to recent research, vitamin D3 can aid in the defense against viral infections, such as Covid-19.

6.1.2. B12

The main source of vitamin B12, also known as cobalamin, is animal-based diets such as meat, fish, cheese, and eggs. This makes B12 difficult for vegetarians or vegans who consume less meat to obtain. The following are signs of a B12 deficiency.

- Chronic fatigue.
- Psychiatric problems.
- Neurological problems.
- Increased risks of Alzheimer's disease.
- Intergenerational neurological disorders.

One reliable source of B_{12} and other B vitamins is meat. The best source found in nature is beef liver, which has 2471% of your RDA per 100 grams. On the other hand, most people find chicken liver to be more palatable and offers 220% per 100 grams. Although animal products are by far the best source of vitamin B_{12}, certain plant foods also include the following amounts:

- Shiitake.
- Nori (seaweed).
- Tempeh (fermented soy).

6.1.3. Vitamin A (Retinol)

Retinol, a fat-soluble form of vitamin A, can only be found in meat and animal products. It is necessary for immune system function, physical development, reproduction, and vision maintenance. You may be aware of the high vitamin A content of certain plants, such as carrots. Carrots do contain vitamin A, but it is found in the form of a carotenoid, which the body must convert to vitamin A. Vitamin A retinal is significantly more bioavailable than its counterpart in plant foods, much like the iron and zinc found in animal foods. According to studies, the absorption efficiency of retinol was approximately 30% in just one hour, while that of vitamin A carotenoids was less than 5%. You will be deficient in vitamin A if you don't get it from your diet, which can lead to infections, dry eyes, hair loss, skin problems, and night blindness.

Retinal is a fat-soluble vitamin that enters the bloodstream and is utilized by the body when it is combined with animal fat. The majority of meats that are high in vitamin A are also high in fat, which is not surprising given that human physiology evolved on a diet consisting primarily of fatty meat for almost two million years. Another piece of evidence supporting the theory that humans are meant to survive on meat is the fact that this is not the case for plant foods high in carotenoids.

The highest retinol foods include:

- Beef Liver—713% DV per serving.
- 1 slice: 6421 mcg (713% DV) 100 grams: 9442 mcg (1049% DV).
- Lamb Liver—236% DV per serving.
- 1 ounce: 2122 mcg (236% DV) 100 grams: 7491 mcg (832% DV).
- Liver Sausage—166% DV per serving.
- 1 slice: 1495 mcg (166% DV) 100 grams: 8384 mcg (923% DV).
- 4. Cod Liver Oil—150% DV per serving.
- 1 teaspoon: 1350 mcg (150% DV) 100 grams: 30,000 mcg (3,333% DV).
- King Mackerel—43% DV per serving.
- Half a fillet: 388 mcg (43% DV) 100 grams: 252 mcg (28% DV).

6.1.4. Creatine

Your muscles and brain cells can easily access energy thanks to a molecule called creatine. Over the past few decades, creatine has become one of the most widely used bodybuilding supplements due to its potential to increase muscle growth, strength, and size. Although your liver can produce small amounts of creatine, muscle tissue requires much larger amounts of this substance to be saturated. Fortunately, red meat contains these higher concentrations. Some proponents of the carnivore diet contend that consuming multiple pounds of meat daily is just as beneficial as taking creatine supplements. In fact, research suggests that meat eaters may already have an adequate supply of creatine and may not benefit from taking supplements. Conversely, it has been demonstrated that eating a vegetarian diet lowers creatine levels. Exercises that deplete creatine reserves may make vegetarians even less successful at preserving their creatine stores.

6.1.5. Carnitine

One amino acid that is mainly present in animal products is L-carnitine. Even though L-carnitine isn't strictly speaking an "essential" amino acid, it still has value. It might activate "fat-burning mode" to a degree that would both improve muscle endurance and slow the progression of Alzheimer's. Although there isn't currently a set RDA or RDV for L-carnitine, many organizations suggest that a dosage of 2–4 grams per day is the most advantageous. Sufficient dosages might have a cardioprotective impact. The highest concentrations of L-carnitine are found in beef and pork. There are approximately 80 milligrams in 3 ounces of beef, 22 milligrams in 3 ounces of pork, and 5 milligrams in 3 ounces of fish.

6.1.6. Carnosine

The muscles and brain contain the vital antioxidant carnosine. Similar to creatine, increased muscle strength and endurance are directly correlated with carnosine. Even though carnosine is regarded as optional, these advantages shouldn't be disregarded. There are two ways that meat eaters obtain carnosine: directly from meat and eggs, or indirectly through beta-alanine, which is also present in meat and eggs. Naturally, omnivores, and carnivores have higher levels of carnosine than vegetarians (Figure 6.1).

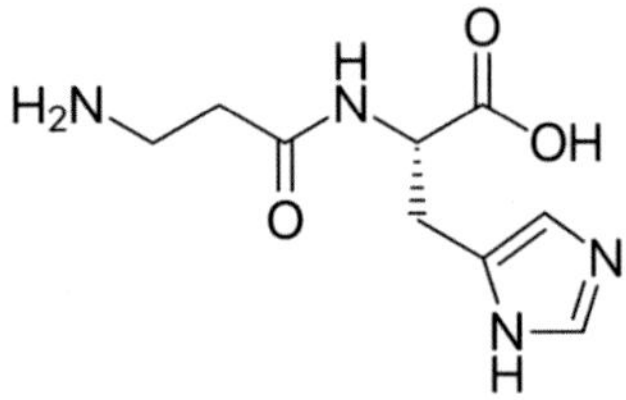

Figure 6.1. *Carnosine.*

Source: By Vaccinationist – Own work using: PubChem, Public Domain, https://commons.wikimedia.org/w/index.php?curid=50903513.

6.1.7. Heme Iron

Heme iron from red meat and non-heme iron from plants are the two forms of iron. It's likely that you've already noticed a trend: heme iron is more readily absorbed. People who eat meat typically have complete iron levels. Conversely, non-heme iron is poorly absorbed, and the plant antinutrients that come with it frequently further reduce its bioavailability. Anemia is more common in vegans and vegetarians due to these factors.

6.1.8. Docosahexaenoic Acid (DHA)

An omega-3 fatty acid called DHA shields the brain from oxidative stress and inflammation. A DHA deficiency can lead to serious mental health issues, particularly in children whose brains develop so quickly. Fish, pastured eggs, flaxseeds, chia seeds, grass-fed beef, and algae are among the best sources of DHA. An animal is more likely to have omega-3s if it was fed and/or raised in a healthier environment.

6.1.9. Eicosapentaenoic Acid (EPA)

Another significant omega-3 fatty acid is EPA, which is primarily present in cold-water seafood such as sturgeon, salmon, herring, shrimp, and eels. Certain foods made from grass-fed animals, such as meats and full-fat dairy, also contain some EPA. The body uses eicosapentaenoic acid (EPA) to make eicosanoids, which are signaling molecules. These molecules have many significant functions, one of which is to lower inflammation, which is the primary cause of many contemporary diseases. Research on the benefits of fish oil rich in EPA and DHA demonstrates a decrease in depressive symptoms. A 2009 study discovered that when EPA and DHA were compared, EPA had stronger antidepressant effects. It has been demonstrated that EPA lessens hot flashes in menopausal women.

6.1.10. Taurine

Sulfur-based taurine is a substance that is present in all body parts and essential organs. Although taurine's exact function is unknown, it appears to be important for the synthesis of bile salts, antioxidants, and muscle function. It also influences the following.

- Eye health.
- Central nervous system function.
- Immune regulation.
- Antioxidant effects.
- Maintaining hydration and balanced electrolytes.

Only animal foods like beef, fish, and dairy contain taurine. The baseline taurine levels in your body are probably influenced by your diet. This implies that vegans have lower taurine levels than omnivores, as one might anticipate.

6.1.11. Other Nutrients in Meat

Meat is a good source of several other vitamins and minerals in addition to the above-mentioned hard-to-find nutrients. These are a few of the standouts.

Vitamins in Meat

- Niacin.
- Riboflavin.
- Vitamin A.
- Vitamin B5.
- Vitamin B6.

6.1.11.1. Niacin

Together with other B vitamins, niacin facilitates the synthesis of intracellular energy. Your body can produce niacin from dietary tryptophan in addition to getting it from food. The best foods include seafood, eggs, cheese, and meats because they are high in tryptophan and niacin, which can raise your body's levels in both directions.

6.1.11.2. Riboflavin

Another B vitamin that promotes metabolism is riboflavin. Its primary functions include encouraging the absorption of iron, preserving good vision, and releasing energy from the food you eat. The majority of animal foods, such as meat, milk, cheese, and yogurt, are excellent providers of riboflavin due to their high riboflavin content.

6.1.11.3. Vitamin A

One of the most crucial nutrients for the health of your skin and eyes is vitamin A. Additionally, degenerative eye issues like these may be avoided with vitamin A.

- Night blindness.
- Vision loss.
- Dry eyes.
- Ulcers.

Foods derived from animals, particularly beef liver, are the best natural source of preformed vitamin A. Carotenoids, which are found in carrots and sweet potatoes, are low in bioavailability because they are "precursors" of vitamin A that your body needs to convert into usable form.

6.1.11.4. Minerals in Meat

- Zinc.
- Selenium.

- Phosphorus.
- Magnesium.
- Potassium.
- **Zinc:** One of the best dietary sources of zinc is red meat. Not only does red meat contain more zinc than almost any other plant food, but it is also four times more readily absorbed. Another nutrient that vegans and vegetarians frequently lack is zinc, as it can be challenging to meet your RDA for zinc only through plants and grains. Maintaining a healthy level of zinc (and copper, with which it closely interacts) enables your body to perform a number of essential processes. Zinc helps develop motor neurons and cognitive function in kids and improves insulin sensitivity and glycemic control in adults. And it's essential for healthy immune system operation at any age.
- **Selenium:** A nutrient that supports the thyroid and indirectly affects many physiological processes is selenium. It's particularly beneficial for your heart. Selenium levels in most beef cuts range from 20 to 30 micrograms per 3 ounces. An even better source of selenium is pork: a cup of roast pork provides 122% of your recommended daily intake. Given the depletion of selenium in our soils, getting your selenium from animal sources is probably more important than it has ever been. Because selenium is so vital to immune function, regions with particularly low soil selenium levels frequently have higher rates of COVID-19 infection.

Meat was a staple of the carnivorous diet that humans evolved to consume. While other non-meat nutrients are poorly absorbed by the body, meat provides a rich combination of vitamins, minerals, and other nutrients, some of which are not available from other food groups. Red meat is a fantastic source of: in particular, organ meats.

- D3.
- B12.
- Taurine.
- Creatine.
- Carnitine.
- Carnosine.
- Heme iron.
- Docosahexaenoic acid (DHA).

No other food group comes close to providing so many nutrients in their most bioavailable forms.

6.2. VITAMIN AND MINERAL CONTENT IN PLANT-BASED ALTERNATIVES

The popularity of plant-based meat substitutes has grown as more people explore methods to consume less meat. Depending on the brand and formulation, their nutritional value can vary greatly, but generally speaking, they try to offer a good balance of vitamins and minerals. An overview of potential contents in these products is provided below:

1. **Protein**: Protein-dense plant-based meat substitutes are frequently made from soy, peas, or other legumes. Although there may be a slight variation in the amino acid profile, the protein content can be similar to that of meat.
2. **Iron**: Iron is frequently added to plant-based meats, usually in the form of ferrous sulfate or other iron salts. But heme iron, which is present in animal products, is more readily absorbed than non-heme iron, which is found in plants. In certain products, vitamin C is also added to improve the absorption of iron.
3. **Calcium**: Certain plant-based meats—especially those made to resemble dairy or meat, which are normally sources of calcium—are fortified with calcium to help meet daily requirements.
4. **B Vitamins**: Since vitamin B12 is not naturally present in plant foods, it is frequently added to meat substitutes made from plants. Depending on the product, additional B vitamins like B6 and folate may also be present.
5. **Zinc**: Meats derived from plants can contain zinc, but plant sources have a lower bioavailability of zinc than animal sources. Zinc may be used to fortify certain products.
6. **Omega-3 Fatty Acids**: Some plant-based meat substitutes include sources of omega-3s, such as flaxseeds or algae oil, to supply these vital fatty acids, even though they aren't always present in significant amounts.
7. **Fiber**: Plant-based meats can have different levels of dietary fiber depending on how they are made, especially if they have whole grains or beans in them.
8. **Sodium**: Since many plant-based meat substitutes include flavorings and preservatives, they may contain high sodium content. To keep an eye on your sodium intake, it's a good idea to read labels.
9. **Other Vitamins and Minerals**: You may also find additional vitamins and minerals, such as magnesium, potassium, and vitamin D, depending on the product.

To obtain exact information, always read the nutritional label of the product you're interested in. Brands and products can have quite different formulations.

6.3. BIOAVAILABILITY AND ABSORPTION DIFFERENCES

The percentage of a nutrient or bioactive compound that is absorbed and used by the body following ingestion is referred to as bioavailability. The process by which these nutrients leave the digestive tract and enter the bloodstream is called absorption. Consumer nutritional outcomes can be greatly impacted by the variations in the bioavailability and absorption of traditional meats and plant-based meat substitutes. There are a few significant distinctions between traditional meats and plant-based meat substitutes when it comes to the bioavailability and absorption of nutrients.

6.3.1. Vitamins in Traditional Meats vs. Plant-Based Alternatives

Many important vitamins, especially B vitamins like B_{12}, niacin (B_3), riboflavin (B_2), and pyridoxine (B_6), are abundant in traditional meats. Although vitamin B_{12} is primarily present in animal products, it is essential for blood formation and nerve function. Because of this, traditional meats are an essential source for people who don't take supplements or fortified foods. In contrast, to replicate the nutritional makeup of conventional meats, plant-based meat substitutes might be enhanced with specific vitamins. For instance, vitamin B_{12} from synthetic or fermentation-derived sources is added to some products. But in the absence of fortification, plant-based substitutes usually don't contain enough of this vitamin. Although they are frequently found in lesser amounts than their animal counterparts, other B vitamins may also be present.

6.3.2. Minerals in Traditional Meats vs. Plant-Based Alternatives

Traditional meats and plant-based substitutes differ significantly in terms of minerals like iron, zinc, calcium, and magnesium:

1. Iron:
   - **Traditional Meats:** The absorption rate of heme iron, which is present in red meat, is higher (15–35%) than that of non-heme iron, which is present in plants (2–20%). Since heme iron does not require additional factors for absorption, the body uses it more effectively.

   - **Plant-Based Alternatives:** Legumes, grains, nuts, and seeds are sources of non-heme iron that are utilized in plant-based products. Phytates, which are present in whole grains, have the ability to prevent non-heme iron absorption. Foods high in vitamin C should be consumed in conjunction with plant sources to improve absorption.

2. Zinc:
   - **Traditional Meats:** Because animal proteins contain less phytate content, which prevent absorption, the zinc they contain is more bioavailable than zinc found in plant sources.
   - **Plant-Based Alternatives:** Because phytates bind zinc ions, reducing their availability for absorption, zinc from legumes and whole grains typically has a lower bioavailability.
3. Calcium:
   - **Traditional Meats:** Certain meats have trace amounts of calcium, which can increase overall intake, even though they are not as important a source as dairy products or leafy greens.
   - **Plant-Based Alternatives:** Plant-based foods fortified with calcium, such as soy milk, can supply enough calcium; on the other hand, oxalates, which are present in some vegetables, such as spinach, can prevent the body from absorbing calcium.
4. Magnesium:
   - Magnesium is found in both conventional meats and plant-based substitutes; however, plant sources such as nuts and seeds tend to contain higher amounts of the mineral as well as potentially harmful compounds that may alter its bioavailability.

6.3.3. Impact on Nutritional Status

A balanced plant-based meat substitute and a range of nutrient-dense plant foods can help guarantee that you are getting enough vitamins and minerals. Optimizing the bioavailability and absorption of nutrients from plant-based sources can be achieved by paying attention to food combinations, processing techniques, and fortification levels. Seeking individualized advice from a healthcare provider or dietitian regarding fulfilling your specific dietary requirements is always a good idea. The nutritional status of an individual is affected by the variations in bioavailability between these two categories.

- People who eat a lot of traditional meats generally have easier access to vital nutrients such as heme iron and vitamin B12.

- People who follow plant-based diets need to plan carefully and be aware of the nutrients they are consuming. They should also make sure to take supplements or foods that are fortified when needed.
- If people don't make up for some of the nutrients that are primarily found in plants that have a lower bioavailability, deficiencies may result.

In general, traditional meats offer more nutrients that are bioavailable, especially in the case of protein, zinc, heme iron, and omega-3 fatty acids. For plant-based substitutes to improve nutrient absorption, fortification or particular food combinations are frequently needed. To address potential deficiencies and improve bioavailability, many plant-based meat substitutes are fortified with vital nutrients like B12, iron, calcium, and zinc. To meet their nutritional needs, people who follow plant-based diets might need to pay closer attention to the sources of their nutrients and might find it helpful to combine foods or use fortified products. Knowing these distinctions can help you, whether you eat a plant-based or an omnivorous diet, make educated food decisions and make sure your nutritional needs are satisfied (Huskisson et al., 2007).

REFERENCES

1. Adzitey, F., & Nurul, H. (2011). Pale soft exudative (PSE) and dark firm dry (DFD) meats: Causes and measures to reduce these incidences—A mini review. *International Food Research Journal*, *18*(1).
2. Ali Hassan, A., Sandanger, T. M., & Brustad, M. (2012). Selected vitamins and essential elements in meat from semi-domesticated reindeer (*Rangifer tarandus* L.) in mid- and northern Norway: Geographical variations and effect of animal population density. *Nutrients*, *4*(7), 724–739.
3. Andersen, H. J., Oksbjerg, N., Young, J. F., & Therkildsen, M. (2005). Feeding and meat quality—A future approach. *Meat Science*, *70*(3), 543–554.
4. Aranceta, J., & Pérez-Rodrigo, C. (2012). Recommended dietary reference intakes, nutritional goals and dietary guidelines for fat and fatty acids: A systematic review. *British Journal of Nutrition*, *107*(S2), S8–S22.
5. Barabási, A. L., Menichetti, G., & Loscalzo, J. (2020). The unmapped chemical complexity of our diet. *Nature Food*, *1*, 33–37. https://doi.org/10.1038/s43016-019-0005-1.
6. Barron-Hoyos, J. M., Archuleta, A. R., del Refugio Falcon-Villa, M., Canett-Romero, R., Cinco-Moroyoqui, F. J., Romero-Barancini, A. L., & Rueda-Puente, E. O. (2013). Protein quality evaluation of animal food proteins by *in-vitro* methodologies. *Food and Nutrition*, *4*, 376–384.
7. Bohrer, B. M. (2019). An investigation of the formulation and nutritional composition of modern meat analogue products. *Food Science and Human Wellness*, *8*, 320–329. https://doi.org/10.1016/j.fshw.2019.11.006.
8. Briguglio, M., et al. (2018). Dietary neurotransmitters: A narrative review on current knowledge. *Nutrients*, *10*, 591. https://doi.org/10.3390/nu10050591.
9. Chan, W., Brown, J., Lee, S., & Buss, D. H. M. (1995). *Poultry and Game*. Fifth supplement to McCance & Widdowson's the composition of foods. London: The Royal Society of Chemistry and the Ministry of Agriculture Fisheries and Food.
10. Chen, J., et al. (2020). Structure-antioxidant activity relationship of methoxy, phenolic hydroxyl, and carboxylic acid groups of phenolic acids. *Scientific Reports*, *10*, 2611. https://doi.org/10.1038/s41598-020-59451-z.
11. Curtain, F., & Grafenauer, S. (2019). Plant-based meat substitutes in the flexitarian age: An audit of products on supermarket shelves. *Nutrients*, *11*, 2603. https://doi.org/10.3390/nu11112603.
12. De Smet, S., & Vossen, E. (2016). Meat: The balance between nutrition and health. A review. *Meat Science*, *120*, 145–156.

13. Droulez, V., Williams, P. G., Levy, G., Stobaus, T., & Sinclair, A. (2006). Composition of Australian red meat 2002. 2. Fatty acid profile. *Food Australia*, *58*, 335–341.

14. Eze, N. M., Maduabum, F. O., Onyeke, N. G., Anyaegunam, N. J., Ayogu, C. A., Ezeanwu, B. A., & Eseadi, C. (2017). Awareness of food nutritive value and eating practices among Nigerian bank workers: Implications for nutritional counseling and education. *Medicine*, *96*(10).

15. Frei, B., England, L., & Ames, B. N. (1989). Ascorbate is an outstanding antioxidant in human blood plasma. *Proceedings of the National Academy of Sciences of the United States of America*, *86*(16), 6377–6381. https://doi.org/10.1073/pnas.86.16.6377.

16. Fricker, R. A., Green, E. L., Jenkins, S. I., & Griffin, S. M. (2018). The influence of nicotinamide on health and disease in the central nervous system. *International Journal of Tryptophan Research*, *11*, 1178646918776658. https://doi.org/10.1177/1178646918776658.

17. Godfray, H. C. J. (2019). Meat: The future series—Alternative proteins. World Economic Forum, Geneva, Switzerland. http://www3.weforum.org/docs/WEF_White_Paper_Alternative_Proteins.pdf (accessed on 22 October 2024).

18. Grunert, K. G., Bredahl, L., & Brunsø, K. (2004). Consumer perception of meat quality and implications for product development in the meat sector—A review. *Meat Science*, *66*(2), 259–272. https://doi.org/10.1016/j.meatsci.2003.10.004.

19. Hocquette, J. F., Gondret, F., Baéza, E., Médale, F., Jurie, C., & Pethick, D. W. (2010). Intramuscular fat content in meat-producing animals: Development, genetic and nutritional control, and identification of putative markers. *Animal*, *4*(2), 303–319.

20. Hu, F. B., Otis, B. O., & McCarthy, G. (2019). Can plant-based meat alternatives be part of a healthy and sustainable diet? *JAMA*, *322*, 1547–1548. https://doi.org/10.1001/jama.2019.13187.

21. Hurrell, R., & Egli, I. (2010). Iron bioavailability and dietary reference values. *The American Journal of Clinical Nutrition*, *91*(5), 1461S–1467S.

22. Huskisson, E., Maggini, S., & Ruf, M. (2007). The role of vitamins and minerals in energy metabolism and well-being. *Journal of International Medical Research*, *35*(3), 277–289.

23. International Food Council. (2020). *A Consumer Survey on Plant Alternatives to Animal Meat*. https://foodinsight.org/wp-content/uploads/2020/01/IFIC-Plant-Alternative-to-Animal-Meat-Survey.pdf (accessed on 22 October 2024).

24. Jensen, J., Rustad, P. I., Kolnes, A. J., & Lai, Y. C. (2011). The role of skeletal muscle glycogen breakdown for regulation of insulin sensitivity by exercise. *Frontiers in Physiology*, *2*, 112.
25. Kabara, J. J. (1984). Antimicrobial agents derived from fatty acids. *Journal of the American Oil Chemists' Society*, *61*, 397–403. https://doi.org/10.1007/BF02678802.
26. Kamruzzaman, M., Makino, Y., & Oshita, S. (2016). Parsimonious model development for real-time monitoring of moisture in red meat using hyperspectral imaging. *Food Chemistry*, *196*, 1084–1091.
27. Kanehisa, M. (2019). Toward understanding the origin and evolution of cellular organisms. *Protein Science*, *28*, 1947–1951. https://doi.org/10.1002/pro.3715.
28. Karunanayaka, D. S., Jayasena, D. D., & Jo, C. (2016). Prevalence of pale, soft, and exudative (PSE) condition in chicken meat used for commercial meat processing and its effect on roasted chicken breast. *Journal of Animal Science and Technology*, *58*(1), 27.
29. Kihara, J., Sileshi, G. W., Nziguheba, G., Kinyua, M., Zingore, S., & Sommer, R. (2017). Application of secondary nutrients and micronutrients increases crop yields in sub-Saharan Africa. *Agronomy for Sustainable Development*, *37*(4), 25.
30. Kohen, R., Yamamoto, Y., Cundy, K. C., & Ames, B. N. (1988). Antioxidant activity of carnosine, homocarnosine, and anserine present in muscle and brain. *Proceedings of the National Academy of Sciences of the United States of America*, *85*(9), 3175–3179.
31. Kumar, M. N. V. R., Muzzarelli, R. A. A., Muzzarelli, C., Sashiwa, H., & Domb, A. J. (2004). Chitosan chemistry and pharmaceutical perspectives. *Chemical Reviews*, *104*, 6017–6084. https://doi.org/10.1021/cr030441b.
32. Lehnen, T. E., da Silva, M. R., Camacho, A., Marcadenti, A., & Lehnen, A. M. (2015). A review on effects of conjugated linoleic fatty acid (CLA) upon body composition and energetic metabolism. *Journal of the International Society of Sports Nutrition*, *12*(1), 36.
33. Leipnitz, G., et al. (2007). *In vitro* evidence for an antioxidant role of 3-hydroxykynurenine and 3-hydroxyanthranilic acid in the brain. *Neurochemistry International*, *50*, 83–94. https://doi.org/10.1016/j.neuint.2006.04.017.
34. Løvaas, E., & Carlin, G. (1991). Spermine: An anti-oxidant and anti-inflammatory agent. *Free Radical Biology and Medicine*, *11*, 455–461. https://doi.org/10.1016/0891-5849(91)90061-7.

35. Lozoff, B., & Georgieff, M. K. (2006). Iron deficiency and brain development. In *Seminars in Pediatric Neurology* (Vol. 13, No. 3, pp. 158–165). Elsevier.

36. Mahan, D. C., & Shields, R. G. Jr. (1998). Essential and nonessential amino acid composition of pigs from birth to 145 kilograms of body weight, and comparison to other studies. *Journal of Animal Science*, *76*(2), 513–521.

37. Marangoni, F., Corsello, G., Cricelli, C., Ferrara, N., Ghiselli, A., Lucchin, L., & Poli, A. (2015). Role of poultry meat in a balanced diet aimed at maintaining health and wellbeing: An Italian consensus document. *Food and Nutrition Research*, *59*(1), 27606.

38. McCance, R. A., & Widdowson, E. M. (1960). *The Composition of Foods* (3rd ed.). London: Her Majesty's Stationery Office.

39. Othman, R. A., & Moghadasian, M. H. (2011). Beyond cholesterol-lowering effects of plant sterols: Clinical and experimental evidence of anti-inflammatory properties. *Nutrition Reviews*, *69*, 371–382.

40. Pathare, P. B., & Roskilly, A. P. (2016). Quality and energy evaluation in meat cooking. *Food Engineering Reviews*, *8*(4), 435–447.

41. Paul, B. D., & Snyder, S. H. (2019). Therapeutic applications of cysteamine and cystamine in neurodegenerative and neuropsychiatric diseases. *Frontiers in Neurology*, *10*, 1–9. https://doi.org/10.3389/fneur.2019.01315.

42. Pereira, P. M., & Vicente, A. F. (2013). Meat nutritional composition and nutritive role in the human diet. *Meat Science*, *93*(3), 586–592.

43. Piper, J. D., & Piper, P. W. (2017). Benzoate and sorbate salts: A systematic review of the potential hazards of these invaluable preservatives and the expanding spectrum of clinical uses for sodium benzoate. *Comprehensive Reviews in Food Science and Food Safety*, *16*, 868–880. https://doi.org/10.1111/1541-4337.12284.

44. Ravisankar, P., Reddy, A. A., Nagalakshmi, B., Koushik, O. S., Kumar, B. V., & Anvith, P. S. (2015). The comprehensive review on fat soluble vitamins. *IOSR Journal of Pharmacy*, *5*(11), 12–28.

45. Reddy, L. H., & Couvreur, P. (2009). Squalene: A natural triterpene for use in disease management and therapy. *Advanced Drug Delivery Reviews*, *61*, 1412–1426. https://doi.org/10.1016/j.addr.2009.09.005.

46. Reiley, L. (2019). Impossible Burger: Here's what's really in it. *The Washington Post*.

47. Ruxton, C. H. S., Reed, S. C., Simpson, M. J. A., & Millington, K. J. (2004). The health benefits of omega-3 polyunsaturated fatty acids: A

review of the evidence. *Journal of Human Nutrition and Dietetics*, *17*, 449–459. https://doi.org/10.1111/j.1365-277X.2004.00552.x.

48. Sakomura, N. K., Ekmay, R. D., Mei, S. J., & Coon, C. N. (2015). Lysine, methionine, phenylalanine, arginine, valine, isoleucine, leucine, and threonine maintenance requirements of broiler breeders. *Poultry Science*, *94*(11), 2715–2721.
49. Schweigert, B. S., & Payne, B. J. (1956). A summary of the nutrient content of meat. Bulletin No. 30, *American Meat Institute Foundation*. Chicago: American Meat Institute.
50. Seehafer, A., & Bartels, M. (2019). Meat 2.0—The regulatory environment of plant-based and cultured meat. *European Food and Feed Law Review*, *14*(4), 323–331.
51. Sha, L., & Xiong, Y. L. (2020). Plant protein-based alternatives of reconstructed meat: Science, technology, and challenges. *Trends in Food Science & Technology*, *20*, S0924–2244. https://doi.org/10.1016/j.tifs.2020.05.022.
52. Shestopalov, A. V., et al. (2006). Biological functions of allantoin. *Biology Bulletin*, *33*, 437–440. https://doi.org/10.1134/S1062359006050037
53. Soetan, K. O., Olaiya, C. O., & Oyewole, O. E. (2010). The importance of mineral elements for humans, domestic animals and plants—A review. *African Journal of Food Science*, *4*(5), 200–222.
54. Soladoye, O. P., Juárez, M. L., Aalhus, J. L., Shand, P., & Estévez, M. (2015). Protein oxidation in processed meat: Mechanisms and potential implications on human health. *Comprehensive Reviews in Food Science and Food Safety*, *14*(2), 106–122. https://doi.org/10.1111/1541-4337.12127.
55. Tawaraya, K., et al. (2014). Metabolite profiling of soybean root exudates under phosphorus deficiency. *Soil Science and Plant Nutrition*, *60*(5), 679–694. https://doi.org/10.1080/00380768.2014.945390.
56. Van Vliet, S., Kronberg, S. L., & Provenza, F. D. (2020). Plant-based meats, human health, and climate change. *Frontiers in Sustainable Food Systems*. https://doi.org/10.3389/fsufs.2020.00128.
57. Wang, X., Guo, M., Song, H., Meng, Q., & Guan, X. (2021). Characterization of key odor-active compounds in commercial high-salt liquid-state soy sauce by switchable GC/GC × GC–olfactometry–MS and sensory evaluation. *Food Chemistry*, *342*, 128224. https://doi.org/10.1016/j.foodchem.2020.128224.
58. Williams, P. (2007). Nutritional composition of red meat. *Nutrition*

& Dietetics, 64(Suppl. 4), S113-S119. https://doi.org/10.1111/j.1747-0080.2007.00197.x.

59. Williams, P., Droulez, V., Levy, G., & Stobaus, T. (2007). Composition of Australian red meat 2002. 3. Nutrient profile. *Food Australia, 59*, 331–341.
60. Wood, J. D., Enser, M., Fisher, A. V., Nute, G. R., Sheard, P. R., Richardson, R. I., Hughes, S. I., & Whittington, F. M. (2008). Fat deposition, fatty acid composition and meat quality: A review. *Meat Science, 78*(4), 343–358.
61. Wood, J. D., Richardson, R. I., Nute, G. R., Fisher, A. V., Campo, M. M., Kasapidou, E., Sheard, P. R., & Enser, M. (2004). Effects of fatty acids on meat quality: A review. *Meat Science, 66*(1), 21–32.
62. Woods, V. B., & Fearon, A. M. (2009). Dietary sources of unsaturated fatty acids for animals and their transfer into meat, milk and eggs: A review. *Livestock Science, 126*(1), 1–20.
63. Wu, G. (2020). Important roles of dietary taurine, creatine, carnosine, anserine and 4-hydroxyproline in human nutrition and health. *Amino Acids, 20*, 329–360. https://doi.org/10.1007/s00726-020-02823-6.
64. Wyness, L., Weichselbaum, E., O'Connor, A., Williams, E. B., Benelam, B., Riley, H., & Stanner, S. (2011). Red meat in the diet: An update. *Nutrition Bulletin, 36*(1), 34–77.
65. Yamabe, N., Loganin, I., & Iridoid Glycoside. (2010). Evaluation of loganin, iridoid glycoside from *Corni Fructus*, on hepatic and renal glucolipotoxicity and inflammation in type 2 diabetic db/db mice. *European Journal of Pharmacology, 648*(1), 179–187. https://doi.org/10.1016/j.ejphar.2010.08.044.
66. Yu, T. Y., Morton, J. D., Clerens, S., & Dyer, J. M. (2017). Cooking-induced protein modifications in meat. *Comprehensive Reviews in Food Science and Food Safety, 16*(1), 141–159.

Chapter 7

Safety and Nutritional Risks with Plant-Based Meat Alternatives

Contents

Introduction 216

7.1. Risks Associated with Plant-Based Meat Alternatives 216

7.2. Food Safety and Nutritional Concerns 224

7.3. Plant-Based Meats, Human Health, and Climate Change 231

References 254

INTRODUCTION

Growing consumer demands for a more ethical, healthful, and sustainable diet along with concerns about global food security brought on by rising global population and climate change have led to a growth in the market for plant-based meat substitutes. Concerns concerning foodborne pathogens, allergenicity, toxicity, and appropriate nutritional composition are among the potential food safety issues that are brought up by the rising consumption of plant-based meat substitutes. Studies of the possible food safety hazards associated with these novel protein sources are poorly documented, and there has been little research conducted from the standpoint of public health on the nutritional and health aspects of plant-based meat products. Numerous studies on the safety and nutritional value of these foods have been commissioned or funded by the companies that make them or by other groups that support them. In this article, the literature is reviewed and the possible health and food safety risks—including dietary, chemical, microbiological, and allergy concerns—of plant-based meat substitutes are analyzed. A number of research gaps have been identified by research, which should be investigated further to help guide future discussions regarding the creation and marketing of plant-based meat alternatives.

The promise of more sustainable, healthier alternatives to real meat is driving a previously unheard-of wave of plant-based meat substitutes into consumer markets. Their marketing is motivated by the need to develop a contemporary food supply that is healthier for the environment, animals, and people. Although the wealthiest countries make up only 17% of the world's population, they could reduce greenhouse gas emissions by approximately 61% and increase carbon sequestration by implementing a more plant-centric diet. Global consumption of alternative proteins was estimated to be 13 million metric tons in 2020. The projected value of the worldwide market for plant-based substitutes is expected to increase from USD $4.6 billion in 2018 to USD $85 billion by 2030.

7.1. RISKS ASSOCIATED WITH PLANT-BASED MEAT ALTERNATIVES

The development of plant-based meat alternatives, or PBMAs, is accelerating due to growing consumer demand for more plant-based diets. PBMAs are usually made of plant proteins and other ingredients. Many plant-based meat alternative (PBMA) products are being developed by the food industry; these products are typically made to resemble real meat products in terms of appearance, texture, mouthfeel, and flavor. Because they are worried about the effects traditional meat has on their health, the environment, and animal welfare, consumers are embracing these products at an increasing rate. Present meat consumption trends have been connected to negative impacts on human health (such as obesity, cancer, and greenhouse gas emissions) as well as unfavorable environmental outcomes (such as pollution, biodiversity loss, and greenhouse gas emissions). A shift toward a diet higher in plants could get rid of or lessen these issues.

PBMAs are being sold as a way to encourage the move away from products derived from animals and to create a modern food system that is good for the environment, animals, and people. Even though they make up only 16% of the global population, the wealthiest countries could cut greenhouse gas emissions by almost 61% and improve carbon sequestration by adopting a plant-centric diet. The market for plant-based foods has grown primarily due to the meat substitute category. The amount of alternative proteins consumed worldwide in 2020 was close to 13 million metric tons. It is anticipated that the PBMA product category will display unique functional, sensory, nutritional, and physical attributes. The food industry needs to be able to identify the best ingredient combinations and manufacturing processes to meet market trends to accomplish this on a large scale at a reasonable cost. The global market for plant-based foods, mainly plant-based milk alternatives and beverages, is expected to grow to a value of USD 162 billion by 2030, a significant increase from the USD 4.6 billion in 2018 and the USD 29.4 billion in 2020. Although the market for alternative proteins is still relatively small and that consumer demand is currently exceeding the capacity of the industry's supply chain, these notable growth figures indicate a positive long-term outlook for investments in this sector.

- There may be new nutritional and food safety hazards as a result of the growing introduction of PBMAs into the food supply chain, which could result in additional health issues. For this reason, it is essential to have a thorough understanding of the physicochemical and molecular characteristics of ingredients derived from plants. As a result, it's critical that the developing plant-based food sector carefully designs its goods takes into account concerns about nutrition and food safety, and encourages healthfulness. These problems must be taken into consideration case-by-case because they depend on the components and manufacturing methods used to make PBMAs, which means that their digestibility, bioaccessibility, and nutrient composition must be controlled. To help consumers make better decisions, researchers have emphasized how important it is to tell them about the nutritional value of PMBAs in comparison to real meat. The purpose of research is to assess the safety and nutritional risks that may arise from the manufacture and use of PBMA products. It draws attention to possible nutritional and safety hazards connected to these products' primary manufacturing steps: (1) protein functionalization and isolation, (2) product formulation, (3) processing, and (4) storage.

7.1.1. How Are Plant-Based Meats Made

The production and consumption of plant-based meat substitutes date back more than a millennium, even though public awareness of them has increased dramatically in recent years. Asian societies were the first to create a variety of meat substitutes in antiquity. They created relatively simple plant-based products like tempeh and tofu or wheat-based products like seitan. It is noteworthy that the initial iterations were not deliberately created to serve as exact substitutes for meat. Extrusion of defatted soy meal, soy protein concentrates, or wheat gluten produced texturized vegetable protein (TVP), which became popular as a meat alternative several decades ago.

Recent years have seen the emergence of a new generation of plant-based meat substitutes that are aimed mainly at omnivores who desire the satisfying texture and flavor of meat. Although there is a growing interest in developing more complex and intricate products, like chicken breast, beefsteak, or pork chop analogs, the majority of PBMA products that are currently available are burger patties, mince, sausages, and nuggets. PBMAs are usually designed to replicate the enticing look, feel, and flavor of real meat products in an environmentally sustainable manner. Some may not place as much emphasis on this feature as others, but some are designed to closely resemble the nutritional makeup of actual meat. Furthermore, since the majority of PBMAs are classified as ultra-processed foods, their rate of absorption and digestion may differ from that of actual meat. This may result in changes to their nutritional value and overall health effects. As plant-based diets gain popularity, worries, and inquiries about the possible risks to nutrition and safety arise when switching from an omnivorous to a more plant-based diet. As previously stated, problems related to safety and nutrition may arise from the components and methods of processing used in the creation of products containing PBMAs.

When it comes to plant-based food, there are a lot of possible issues with nutrition and food safety. These difficulties cover a wide range of topics, such as the existence of diverse chemical and microbiological contaminants in the ingredients used, worries about food adulteration, high concentrations of food additives, the use of genetically modified ingredients, incorrect labeling, the introduction of new allergen sources, possible vitamin or mineral deficiencies, and modifications to the macronutrient composition (such as the amount of protein, carbohydrate, or fat). Therefore, it is crucial to take these concerns into account when creating the upcoming line of plant-based foods. This guarantees that they are safe, nourishing, and advantageous for both consumers and the environment and that they also meet environmental sustainability goals.

The production of PBMAs consists of four main stages:

- **Protein Isolation and Functionalization:** To make flours, concentrates, or isolates, plant proteins are extracted and purified. To improve their functionality, they may occasionally go through additional processing techniques like denaturation, hydrolysis, or conjugation.
- **Product Formulation:** To create plant-based meat substitutes that mimic the appearance, feel, texture, flavor, and functionality of real meat products, the plant proteins are combined with a range of other ingredients, such as carbohydrates, lipids, salts, flavors, and colors. Nutraceuticals, vitamins, minerals, and dietary fibers are some examples of nutrients that can be added to match or even exceed the nutritional value of actual meat.
- **Processing:** The combination of ingredients undergoes a series of processing steps that encourage the development of structures and characteristics similar to those of meat. These processes frequently entail cutting, molding, shearing, mixing, and extrusion.
- **Storage:** To guarantee the safety and quality of PBMAs throughout storage, transportation, and distribution, the product, packaging materials, and environmental factors must be carefully designed. The processes of microbiological, chemical, and physical deterioration must be carefully managed and controlled to achieve this.

Ensuring the safety and nutritional value of the final product while producing a faithful replication of the desired quality and sensory attributes of real meat is the ultimate goal of this process. The most popular technique for producing PBMAs that closely resemble real meat is currently extrusion technology. Other alternative technologies, such as shear cells, 3D printing, recombinant proteins, and mycelium, are still being developed, though (van Vliet et al., 2020).

7.1.2. Protein Isolation and Functionalization

It is possible to think of a protein isolation process as a sequence of linked steps where the purity of the protein rises gradually at each stage. These consist of (a) finding an appropriate protein and obtaining the material; (b) extracting from the source; (c) separating from non-protein components, such as lipids and nucleic acids; (d) condensing the main protein and dividing it into fractions containing different proteins; and (e) finishing with sophisticated techniques, such as chromatography, that produce the final product. The physicochemical, technological, and sensory properties of plant-based food products are significantly shaped by proteins. For example, their influence on the structure, texture, look, chewiness, water retention, and nutritional value of PBMAs is significant. Selecting the right plant protein is essential because it can give the finished product a variety of desired functional qualities, such as structuring, foaming, thickening, gelling, emulsifying, and fluid retention.

Typically, plant proteins are made into flours, concentrates, or isolates from sources like corn, soybeans, or peas. There are many different ways to achieve extraction and purification. These methods include traditional approaches that use harsh chemicals such as acids, bases, and/or organic solvents; environmentally friendly green extraction methods that use one or more enzymes; and advanced physical extraction methods that use ultrasound, pulse electric field, microwave, and high-pressure assistance. The fact that a large portion of the methods used to extract proteins were not designed with the goal of improving and optimizing their functionality should be noted. Rather, these techniques were designed to extract starch or oil from plant materials. As a result, during the isolation process, proteins may become denatured or aggregate, which could impair their functionality.

7.1.3. Product Formulations

Comprehending the interplay between materials is crucial for formulating products that have improved qualities, maximize processing effectiveness, and efficiently distribute active ingredients. When it comes to food, the formulation is the process of creating, arranging, or modifying food products to add particular features. These features can range from adding extra nutritional benefits to enhancing food items. Products that are carefully designed to mimic the color, flavor, taste, consistency, texture, aroma, and visual attributes of animal-derived products to replicate or at least closely resemble the sensory experience of consuming meat products are known as PBMAs. The physicochemical and organoleptic properties of conventional meat products are mimicked in the formulation of PBMAs.

Plant proteins are mixed with a variety of other useful ingredients at this stage of production to give the finished product the right look, feel, texture, and flavor. These extra ingredients include binding agents, gelling agents, emulsifiers, flavorings, and colorings. Because plant-based meat substitutes are made up of a variety of ingredients, including proteins, water, fats, carbohydrates, flavorings, binding agents, and colorants, their functional properties play a critical role in determining their characteristics. Concerns about PMBAs' heavy additive use have been raised by some customers, who see them as highly processed foods. Saturated fats, highly refined flours, and high salt content are just a few of the ingredients that have sparked nutritional concerns about these carefully constructed food products. Since different products use different fortification techniques, each product should be evaluated individually.

The presence of leghemoglobin in some PBMAs has also prompted concerns from certain researchers. Leghemoglobin is a hemeprotein that contains iron and is obtained from soybean root nodules. It is used to mimic the desired red color and meaty flavors of hemoglobin found in actual meat. Owing to the difficulties in extracting enough protein from soybeans, genetically modified yeast cultures are frequently used to generate this protein. Researchers have found links between elevated body iron stores, increased heme iron intake, and a higher risk of type 2 diabetes. However, the amounts of these proteins found in plant-based diets do not appear to be harmful to health, according to the scant scientific data. There are also worries about the possible negative effects on health that can arise from consuming PBMAs, which contain a variety of additives, such as flavorings, colorings, binding agents, preservatives, and sweeteners. Nevertheless, there is little evidence to support the notion that this presents a serious health risk. It is obvious that additional research is needed to determine the safety and nutritional value of PBMAs.

Some PBMAs have more saturated fat than other plant-based protein sources, such as beans and lentils, or traditional meat products. In fact, some scientists contend that PBMAs have roughly the same caloric and saturated fat content as livestock meat, surpassing the recommended daily allowance for saturated fat. Consequently, increasing the consumption of saturated fats in the diet has raised questions regarding the possible health effects. However, it's important to remember that plant-based meat substitutes can be modified to have less saturated fat. Furthermore, nutrition scientists continue to disagree about the detrimental effects of saturated fats on nutrition. There may be health concerns associated with the noticeable elevated salt levels found in some PBMAs. Elevated levels of salt in the diet may increase the chance of developing diseases like kidney disease, osteoporosis, high blood pressure, heart disease, and stomach cancer. High sodium content is considered nutritionally undesirable and may eventually raise the long-term risk of cardiovascular issues for those who consume excessive amounts of it over an extended period of time. Therefore, it might be imperative that producers take into account reducing the salt content of their PBMA products.

PBMAs often have a long list of ingredients that include plant-based extracts, flavoring agents, colors, emulsifiers, isolated macronutrients (proteins, fats, and carbohydrates), and micronutrients (vitamins and minerals). Many of these ingredients, as well as the combinations they form in PBMAs, have unknown and uncertain potential health effects. The nutritional makeup of PBMAs varies greatly amongst products and is contingent upon the ingredients utilized by the manufacturer. Because plant proteins are less digestible and lack some essential amino acids, they are frequently regarded as having a lower nutritional quality than animal proteins. In addition, important micronutrient levels in plant-based diets are frequently lower (e.g., iron, and vitamin B12) compared to the foods made from animals that they are intended to replace. Calcium, potassium, magnesium, zinc, and vitamin B12 levels are reduced in PBMAs. Furthermore, compared to meat, a lot of plant-based meat substitutes have higher sodium and saturated fat content. Consequently, converting from an omnivore to a plant-based diet may raise certain nutritional issues. However, these issues can frequently be resolved by appropriately fortifying plant-based foods with nutrients, as some of the major producers of these products are currently doing.

7.1.4. Nutritional Profile

Due to variations in their compositions, structures, and processing, moving from animal-based to plant-based diets may have an effect on human health and nutrition. Numerous studies that were part of research stated that PBMAs are highly processed foods with a longer ingredient list than comparable animal-based products, which may have different nutritional effects. As previously mentioned, there can be significant differences in the composition of PBMAs among products, which could result in varying health effects. Additionally, there might be variations in the absorption and digestibility of meat products and PBMA. PBMAs are produced by a range of processing techniques, such as dehulling, soaking, blanching, pH adjustment, enzyme treatments, shearing, thermal processing, and size reduction, which all affect the final product's composition and digestibility. Due to their high level of processing, food manufacturers can regulate the nutritional composition and digestibility of PBMAs, which in turn affects how they affect health. For example, their levels of macro- and micronutrients can be regulated, and they can be supplemented with additional ingredients that promote health. Reducing the salt content of PBMAs could enhance their nutritional profile because consuming too much sodium is linked to a higher risk of cardiovascular diseases. Generally speaking, though, the amount of salt content will probably vary depending on the specific plant-based food type under consideration as well as the product formulation used by the producers.

Phytoestrogens, one type of soy isoflavone compound that has been connected to certain health issues, may be present in PBMAs. Estrogens are generally thought to be helpful in preventing menopausal symptoms, osteoporosis, cardiovascular disease, and breast cancer. Whether consumed phytoestrogens function similarly to endogenous estrogens in the human body is unknown, though. According to certain studies, consuming phytoestrogens in excess can have a negative impact on reproductive health. Compared to meat proteins, plant proteins have different amino acid profiles, which may have an impact on their nutritional value. Some of them might be lacking in one or more essential amino acids (tryptophan, lysine, or methionine, for example) that are necessary for human health but that the body is unable to produce on its own. While this may theoretically give rise to health issues, most vegan or vegetarian diets include a wide range of different protein sources that typically supply sufficient amounts of these vital amino acids, making the issue essentially insignificant.

Starches, flours, and binding agents are not the only carbohydrates found in PBMA products. While starchy foods are an important component of a healthy diet because they provide you energy and fiber, they can also be harmful to your health (e.g., for those suffering from illnesses like diabetes). Furthermore, in high-temperature cooking e.g., starchy, plant-based foods (such as those made by boiling, roasting, and baking) can release some potentially hazardous chemicals, like acrylamide, which may be the case with cooked PBMAs. A cool, dry place to store food and refraining from high-temperature cooking are two ways to reduce the amount of acrylamide that consumers are exposed to. It is possible to create and formulate meatless products with liquid smoke flavorings, which have been shown to have carcinogenic potential when ingested frequently and in large enough amounts. Additionally, it has been noted that plant-based meat substitutes are lacking in a few particular amino acids and their derivatives, such as taurine, anserine, and creatine. Since these substances can affect both muscle and brain function, it is thought that they have a significant and meaningful impact on human health.

Various additives were used by food manufacturers to enhance the appearance, texture, and flavor of their products. These include gums or binding ingredients like methylcellulose, acacia gum, xanthan gum, carrageenan, and others that act as thickeners, binders, stabilizers, and emulsifiers. Customers are calling for clearer labeling, and there have been concerns raised about the safety and health of these products. Food processors are already working to reduce or completely remove these ingredients from a variety of processed food products, even though there are currently no known health risks or concerns. It is possible to improve the nutritional profiles of PBMAs by fortifying them with desired nutritional components. Some of the ingredients in these alternatives may have lower bioavailability of certain essential minerals, such as iron, zinc, magnesium, and calcium. As a result, it's critical to create plans to boost bioavailability and to have techniques for measuring the bioavailability of minerals and other nutrients in PBMAs both *in vitro* and *in vivo*. Numerous varieties of natural colorants (e.g., red beet, red cabbage, leghemoglobin, etc.) and flavorings (e.g., herbs, and spices) to mimic the ideal color and flavor of real meat in PBMAs. These ingredients may chemically break down during food processing, resulting in undesirable changes to quality attributes. They are also frequently less stable than synthetic substitutes. The possible safety implications of the chemical deterioration of natural flavorings or colorants in plant-based diets are not well studied at this time.

7.2. FOOD SAFETY AND NUTRITIONAL CONCERNS

This stage of PBMA creation is associated with many food safety concerns, which are covered in detail in the following sections. These concerns include the presence of allergens, bacteria, mycotoxins, anti-nutrients, thermally induced carcinogens, and natural toxins.

7.2.1. Allergens

The possibility of allergens in PBMAs poses a serious risk to food safety. Given the recent global surge in food-related allergies, this is especially crucial. Rather than being inherited, the development of food allergies is thought to be dependent on an individual's response to food. Food allergies and potentially fatal food-related reactions can develop at any age and either go away or persist for the duration of a person's life. Therefore, it is essential to assess new foods for allergens to make sure there are no risks involved. Certain food ingredients can cause symptoms ranging from minor (like itching, swelling, and stomachache) to severe (like anaphylaxis) in even small amounts. Some people may experience allergic reactions to common PBMA components. For example, a wide range of plant proteins, including those from legumes and cereals like soy, peas, wheat, rye, barley, and lupin, are known to cause allergies. The amount of the product used affects how severe the allergic reactions to food are. Certain plant proteins derived from genetically modified (GM) sources have some consumers worried that they could increase the risk of allergies. Thus, before the foods are released for general consumption, these novel proteins should be thoroughly evaluated for possible allergenicity. Proteins that are resistant to being broken down by the gastrointestinal tract's proteases are more likely to cause an allergic reaction. A clear labeling of such genetically modified food ingredients used in a product formulation is required for transparency and to allay consumer concerns. However, there is still disagreement over the health and environmental risks associated with eating genetically modified foods, which calls for more research.

Moreover, increased use of products that contain significant amounts of soy, pea, wheat, and other plant proteins may cause and exacerbate allergic reactions in people who have never had problems with these foods before. The use of high-protein pea components in PBMA formulations, such as concentrated pea protein, may raise the possibility of allergies. Like peanuts and lentils, peas are a legume and can also cause allergies. Allergic reactions may result from the continued practice of adding pea protein isolates and concentrates to a variety of foods to boost their protein content and add bulk. The similarity between the proteins in peas and peanuts may cause cross-reactions, as reported by people with peanut allergies who have indicated and reported post-consumption allergic symptoms. Peas may become a more common allergenic food as a result of their increased exposure.

Another common allergen that can cause both less serious but still undesired reactions and potentially fatal severe anaphylaxis reactions is wheat proteins, which are used in PBMAs. These proteins are frequently used as binders in different PBMA products. A negative allergic response to gluten, a protein present in grains like wheat, barley, and rye, can also occur in consumers with celiac disease and wheat allergies. When creating plant-based meals, it is crucial to choose proteins with low allergenicity and to take extra care during processing and labeling. One of the Codex Alimentarius Commission's General Standards for the Labeling of Prepackaged Foods is a list of priority allergens. This list is created and constructed using preset standards that consider the prevalence of allergens worldwide as well as other known variables.

7.2.2. Bacteria

Bacteria have caused a disproportionate amount of human illnesses and deaths throughout history. It is crucial to comprehend the different kinds of bacteria that are connected to food to guarantee food safety. PBMA products may include pathogenic bacteria that originated from the raw ingredients. Studies have revealed that certain endospore-forming bacteria, (e.g., those that form during food processing procedures like extrusion and cooking), are normally rendered inactive. Clostridium species or Bacillus species and other bacteria (e.g., *Lactobacillus sakei* and *Enterococcus faecium*) can survive the heating process and/or re-contaminate the product after it has been extracted. In the field of food microbiology, bacteria can include not only infectious pathogens but also strains that produce toxins, which also pose serious safety concerns.

7.2.3. Toxins

Substances, whether they are of natural or artificial origin, can pose risks when the level of exposure reaches a certain threshold.

7.2.3.1. Mycotoxins

Mycotoxins are dangerous, toxic substances that are made by different types of fungi, or mold. The most common mycotoxins found in food commodities are ochratoxins and aflatoxins, which are produced by Aspergillus species, and patulin and ochratoxins, which are synthesized by Penicillium. Fusarium species also produce zearalenone, deoxynivalenol, and fumonisins. Globally, a wide range of acute and chronic human illnesses are caused by mycotoxins, including fumonisins, patulin, aflatoxins, and ochratoxins. Mycotoxins, which are toxic to humans if consumed in large enough doses, can contaminate a wide variety of edible plants. In fact, it has been reported that mycotoxins have contaminated many significant agricultural crops. They can be found in some of the raw materials used to make PBMAs, including soy, nuts, and grains like rice and oats. Mycotoxins are strong toxins that have negative effects on people's health because they are "mutagenic, teratogenic, and carcinogenic." The duration of exposure, the concentration of mycotoxin, and the sensitivity of the consumer determine how toxic the product is. Mycotoxins, like bacteria, are not entirely destroyed by heat and can withstand temperatures within the typical range for food processing. The discovery of ochratoxin A in plant-based foods and ingredients raises the possibility that PBMAs may be contaminated with mycotoxin. Fumonisin FB1, another mycotoxin, is primarily present in beer, sorghum, soybeans, corn, rice, cowpea seeds, and beans. Establishing an appropriate regulatory framework for mycotoxins in PBMAs is imperative to reduce mycotoxin contaminations and their detrimental effects on health, including the emergence of life-threatening illnesses (e.g., cancer of the liver).

7.2.3.2. Natural Toxins

There's a chance that some of the ingredients that go into making plant-based foods are naturally toxic. Usually, plants produce these compounds as metabolic byproducts to protect themselves from various threats like bacteria, fungi, insects, and predators. Several common plant toxins include lectins, which are present in green, red, and white kidney beans; cyanogenic glycosides, which are found in bitter apricot seeds, bamboo shoots, cassava, and flaxseeds; glycoalkaloids, which are found in potatoes; 4'-methoxypyridoxine, which is obtained from ginkgo seeds; colchicine, which is found in fresh lily flowers; and muscarine, which is found in certain wild mushrooms. Consequently, all plant-based components used in product formulation must go through careful selection and processing steps to stop, get rid of, or neutralize these toxins. A few academics have made people aware of the use of carrageenan. Seaweed is the source of carrageenan, which is a food product ingredient. Sometimes used as a thickener, gelling agent, stabilizer, or binder in PBMAs, carrageenan is a polysaccharide.

While carrageenan of food grade is deemed safe for ingestion, there have been indications that it may aggravate gastrointestinal disorders, alter intestinal microbiota, cause irritable bowel syndrome, and increase the risk of colon cancer and other health problems. Furthermore, if carrageenan is derived from tainted or polluted seawater, it may absorb high levels of heavy metals, which could be harmful to human health. However, there is currently little agreement among scientists regarding the possible health risks connected to carrageenan. In spite of this, the industry of packaged food manufacturers (PBMAs) is being encouraged to lower the number of ingredients that raise red flags in their products due to consumer demand for cleaner labels. Furthermore, to reduce the risks to food safety, all current and future products containing PBMAs must have updated natural toxins regulations and risk assessments.

7.2.3.3. Synthetic Toxins

Pesticide residues and other synthetic toxins may be present in some of the ingredients used to make PBMAs. Using organic solvents (such as hexane) for protein extraction can have negative effects on the environment and human health, particularly if the final product contains comparatively large residual amounts.

The use of the chemical hexane in the processing of soy protein isolates has sparked concerns because it is a neurotoxin. However, accurate information about the amounts of hexane utilized in the soy and pea protein isolate production process for plant-based substitutes, as well as any residual amounts that might remain in the finished product, is currently lacking. This emphasizes how crucial it is to carry out more research in this specific field.

7.2.4. Thermally Induced Carcinogens

As with real meat products, cooked PBMAs can produce carcinogens through thermal processing, which is used to lower microbial contamination or cook food. Little is known at this time regarding the creation and consequences of dangerous chemicals and thermally induced substances in PBMAs. Obviously, additional research is required to confirm the probable safety risk associated with the chemicals created during the high-temperature processing of plant-based meat substitutes (van Vliet et al., 2020)

7.2.5. Antinutrients

Legumes and other plant-based ingredients like phytic acid, lectins, oxalates, goitrogens, saponins, phytoestrogens, phytates, and tannins are examples of antinutrients that can negatively impact human health. These antinutrients can limit the digestion of proteins or the absorption of minerals, which lowers the bioavailability of important nutrients. Antinutrients can be found in certain plants by nature and may not be completely eliminated or rendered inactive during the extraction and extrusion procedures. Research has indicated, for example, that some might be found in the starch ingredients of cassava, soybean, and maize. Lectins are a problem when making rice and soybean milk substitutes, but they could also be a problem when making PBMAs. Certain protein conformations, antinutrients, and proteolysis-resistant structures are among the elements that some researchers have found in plant-based proteins that may reduce nutrient bioavailability. Testing for anti-nutrients is therefore a crucial step in the development of plant-based proteins.

7.2.6. Processing

Food processing affects how a food's chemical, physical, and nutritional qualities change. Plant proteins can be made with a variety of technological and food engineering techniques, but it is still challenging to balance processing techniques so that all required mechanical properties are achieved while maintaining the nutritional value of the finished product. However, there remains a great deal of work to be done to improve their functionality, sensory aspects, safety, and component selection. Generally, water, carbohydrates, salts, flavorings, and edible fats are mixed with plant proteins, which are typically defatted. After that, this mixture is run through a twin-screw extruder that runs at high temperatures and different moisture contents. By encouraging the proteins to form a fibrous structure resembling meat, this process produces a meat substitute that can be used in a variety of culinary applications.

"Formulations made mostly or entirely from substances derived from foods and additives, with little if any intact (whole) food," is how ultra-processed PBMA products are described, and they perfectly fit the description. Consuming ultra-processed foods has been linked to many detrimental health effects, including type 2 diabetes, obesity, cancer, heart disease, and all-cause death. Higher processing levels aside, PBMAs typically have a wider variety of additional ingredients to replicate the features and sensory aspects of traditional animal-based products. All of this creates an environment where many different sources can emerge and turn into potential hazards. A key component of the conversation surrounding the critical role these substitute products play in the shift to a more sustainable diet is the argument regarding the ultra-processed nature of PBMAs and its link to a possible elevation risk of health-related harm, which could affect the consumer motivation for their regular consumption.

Because so many vitamins, minerals, and nutrients can be lost or eliminated during processing, varying levels of processing have a negative effect on nutrition and health. More research into the potential effects is required because, in some cases, adding particular ingredients intended to improve nutritional quality (e.g., consumers may become less interested in a product as a result of, fortification). Certain foods and ingredients can have their bioavailability, digestibility, nutritional value, and functional qualities increased or decreased through processing. This should be supplemented with the lengthy list of ingredients and the allowance for differences between the nutritional values listed on the PBMA food labels and those that differ from international food label laws. Similar to meat products, it's critical to guarantee the microbiological safety of plant-based meat substitutes by subjecting them to sufficient and adequate thermal processing before consumption. A factory, restaurant, or even a person's house can be the location of this thermal treatment. Special consideration must also be given to the possibility of microbial proliferation and the presence of yeast and Enterobacteriaceae following the preparation of hot meals using PBMAs, as some researchers have found that this risk is marginally higher than that of food that contains meat. Due to the possibility of pathogenic bacteria contamination, consumers must follow the manufacturer's food preparation instructions as listed on food labels when it comes to legumes, grains, and vegetables. By following good food safety procedures, it is ensured that these foods won't hurt customers when they are prepared and eaten as directed.

The majority of PBMAs are processed through extrusion techniques, including power heated, wet or high moisture, and dry extrusion. The well-known processing method of conventional dry extrusion is ideal for creating meat substitutes that resemble minced meat. However, new developments in high-moisture extrusion technology make it possible to produce visually appealing whole portions of non-meat substitutes. Extrusion is a high-temperature, high-pressure process that helps reduce the microbial load while simultaneously giving products the form and texture they want. The exact temperature and time parameters are taken into account and are thought to be crucial control points. To address concerns related to food safety and quality as well as the continuous expansion and diversification of protein offerings in the market, more research into non-protein constituents, technological advancements and innovations in the production of alternative protein products, improvements, and enhancements in overall appearance and flavor, strict control over biological and chemical safety, and careful selection of protein sources are all necessary. Processing PBMAs has the potential to result in several chemical hazards. As with actual meat products, known carcinogens like polycyclic aromatic hydrocarbons, nitrosamines, and heterocyclic aromatic amines may be created during the thermal processing of PBMAs. Similarly, other heat-induced contaminants, like glycidyl fatty acid esters, 2-monochloropropanediol (2-MCPD), and 3-monochloropropanediol (3-MCPD), may be produced in PMBAs during thermal processing. However, more research is needed to understand how these harmful compounds are produced and what impact they have on PBMA.

PMBAs' health may also be negatively impacted by the use of lipid sources that contain trans-fatty acids, which are produced when vegetable oils are partially hydrogenated. On the other hand, artificial trans-fatty acids made in factories are prohibited from being used in food products by law in a large number of nations. The potential health benefits of plant-based foods may be diminished if certain essential nutrients and phytochemicals are lost during food processing. Alternatives made of plants are typically just as nutrient-dense as their corresponding counterparts made of animals. The results of their metabolomics study suggest that, in terms of supplying advantageous nutrients, the PBMAs and the animal-based product (beef) are more likely to be complementary than identical or interchangeable. The bioavailability and digestibility of PBMA products are also concerning issues. In light of the potential for positive changes in nutrient bioavailability, human gut microbiome, and protein digestibility, some researchers think that industrial processing of plant-derived ingredients to form meat substitutes may not always be detrimental. According to some, there is an unquestionable need for a comprehensive risk assessment and a multidisciplinary approach because the novel protein sources may cause unfavorable reactions related to allergies and other triggers.

7.2.7. Storage

Another matter of concern is the impact of microbial contamination on the safety of PBMAs. Similar to meat products, PBMA products need to be kept in the right storage environment (e.g.temperature, humidity, and packaging) to hinder the development of harmful microbiological contaminants. In general, PBMAs have little to do with worries about infectious illnesses. They may, nevertheless, result in foodborne infections. For example, they could come into contact with sources such as water, other foods, or animal dung and get contaminated with pathogens. Because of their neutral pH, high moisture content, and advantageous nutritional profile, PBMAs are generally very vulnerable to microbial development and spoiling. Research on the safety and microbial contamination of plant-based foods is currently lacking, and it is obvious that more research is required. Meta-genetics has been utilized by researchers to investigate changes in the types and numbers of microorganisms in plant-based products as they are stored. Studies have shown that some microbes, such as *Lactobacillus sakei*, *Enterococcus faecium*, *Enterobacteriaceae*, and yeasts, are more common as a product nears the end of its shelf life. These microorganisms could be there because heat-treated PBMAs were contaminated after processing. To guarantee the safety of PBMAs, a better understanding of the different kinds of microorganisms present is essential (van Vliet et al., 2020).

7.3. PLANT-BASED MEATS, HUMAN HEALTH, AND CLIMATE CHANGE

Plant-based meats have emerged as a significant alternative to traditional animal products, offering potential benefits for both human health and climate change mitigation. These products are typically lower in saturated fats and cholesterol, which may reduce the risk of heart disease and improve overall health when incorporated into a balanced diet. Furthermore, plant-based meats often contain higher levels of fiber, vitamins, and minerals, promoting better digestive health and nutrition. From an environmental perspective, the production of plant-based meats generates significantly fewer greenhouse gas emissions compared to conventional livestock farming, reducing the carbon footprint associated with meat consumption. Additionally, plant-based protein sources require less land and water, making them more sustainable. By shifting dietary patterns toward plant-based options, individuals can contribute to improved health outcomes while also addressing urgent climate challenges, making plant-based meats a compelling choice for a more sustainable future.

Plant-based meat substitutes come in different varieties, but the latest ones are designed to replicate the texture and nutritional value of meat through the use of plant-based proteins. (e.g., rice, wheat, soybean, pea, potato, rice, and/or mycoprotein), fats (e.g., coconut, soybean, canola, and/or sunflower oil), as well as additional cutting-edge components (e.g., red-colored vegetable extracts, soy leghemoglobin, and/or flavorings). Furthermore, meat contains a variety of naturally occurring vitamins and minerals, (such as iron, zinc, and B-vitamins) that are progressively added to meat substitutes made from plants. This allows plant-based meat substitutes to closely resemble meat (Figure 7.1). Furthermore, those who are omnivores but want to consume fewer animal products—flexitarians—are the target market for innovative meat substitutes. We will talk about the nutritional and ecological effects of consuming plant-based meat alternatives vs. meat because novel plant-based meat substitutes are quite similar to meat, and animal meats, in addition to offering a more thorough analysis of the environmental and health implications of substituting plant-based diets for animal products.

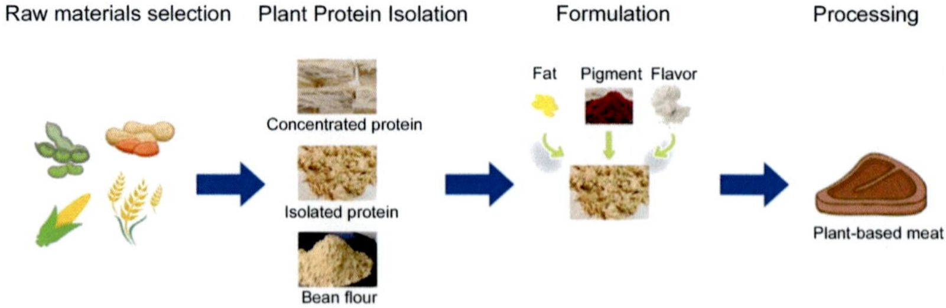

Figure 7.1. Alternatives to meat.

Source: https://ars.els-cdn.com/content/image/1-s2.0-S2214799321001429-gr2.jpg

7.3.1. Omnivory or Optionality

Omnivory refers to the dietary behavior of organisms, particularly humans, that consume both plant and animal sources for nutrients. As omnivores, humans have historically adapted to a flexible diet that includes a wide range of foods, from meat to grains, vegetables, and fruits. This dietary adaptability, known as omnivory, has contributed to human survival in various environments by allowing people to take advantage of diverse food sources. However, as concerns about health, environmental sustainability, and animal welfare grow, there has been a shift toward "optionality," or the conscious choice to include plant-based alternatives in diets traditionally centered around meat. This shift emphasizes the idea that modern humans have the option to fulfill nutritional needs without relying exclusively on animal products, highlighting a broader range of dietary choices available today.

The term "optionality" in this context refers to the freedom and flexibility to replace animal-based foods with plant-based or alternative protein sources without sacrificing taste, texture, or nutrition. Plant-based meat alternatives are designed to mimic the flavor and experience of traditional meat products, making them attractive to consumers who may not want to eliminate meat from their diet entirely but wish to reduce their intake for various reasons. These alternatives, such as soy-based or pea protein burgers, sausages, and nuggets, have become widely available and popular. By offering taste and mouthfeel similar to conventional meat, these products are closing the gap between traditional omnivorous diets and plant-based eating. This flexibility allows people to maintain a mixed diet while reducing their environmental impact and addressing ethical considerations surrounding animal welfare.

Plant-based meat alternatives fit into the concept of optionality by giving consumers the choice to diversify their diets. Many consumers are interested in reducing meat consumption due to evidence linking high red and processed meat intake with various health risks, such as heart disease, certain cancers, and diabetes. Furthermore, animal agriculture is resource-intensive and contributes significantly to greenhouse gas emissions, deforestation, and water use. Plant-based meat alternatives, by comparison, generally have a smaller environmental footprint. For instance, producing plant-based protein requires fewer resources like water and land than raising livestock, and emits lower levels of carbon dioxide and methane. For consumers who are concerned about these issues, plant-based options offer a way to reduce personal environmental impact without committing to a vegetarian or vegan diet.

Additionally, plant-based meat alternatives contribute to dietary optionality by appealing to a range of dietary preferences, including those with religious or ethical restrictions. Certain groups avoid animal products due to religious beliefs, personal values, or ethical stances on animal rights. By providing products that replicate the meat experience, plant-based alternatives enable these individuals to enjoy familiar flavors and textures without compromising their values. Moreover, the concept of optionality recognizes that dietary choices do not need to be all-or-nothing; people can incorporate plant-based meals some of the time, creating a flexible, blended approach known as "flexitarianism." Flexitarians often choose to reduce, rather than eliminate, animal products, which can improve health and reduce environmental impact while still enjoying occasional meat-based dishes.

The rise of plant-based alternatives is not only reshaping diets but also challenging the food industry to innovate. Companies are investing in technology and research to make plant-based meats more appealing, nutritious, and accessible to a broader audience. Techniques such as fermentation, extrusion, and the use of legume-based proteins are advancing, with companies striving to improve texture, flavor, and nutritional value. For instance, innovations like heme from soy plants used by some companies are intended to create a meat-like juiciness and taste in plant-based burgers. This technological advancement in food science supports omnivory's optionality by making alternatives so similar to meat that consumers do not feel they are making a significant sacrifice by choosing them. Additionally, plant-based meats are being made healthier, often containing added vitamins, minerals, and dietary fiber to improve their nutritional profile, although they sometimes contain added sodium or processed ingredients, which consumers need to be mindful of.

However, some criticisms arise with the concept of plant-based meat in the context of omnivory and optionality. While plant-based alternatives are intended to provide similar nutritional benefits to meat, they are sometimes criticized for being overly processed. Critics argue that whole plant foods, such as beans, lentils, and vegetables, may offer a healthier alternative, especially for those concerned about processed food consumption. There is also concern about the impact of these alternatives on agricultural economies, particularly those dependent on animal farming. As consumer interest grows in plant-based options, the market dynamics of animal farming and food production are changing, affecting jobs and economies tied to traditional meat production. Furthermore, despite the promise of lower environmental impact, plant-based meat production still involves agricultural resources, such as land, water, and energy. While more sustainable than animal agriculture, the production of plant-based meats is not entirely free of environmental consequences.

These individuals mainly rely on eating fruits and vegetables that contain provitamin A (mainly β-carotene) to meet their vitamin A requirements, but many of them do not get enough of it. Naturally, genetic variations about the metabolism of nutrients account for why certain people can thrive on plant-based diets, while others who follow a vegan or vegetarian diet experience health issues linked to dietary deficiencies. In the US, for instance, there are five times as many former vegans and vegetarians as there are present ones, with 53% of the former group stating that they followed the diet for less than a year. It is extremely unlikely that everyone can thrive on a plant-based diet due to intra-individual differences in nutrient metabolism, even though a variety of factors, such as social factors and food options, contribute to the challenges associated with sticking to plant-based diets. If a person follows a "carnivorous" diet, which consists mostly or only of animal products, the same is probably true, though further research is required to support this theory.

In contrast, minerals like iron and zinc, long-chain polyunsaturated fatty acids (PUFAs) (e.g., retinol), vitamins D (cholecalciferol), B_{12} (adenosyl- and hydroxocobalamin), and K2 (menaquinone-4). G. animal sources are the more convenient way to obtain docosahexaenoic acid (DHA) and eicosapentaenoic acid (EPA); plant sources are less convenient. The development and regeneration of tissue depend on these nutrients. Although the precursors to these nutrients are frequently found in plant foods, a significant portion of the population has poor *in vivo* enzymatic conversion of plant precursors into forms that the human body can use. For instance, the proportion of carotenoids (provitamin A) that converts to retinol (vitamin A) varies from approximately 30.5 to 28%, contingent on individual genetic variability. This indicates that individuals classified as "poor converters" cannot obtain adequate retinol from plant-based diets alone. In the developing world, retinal deficiency is especially common, especially in young children and women who are pregnant or nursing. These individuals mainly rely on eating fruits and vegetables that contain provitamin A (mainly β-carotene) to meet their vitamin A requirements, but many of them do not. Naturally, genetic variations pertaining to nutrient metabolism also account for the fact that while some people thriving on plant-based diets, others adhering to a vegan or vegetarian diet report health issues linked to nutrient deficiencies.

It can be argued that human omnivory does not truly imply a choice between plant-based and animal-based foods. However, there is growing consensus that replacing traditional animal foods with plant-based foods would suffice to meet most of our nutritional needs, due to the detrimental effects of animal foods on human and environmental health. In addition, the use of isolated plant proteins, bioengineered ingredients, and/or synthetic vitamins and minerals to produce plant-sourced foods that can match the macronutrient, vitamin, and mineral content of animal foods has made it possible for the shift toward replacing animal foods with plant substitutes. The fact that novel plant-based meats can better satiate people's "intrinsic desire" for meat consumption could also be a factor in consumers' interest in these meat-like, tactile, and tasting plant versions. The intrinsic motivational salience, or the desire for meat, for instance, was retained in vegetarians on a brain level comparable to that of omnivores, despite their subjective dislike of animal products. The fact that this "subconscious motivation for eating meat" was noticed after just one overnight fast—far from being a condition akin to starvation—is noteworthy. The question of whether plant-based meat substitutes can satisfy the nutritional needs typically met by consuming animal products is addressed in the section that follows because novel plant-based meat substitutes resemble meat quite a bit.

7.3.2. Protein

Protein is a fundamental macromolecule that plays critical roles in the body, involved in almost every biological process. Comprised of amino acids, proteins are the building blocks for nearly all tissues and structures, from muscles and organs to enzymes and hormones. Proteins are essential for growth, repair, and maintaining cell structure and function. They support the immune system, help transport molecules, and catalyze biochemical reactions that keep the body functioning. Given these versatile roles, proteins are often regarded as the primary component of living organisms, second only to water in terms of volume within cells.

Structurally, proteins are made up of long chains of amino acids linked by peptide bonds. These amino acids, 20 in total, are arranged in specific sequences to form thousands of different proteins, each with a unique shape and function. The amino acid sequence is determined by genetic information, and variations in sequence can result in different protein shapes, which, in turn, define a protein's function. For example, the protein hemoglobin has a specific shape that allows it to carry oxygen through the bloodstream, while enzymes like lactase have active sites that bind to specific molecules, facilitating biochemical reactions. The body's ability to create complex proteins from basic amino acids is a testament to its intricate biochemistry.

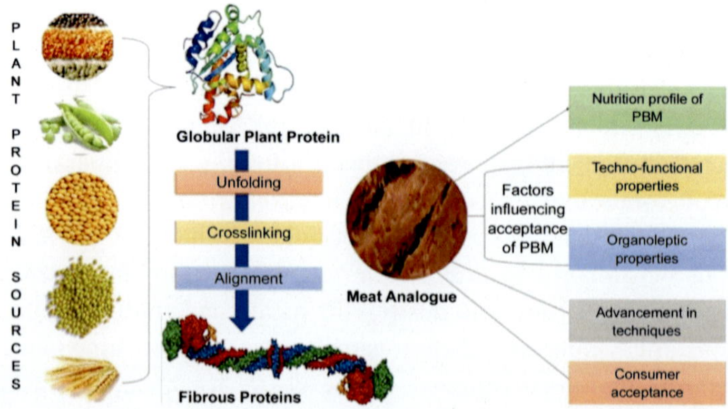

Figure 7.2. Transforming plant proteins into plant-based meat alternatives: challenges and future scope.

Source: https://media.springernature.com/lw685/springer-static/image/art%3A10.1007%2Fs10068-024-01683-0/MediaObjects/10068_2024_1683_Figa_HTML.png?as=webp

Dietary proteins provide essential amino acids that the body cannot produce on its own. These amino acids, called "essential amino acids," must be obtained through food. Protein sources are classified as either complete or incomplete, depending on their amino acid profile. Complete proteins, such as those found in meat, eggs, and dairy, contain all nine essential amino acids in adequate amounts. In contrast, plant-based sources like grains, legumes, and nuts may lack one or more essential amino acids, though combining different plant foods (like rice and beans) can provide a complete amino acid profile. This is particularly important for vegetarians and vegans, who rely on plant-based foods to meet their protein needs.

The daily recommended protein intake varies based on factors like age, gender, activity level, and health status. Generally, adults are advised to consume around 0.8 grams of protein per kilogram of body weight per day. For athletes or individuals engaging in intense physical activities, protein requirements are typically higher, as protein supports muscle recovery and growth. Likewise, growing children, pregnant or breastfeeding women, and elderly individuals may also require additional protein to support bodily needs. Inadequate protein intake can lead to various health issues, including muscle wasting, weakened immunity, and impaired wound healing.

Proteins are also a critical focus in diet and nutrition, especially with the recent popularity of plant-based and alternative protein sources. Beyond traditional sources like meat, eggs, and dairy, protein-rich plant foods such as soy, lentils, and quinoa are gaining attention for their health benefits and environmental sustainability. The development of plant-based meat alternatives, made with ingredients like pea and soy protein, is providing people with more options to diversify their protein intake, particularly appealing to those seeking to reduce meat consumption for health or ethical reasons.

Beyond physical health, protein intake influences mental function, energy levels, and mood. Certain amino acids act as precursors to neurotransmitters, impacting cognitive processes and emotional regulation. For instance, the amino acid tryptophan is a precursor to serotonin, often dubbed the "happiness hormone," and can affect mood and sleep. Therefore, adequate protein intake not only supports bodily functions but can also contribute to overall well-being.

7.3.3. Vitamin B_{12}

Vitamin B12, also known as cobalamin, is an essential water-soluble vitamin required for various physiological processes, including red blood cell formation, DNA synthesis, and neurological function. It is unique among vitamins in that it is found naturally almost exclusively in animal-derived foods, such as meat, dairy, eggs, and fish. For those who follow vegetarian or vegan diets, or rely heavily on plant-based meat alternatives, obtaining adequate B12 can pose a challenge, as plant-based foods generally do not naturally contain significant amounts of this vitamin. Consequently, understanding the importance of B12 and strategies for obtaining it within plant-based diets is essential for individuals seeking to avoid potential deficiencies.

The biological importance of vitamin B12 lies in its role in producing red blood cells and maintaining the nervous system. It is crucial for converting homocysteine, an amino acid, into methionine, another amino acid, a reaction that plays a part in DNA synthesis and repair. Additionally, B12 supports the production of myelin, a fatty sheath around nerves that ensures proper neural transmission. Deficiency in B12 can lead to anemia, fatigue, and neurological issues, including memory problems, mood changes, and, in severe cases, irreversible nerve damage. This makes B12 a vital nutrient for individuals across all ages, but it is particularly significant for older adults, as the ability to absorb B12 tends to decline with age. For people consuming plant-based meat alternatives, B12 intake must be carefully considered since it is not naturally present in plant foods. Some individuals mistakenly assume that a plant-based diet inherently provides all necessary nutrients, but without careful planning or supplementation, B12 intake may be insufficient, potentially leading to deficiency over time. Symptoms of B12 deficiency may develop slowly and can initially be subtle, including fatigue, muscle weakness, and tingling in extremities, which can be misattributed to other health issues. As deficiency progresses, more serious complications can arise, making it important for those on plant-based diets to proactively address their B12 needs.

Many plant-based meat alternatives are fortified with vitamin B12 to help bridge this nutritional gap. Fortification involves adding B12 to products like plant-based burgers, sausages, and dairy substitutes, providing a convenient source of this essential vitamin. Fortified plant-based products are now widely available, making it easier for individuals following plant-based diets to meet their B12 needs without relying on animal-derived foods. Common sources of fortified B12 include fortified breakfast cereals, nutritional yeast, plant-based milks, and meat substitutes, all of which can contribute to daily B12 intake. However, it is important to check food labels, as not all plant-based products are fortified with B12, and the fortification levels can vary significantly.

In addition to fortified foods, B12 supplements are widely recommended for those on plant-based diets, especially for individuals who do not consume fortified foods regularly. B12 supplements come in various forms, including tablets, sublingual (under-the-tongue) drops, and even B12-enriched sprays, which can be absorbed through the lining of the mouth. Dosage recommendations for B12 supplements can vary based on individual needs, but daily intake generally ranges from 2.4 to 6 micrograms for adults. Larger doses are often recommended in supplements since B12 absorption can be limited, with only a small percentage being absorbed per dose. Individuals who have difficulty absorbing B12, such as the elderly or those with specific gastrointestinal conditions, may require higher doses or periodic injections for effective intake.

In recent years, there has been interest in finding natural, plant-based sources of B12. Some types of algae, such as chlorella and certain seaweeds, have been found to contain forms of B12, but the bioavailability and effectiveness of B12 from these sources remain controversial. Studies have indicated that not all B12 in these sources is in a form that the human body can readily use, meaning it may not effectively prevent deficiency. Likewise, fermented foods like tempeh and kimchi, once believed to contain B12 due to bacterial fermentation, typically contain very minimal or inconsistent levels of B12 and cannot be relied upon as primary sources. As a result, for individuals on plant-based diets, obtaining reliable B12 from fortified foods or supplements is still the most effective strategy.

The importance of ensuring adequate B12 intake for those on plant-based diets, including those who consume plant-based meat alternatives, is increasingly recognized in the nutrition and health community. Regular blood testing can help monitor B12 levels, particularly for individuals who have adopted a vegan diet or who may be at risk of deficiency. Monitoring B12 status is especially important for pregnant or breastfeeding women on plant-based diets, as B12 is crucial for fetal brain and nervous system development. Deficiency in pregnant women can lead to serious developmental issues for the baby, emphasizing the need for careful planning and, in many cases, supplementation.

As the popularity of plant-based diets continues to rise, food producers are increasingly fortifying plant-based products with B12 to cater to this growing market. This trend is beneficial for consumers as it makes meeting dietary needs more convenient. However, relying on fortified products does not eliminate the need for awareness and proactive health management. Those adopting a plant-based lifestyle should educate themselves on the essential nutrients that might be less accessible in their diet, with B12 being one of the most critical.

7.3.4. Iron

Iron is a critical mineral that plays essential roles in transporting oxygen through the blood, supporting immune function, and aiding in energy metabolism. It is a fundamental component of hemoglobin, the protein in red blood cells responsible for carrying oxygen from the lungs to tissues throughout the body. Iron is also a key part of myoglobin, a protein in muscles that supplies oxygen for muscle contractions, and numerous enzymes that drive biochemical reactions within cells. Without adequate iron, the body cannot produce enough healthy red blood cells, leading to iron-deficiency anemia, which causes symptoms like fatigue, weakness, shortness of breath, and impaired cognitive function. Ensuring adequate iron intake is therefore essential for maintaining health and vitality, and it is especially important for those relying on plant-based meat alternatives, as plant foods differ significantly in iron content and type compared to animal-based foods.

Iron exists in two dietary forms: heme and non-heme iron. Heme iron, found in animal-based foods like red meat, poultry, and fish, is more readily absorbed by the body. Non-heme iron, on the other hand, is found in plant-based foods, such as legumes, whole grains, nuts, seeds, and leafy greens. While non-heme iron can also support health, it is less bioavailable, meaning the body absorbs it less efficiently than heme iron. This reduced bioavailability makes iron intake a particular consideration for those following plant-based diets, as they must often consume more iron to compensate for lower absorption rates. Studies indicate that individuals who consume plant-based diets generally need nearly twice as much iron as those who consume meat to meet the same physiological needs due to the lower absorption rate of non-heme iron.

For people who rely on plant-based meat alternatives, understanding iron bioavailability and enhancing iron absorption is essential. Certain factors can inhibit or enhance the body's ability to absorb non-heme iron. For instance, plant compounds like phytates (found in grains and legumes), oxalates (found in leafy greens), and polyphenols (found in tea and coffee) can bind to non-heme iron, reducing its absorption in the gut. On the other hand, vitamin C significantly boosts non-heme iron absorption by converting it into a more absorbable form. Incorporating vitamin C-rich foods, like citrus fruits, bell peppers, tomatoes, or berries, into meals with iron-rich plant foods can help improve iron absorption. Additionally, cooking techniques such as soaking, fermenting, or sprouting grains and legumes can reduce phytate levels and enhance iron availability. These strategies can help individuals following plant-based diets improve their iron intake from natural sources.

In response to the growing demand for plant-based meat alternatives, many manufacturers are fortifying these products with iron to help consumers meet their nutritional needs. Fortification involves adding iron to plant-based foods, especially popular meat substitutes like burgers, sausages, and nuggets made from soy, pea, or other legume proteins. Fortified iron can help bridge the gap between the naturally lower iron content of plant-based diets and the body's needs. However, the type of iron used in fortification is typically non-heme iron, which still has lower bioavailability compared to heme iron in meat. Despite this limitation, fortified plant-based products can be a valuable tool for increasing iron intake, especially when paired with iron-absorption-enhancing foods.

Certain populations, such as women of childbearing age, pregnant women, children, and athletes, have higher iron requirements and are at greater risk of iron deficiency. This is particularly relevant for those on plant-based diets, as iron deficiency can have significant consequences for health, development, and athletic performance. For example, iron is essential during pregnancy to support both the mother's increased blood volume and the developing baby's growth. Similarly, athletes require more iron to support muscle oxygenation and endurance. Therefore, individuals in these groups may need to pay extra attention to their iron intake when consuming plant-based meat alternatives and consider strategies such as fortified products, combining iron-rich foods with vitamin C sources, and possibly supplements if recommended by a healthcare provider.

Despite the benefits of plant-based iron fortification, some individuals express concern over potential side effects or risks associated with excessive iron intake, such as gastrointestinal discomfort. Generally, the risk of excessive iron intake from plant-based sources is low, as the body regulates non-heme iron absorption according to its needs. Iron toxicity is rare in healthy individuals consuming iron from plant-based foods. However, those taking iron supplements should do so under medical supervision to avoid the risk of excessive intake, as supplements provide higher and more concentrated doses of iron than foods.

For those looking to naturally increase iron intake through whole foods within a plant-based framework, certain foods are particularly iron-rich. Lentils, chickpeas, beans, tofu, quinoa, chia seeds, hemp seeds, pumpkin seeds, spinach, and kale are all good sources of non-heme iron. By including these foods in a varied and balanced plant-based diet, individuals can work towards meeting their iron needs, especially when prepared in ways that maximize absorption. For instance, a meal consisting of lentil stew with bell peppers and a side of leafy greens would combine iron with vitamin C for optimal absorption, and sprouted grains or fermented foods could further support iron uptake. Nutritional yeast, often fortified with B vitamins and iron, is another versatile addition that can help boost iron intake while adding flavor to plant-based dishes.

The increasing popularity of plant-based meat alternatives and the trend towards plant-forward eating are encouraging more individuals to consider iron intake in their diets. With thoughtful planning, individuals following plant-based diets can successfully meet their iron requirements by combining fortified products, whole plant foods rich in iron, and absorption-enhancing strategies. Health professionals and dietitians play a crucial role in educating people about these practices, ensuring that individuals following plant-based diets are informed about managing their iron intake effectively.

7.3.5. Zinc

Zinc is an essential trace mineral required for numerous physiological processes, including immune function, DNA synthesis, wound healing, and cell division. It also plays an important role in growth and development, particularly during childhood, adolescence, and pregnancy. Zinc acts as a cofactor for over 300 enzymes, which means it is involved in countless reactions throughout the body that regulate everything from protein synthesis to hormone production and cognitive function. The body cannot produce or store zinc, so it must be regularly obtained from the diet. This requirement makes adequate zinc intake critical for health, especially for individuals following plant-based diets who may rely on plant-based meat alternatives, as plant foods generally contain less zinc or contain zinc in forms that are less bioavailable than those found in animal-based foods. Zinc exists in both animal and plant foods, but animal sources such as meat, dairy, eggs, and seafood provide a highly absorbable form known as "heme zinc." In contrast, the zinc found in plant foods is non-heme, which the body absorbs less efficiently. Additionally, certain compounds in plants, such as phytates, bind to zinc and reduce its absorption. These phytates are present in legumes, whole grains, seeds, and nuts—foods commonly relied upon in plant-based diets for protein and other nutrients. Phytates form complexes with zinc that inhibit its absorption in the gut, meaning that individuals consuming plant-based diets may need to consume more zinc to meet their needs than those who consume meat, which naturally contains readily bioavailable zinc.

To address these challenges, many producers of plant-based meat alternatives fortify their products with zinc. Fortification involves adding zinc in a form that the body can absorb, such as zinc sulfate or zinc gluconate, to popular plant-based meat substitutes like veggie burgers, sausages, and nuggets. This is especially useful for individuals who do not consume animal products, as it provides a more reliable zinc source within a plant-based dietary framework. Although fortified zinc in plant-based products may not achieve the same bioavailability as zinc from meat, it can help bridge the dietary gap and contribute to overall zinc intake.

For those following plant-based diets, certain strategies can further enhance zinc absorption from plant foods. Cooking, soaking, sprouting, and fermenting foods containing phytates can reduce these compounds, making zinc more accessible for absorption. For instance, soaking beans or fermenting grains for products like sourdough bread helps reduce phytate levels, increasing the bioavailability of zinc. Additionally, pairing zinc-rich plant foods with foods high in sulfur-containing amino acids (such as garlic and onions) or certain organic acids (such as citric acid) may improve absorption. Although these strategies are effective, they require intentional planning and food preparation, which is why fortified foods can be particularly convenient and beneficial for those on plant-based diets.

Zinc requirements vary based on age, gender, and life stage. Adult men and women are generally recommended to consume around 11 mg and 8 mg of zinc daily, respectively. However, individuals on plant-based diets may need to aim for slightly higher intakes due to reduced absorption rates from plant sources. Children, adolescents, pregnant, and lactating women also have higher zinc requirements to support growth, development, and increased metabolic demands. Zinc deficiency is a global health concern, particularly in populations that consume limited animal-based foods. Deficiency can lead to symptoms such as weakened immunity, delayed wound healing, impaired taste and smell, hair loss, and, in severe cases, growth retardation and developmental delays in children. Zinc-rich plant foods include legumes (such as chickpeas, lentils, and beans), whole grains, nuts, seeds, and certain vegetables. For example, pumpkin seeds and hemp seeds are relatively high in zinc compared to other plant sources, making them valuable additions to a plant-based diet. Tofu, tempeh, quinoa, and fortified cereals are also good sources of zinc. By consuming a variety of these foods regularly, individuals on plant-based diets can work towards meeting their zinc needs, especially when combined with cooking and preparation techniques that improve zinc availability.

Despite the benefits of fortified plant-based products and whole foods for zinc intake, certain populations following plant-based diets may benefit from zinc supplements, particularly if they have limited access to fortified foods or face increased zinc needs. Zinc supplements are available in various forms, including zinc gluconate, zinc sulfate, and zinc citrate, each with slightly different bioavailability levels. However, supplements should be taken with caution, as excessive zinc intake can cause side effects, such as gastrointestinal discomfort, and may interfere with copper absorption, leading to copper deficiency over time. For this reason, it is generally best to obtain zinc primarily from food sources and only supplement as needed under the guidance of a healthcare provider.

The growing popularity of plant-based diets has prompted a wider discussion about essential nutrients like zinc, with an emphasis on ensuring that individuals following these diets can meet their nutritional needs. Fortification of plant-based meat alternatives with zinc provides a convenient solution for increasing zinc intake and supporting optimal health among those reducing or eliminating animal products. By offering fortified products and encouraging diverse sources of zinc-rich plant foods, the plant-based food industry is helping to make zinc more accessible to a broader population, including vegans, vegetarians, and flexitarians.

7.3.6. Essential Fatty Acids

Essential fatty acids (EFAs) are types of polyunsaturated fats that are crucial for various bodily functions, including brain development, cell membrane structure, inflammation regulation, and hormone production. They are termed "essential" because the body cannot produce them independently and must obtain them from the diet. There are two primary types of essential fatty acids: omega-3 and omega-6. Within these categories, two specific fatty acids are vital: alpha-linolenic acid (ALA), an omega-3 fatty acid, and linoleic acid (LA), an omega-6 fatty acid. Both ALA and LA play significant roles in health, supporting everything from heart health to immune function. However, obtaining sufficient EFAs, especially omega-3s, can be more challenging for those relying on plant-based diets and plant-based meat alternatives, as these essential fats are abundant in certain animal sources.

Omega-3 fatty acids are particularly known for their role in reducing inflammation, supporting heart health, and promoting brain function. While ALA is the primary omega-3 found in plant foods, other omega-3s—eicosapentaenoic acid (EPA) and docosahexaenoic acid (DHA)—are often cited for their potent health benefits. EPA and DHA are primarily found in fatty fish and algae, and while ALA can be converted into EPA and DHA within the body, the conversion rate is low, especially for DHA, which is crucial for brain and eye health. Plant-based sources of ALA include flaxseeds, chia seeds, walnuts, hemp seeds, and canola oil. Although these foods provide beneficial omega-3s, individuals on plant-based diets or those who consume plant-based meat alternatives exclusively may still fall short of recommended EPA and DHA levels without supplementation or specific foods like algae-based oils that contain preformed EPA and DHA.

Omega-6 fatty acids, especially linoleic acid, are also essential and are more readily available in plant-based diets, as they are abundant in vegetable oils (such as soybean, sunflower, and corn oil), nuts, seeds, and many processed foods. Omega-6 fats play essential roles in cell function, and, like omega-3s, they contribute to the body's inflammatory and immune responses. However, balancing omega-6 and omega-3 intake is crucial, as excessive omega-6 intake relative to omega-3 can promote inflammation. Western diets often include high amounts of omega-6 fats from processed foods, which can create an imbalance that may counteract some of the anti-inflammatory benefits of omega-3s. For those on plant-based diets, paying attention to the ratio of omega-6 to omega-3 is particularly important to maintain optimal health.

The popularity of plant-based diets has led to increased interest in plant-based sources of EFAs and the fortification of plant-based meat alternatives with omega-3s. Some plant-based meat alternatives are now fortified with ALA-rich oils, such as flaxseed or chia, to increase their omega-3 content. However, EPA and DHA remain more challenging to incorporate into plant-based foods because they typically come from marine sources, primarily fish and algae. To meet EPA and DHA needs, those on plant-based diets may consider algae oil supplements, which are vegan and provide a direct source of these omega-3s without relying on fish. Algae-derived DHA supplements are widely available and can help ensure that individuals consuming plant-based diets meet their essential fatty acid requirements.

In addition to fortified foods and supplements, certain dietary strategies can improve the body's utilization of ALA. For example, reducing the intake of omega-6-rich oils and prioritizing omega-3-rich plant foods may help increase the conversion of ALA to EPA and DHA. While this conversion is limited, it is possible to improve it slightly by maintaining a balanced intake of these fatty acids. Consuming a varied diet that includes plenty of whole plant foods rich in ALA can contribute to an overall healthier fatty acid profile.

For those following a plant-based diet, achieving a balanced intake of omega-3 and omega-6 fatty acids can support cardiovascular health, reduce inflammation, and promote overall well-being. Although some people may still benefit from taking an algae-based EPA/DHA supplement to ensure adequate levels, fortifying plant-based meat alternatives with ALA is a positive step towards making EFAs more accessible to those who choose not to consume animal products.

Two essential fatty acids that humans cannot synthesize *in vivo* are linoleic acid (C18:2, LNA) and alpha-linolenic acid (C18:3, ALA), both of which are ω-6 fatty acids that must be obtained through diet. Eicosapentaenoic acid (C20:5 n-3, EPA) and docosahexaenoic acid (C22:6 n-3, DHA), two long-chain polyunsaturated fatty acids (LCPUFA), are derived from ALA. DHA and EPA can only be found in animal foods and some types of algae, whereas ALA and LNA are frequently found in plant foods and can also be found in limited amounts in animal foods. A series of elongation and desaturation steps can convert ALA to DHA and EPA, but the conversion rate is often less than 1%. Furthermore, co-factors like zinc, iron, selenium, and vitamin B_6, which are less bioavailable from plant-based diets, are necessary for this conversion efficiency to occur. These explanations explain why, in comparison to omnivores, vegetarians may have lower DHA and EPA levels.

LNA and other ω-6 fatty acids directly compete with ALA for the enzymes responsible for elongation and desaturation, which further reduces ALA's capacity to yield DHA and EPA. This is another possible problem. The issue arises especially when one considers that the modern Western diet's increased intake of high-LNA seed oils has led to a 16:1 ω-6-to-ω-3 fatty acid ratio, while historical intakes place this ratio closer to 1:1. In Western countries, the rising incidence of metabolic disease and all-cause mortality can be attributed in part to this high ω-6-to-ω-3 fatty acid ratio. A potential mechanism for the potential cardioprotective effects of consuming ω-3-rich ALA oils is demonstrated by the reduction in inflammation that occurs when ω-6-rich LNA oils are experimentally substituted with ω-3-rich ALA oils. Therefore, using high ALA oils, rich in ω-3, rather than high LNA oils, rich in ω-6 fatty acids, could be a suitable improvement to the novel plant-based meat substitutes. Although it's important to take into account that high ALA oils are more likely to be oxidized by lipids, which may be the reason why high LNA oils are typically used in plant-based meat substitutes, producers of plant-based meat substitutes that take into account the use of high ALA oils in their products should consider exploring the addition of natural anti-oxidants as well as entrapment of high ALA oils with isolated plant proteins as worthwhile opportunities to explore.

7.3.7. Secondary Nutrients

As we've covered so far, the essential fatty acid, amino acid, vitamin, and mineral content of foods indicates much less about their natural state than does the complexity of these foods. Biochemicals that are vital to human metabolism can be found in food sources in quantities ranging from hundreds to thousands. It is important to remember that these nutrients have an impact on human metabolism even though many of them are regarded as conditionally or non-essential depending on life stages and are frequently overlooked in discussions of the nutritional needs of humans. To improve athletic performance, for instance, creatine has been the subject of much research; however, creatine is also crucial for cognitive function. Since vegetarians and vegans can only obtain creatine from animal sources, their bodies typically contain less of it. However, when vegetarians were given additional creatine, their memory tests significantly improved. Likewise, animal-based foods are the only source of the antioxidants taurine, carnosine, and anserine (virtually). Neurocognitive protection in humans is provided by increased anserine and carnosine intake.

Although trace amounts of taurine—a thousand times less than in animal foods—may be found in some plant foods, such as cereals, legumes, and grains, they are insufficient to meet human needs. Taurine is an amino acid that is almost exclusively found in animal foods. The argument that taurine needs can be satisfied by consuming plant proteins rich in methionine and cysteine—which are present in sufficient amounts in many grains and legumes—is frequently made because taurine can be synthesized *in vivo* from methionine and cysteine via cysteinesulfinic acid decarboxylase (CSD). To sustain tissue concentrations over time, the human body's levels of CSD, which permits taurine to be produced from methionine and cysteine, are limited. Taurine is essential for almost all bodily systems and plays a major part in mitochondrial processes, cardiovascular health, brain function, skeletal muscle cell differentiation, and eye health. To gain a deeper understanding of the effects of these variations in secondary nutrients on both short- and long-term health, more research is required on plant-based meat alternatives and meat.

7.3.8. Fortifying Foods to Mimic the Natural Food Source

It is a common concern that the reductionist approach of attempting to "mimic" whole food sources (meat or other foods) by combining multiple isolated nutrients likely underestimates the true complexity and health benefits of consuming whole foods. This is because natural whole foods are extremely complex. In instance, adding zinc and other minerals from meat to a low-meat diet did not improve zinc levels in the same way as if these minerals were included in the diet naturally as part of the meat's matrix. Further, a recent large population-based study found that adequate intakes of zinc, copper, and vitamins A and D when derived from foods, but not from supplements, were associated with decreased risk of cardiovascular disease and all-cause mortality.

In a similar vein, foods high in carotenoids are linked to a lower risk of heart disease, retinopathies, and many cancers. However, the findings of epidemiological and interventional research indicate that taking supplements containing carotenoids and/or vitamin A may even increase the risk of certain subpopulations of cancer or cardiovascular disease.

7.3.9. The Ecological Impacts of Plant-Based Meat vs. Meat Consumption

The plant vs. meat controversy takes on other dimensions when assessing environmental degradation and climate change, both of which adversely affect human health and are crucial considerations when making recommendations on diets for livestock and humans. Fulfilling nutrient needs with plant-based foods (e.g., manganese, thiamin, copper, β-carotene, and folate) could have a smaller environmental impact (i.e., fewer greenhouse gas emissions) in comparison to eating animal products to obtain these nutrients. However, such comparisons do not account for the reduced bioaccessibility and bioavailability of plant sources for these nutrients. Nevertheless, it has been suggested that similar amounts of protein, iron, and vitamin A can be obtained from carefully chosen plant-based diets at a lower carbon footprint when compared to omnivorous diets/animal foods. Furthermore, amino acid content and nutrient density are taken into account when footprints—land used for production as well as greenhouse gas emissions (GHGE)—are computed. G. because animal foods can more easily satisfy our needs for these particular nutrients (iron, vitamin B_{12}, zinc, retinol, and amino acids), the nutritional footprint of animal foods may resemble that of plant foods (van Vliet et al., 2020).

Plant-based meat substitutes are often preferred over beef because of their lower carbon footprint. Compared to US beef finished on total-mixed rations in feedlots, the Beyond Burger® and the ImpossibleTM Burger have less carbon footprints (+3.2 and 3.5 kg CO_2-eq emissions/per kg product, respectively). The difference in carbon footprint between the two products depends on the model used, the location of the cattle's raising, the inclusion of GHGE potential of retail, distribution, restaurant or at-home use, and end-of-life stages (Figure 7.3).

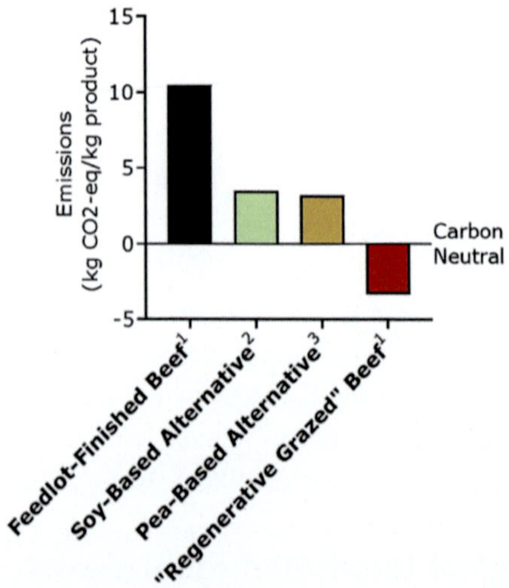

Figure 7.3. Comparison of possible greenhouse gas emissions impacts of various beef production systems and meat alternatives. All values include cradle-to-distribution LCA but exclude GHGE potential of retail, restaurant or at home use, and end-of-life stages.

Source:: van Vliet S, Kronberg SL and Provenza FD (2020) Plant-Based Meats, Human Health, and Climate Change. Front. Sustain. Food Syst. 4:128. doi: 10.3389/fsufs.2020.00128.

Even after accounting for every facet of the production process, well-managed pasture-based livestock systems fix at least all the GHG they emit, and sometimes more. Meat substitutes may have less of an environmental impact than feedlot-finished beef. The amounts of carbon sequestered in the soil more than offset the ruminants' GHGE, resulting in a net negative carbon footprint, according to pastured beef systems that use land management techniques like cover crop grazing and/or rotational grazing, which allows lands to properly recover after a grazing period. Such production systems have the potential to help mitigate climate change, or at the very least not exacerbate it further, while ensuring a certain level of food security. Livestock contribute to carbon cycling by spending their lives on well-managed pastures, which groom and fertilize vegetation and soil. In many different parts of the world, well-managed grasslands—particularly those with more mesic areas—can function as carbon sinks. Depending on the location, these grasslands may even be more dependable than forests as carbon sinks. Notably, silvopastoralism—a form of agroforestry that integrates trees, forage, and livestock—is successfully implemented in forested areas worldwide as a tactic to improve carbon sequestration, soil health, and food security for those residing in such areas, indicating that the two (forests and livestock) are not mutually exclusive to begin with. To avoid trying to alter the environment to fit the production system, often at great expense, considerations regarding livestock-production systems should be tailored to fit the geophysical landscape. For instance, in environments like the Amazon, it would be appropriate to use silvopasture techniques with locally adapted animals instead of trying to turn its forests into pasturelands.

However, it should be remembered that not all pasture-based (grass-fed) operations are inherently regenerative or neutral and that the carbon footprint of grass-fed beef systems may be higher than that of some feedlot systems, depending on management practices. The fact that properly managed livestock grazing on carbon-depleted soils initially sequesters more carbon and gradually reduces it as soil health improves is also important to note. This is to be expected once ecological systems have reached equilibrium. The following discussions should take this idea into account.

By performing an ISO-compliant partial LCA of pasture-raised (grass-fed) beef in the Midwest US, Stanley et al. (2018) found a net negative carbon footprint of ~−3.5. CO_2-eq/per kg beef (assuming a 55% edible yield of hot carcass weight). We note that the value from the pasture-finished beef LCA assessment excludes the GHGE of getting the product case-ready, which is included in the plant-based meat LCAs. This is expected to add +0.3 kg CO_2-eq/kg product which would put the LCA analysis from so-called regenerative, grass-fed beef at −3.2. CO_2-eq/per kg beef. This means that over the lifecycle of the animal more carbon was sequestered than emitted.

Remarkably, Quantis International, the same company that proved a +3.5. CO_2-eq emissions/per kg product in the Impossible BurgerTM LCA analysis, also proved a −3.5. CO_2-eq/per kg beef made with livestock grazing techniques. It is encouraging that the values reported by Quantis International are comparable to those reported in peer-reviewed research on so-called regenerative beef and in academic scientists' work on plant-based meat substitutes, even though these reports are not peer-reviewed. However, the life cycle assessments conducted on meat substitutes and pasture-finished beef do not account for greenhouse gas emissions from retail, dining establishments, home cooking, end-of-life situations, or other localized or indirect effects. Given the challenge of evaluating every element of environmental footprints, future research must validate these LCA analyses by fully accounting for all GHGE to enable impartial evaluations.

Many times, greenhouse gases are grouped together under the CO_2-eq acronym, which compares various GHGs to carbon dioxide (CO_2). However, the global warming potentials (GWP) of various gases vary, and their precise values rely on the CO_2-eq metrics and the timescale (e.g., 20 or 100 years) that is employed to convey its contribution to global warming. While there are significant emissions of CH_4 (methane) associated with livestock production, the majority of emissions from plant-based meat come from the production of energy. Globally, livestock contribute 14.5% to GHGE. of that, 95% is used to produce feed for animals, process, and transport, with the remaining 5% coming from the fermentation of manure and rumen (enteric stomach). Although methane is a strong greenhouse gas, it is only present for 10 years before it decomposes into CO_2, which can be stored in soil. Cattle's methane belches are not expected to contribute additional carbon to the atmosphere because the number of cattle in the US is stable or slightly declining, albeit not worldwide. However, carbon dioxide stays in the atmosphere for thousands of years after it is released by burning fossil fuels for energy or food production. Understanding these subtleties is crucial when debating the relative carbon footprints of various foods and eating habits.

Since most beef consumed in the US comes from feedlots that use fossil fuels heavily to produce it (roughly 96% of all beef in the US is finished in feedlots), meat alternatives probably have a lower carbon footprint than most beef. Some predict that as US beef production systems shift more toward pasture-based practices, domestic beef consumption will have to drop by roughly 40% because there won't be as much land available for the production of roughage feed, which is used to supplement cattle on pasture. The potential for higher carrying capacity from multi-species grazing with little dietary overlap, such as combining cattle with sheep or goats, is not factored into these estimations. This type of grazing increases the productivity of both animals and vegetation when compared to grazing by either animal alone. Additionally, with proper management, multi-species grazing can preserve plant diversity, enhancing pasture health and ecological resilience. This would obviously require us to include products from other livestock, such as sheep, goats, and possibly smaller mammals like ducks and rabbits, in addition to meat and milk in our diet. Adding strategically edible by-products to livestock on pasture is another way to further increase the carrying capacity of pasture-based livestock, something that is frequently overlooked in discussions on the carrying capacity of pasture-based production systems. Ruminants are uniquely suited to recycle agricultural and industrial waste products. For instance, only the starch portion of corn is used to produce ethanol; the remaining byproducts, which include the oil, germ, and outer shell, can be combined to create a high-protein, high-fat cake that is fed to cattle. Crop wastes like stover, straws, and sugarcane tops as well as fruit and vegetable processing byproducts rich in phytochemicals like leaves, pomace, peels, rinds, pulp, seeds, stems, etc. livestock offer comparable chances for recycling these nutrients. By providing these byproducts to cattle while they are on pasture, it may be possible to improve the utilization of undesirable vegetation, lessen the chance of overgrazing, and address problems related to a shortage of available land (van Vliet et al., 2020). Encouraging cattle to graze on by-products instead of feeding them in feedlots would also help alleviate some of the negative effects feedlots have on animal welfare, like strange surroundings, the incapacity to choose one's own food, and the loss of natural behavior. While still retaining the ability to upcycle byproducts to satisfy consumer demand, providing by-products to cattle on pasture may prove to be a valuable opportunity for the livestock industry to enhance consumer perception. As opposed to cultivating feed specifically for livestock, it will be crucial to use only industrial byproducts that would have been produced anyhow. According to recent research, animals' time to reach slaughter weight is also influenced by the variety of forages they eat while on pasture. Cattle eating a variety of plant mixtures—some of which contain tannins—gain weight more efficiently and can reach finish body condition almost as quickly as those in feedlots, all while consuming less GHGE, when compared to grazing a monoculture of grass. If ruminants are fed forages rich in secondary compounds like tannins and terpenes, their urine will contain less nitrogen and their manure, which is rich in tannins and increases soil organic matter, will contain more nitrogen.

Lastly, the question of whether pasture-based production systems can support the consumption of meat centers on the system's capacity to sustain the consumption of common retail cuts (i.e., beef, roasts, and steaks). For example, in their estimates of the carrying capacity of pasture-based beef production systems to support US consumption, they assume a yield that is 60% edible. Although this percentage makes sense, an additional 20% of the animal, which consists of organs, bones, and tallow, is perfectly fit for human consumption. For instance, consuming more organ meats, which are frequently far denser in vitamins and minerals (e.g., 10–1000 times higher in retinol, iron, copper, and vitamins B_6, B_{12}, and K_2) than muscle meat—was discovered to slash GHGE linked to meat consumption by 14%. Agricultural techniques that combine regenerative livestock grazing with plant farming are a significant step in the right direction to lessen the carbon footprint and land use of animal agriculture, even though they are not a miracle cure for climate change. Regenerative techniques—farmland restoration, conservation agriculture, agroforestry, silvopastoralism, and managed grazing—collectively rank first among the 80 methods of mitigating climate change that Project Drawdown evaluated. Moreover, crop yield and soil fertility can be raised by combining plant farming with livestock grazing. It is crucial to recognize the mutually beneficial relationship between plants and herbivores when debating whether or not to replace livestock production. An estimated 80% increase in GHGE results from the predicted rise to nearly 10 billion people and a 32% increase in emissions per person worldwide due to the shift to ultra-processed diets by 2050. Conversely, if a person followed a vegetarian, pescatarian, or Mediterranean diet that included whole foods like fruits, vegetables, seafood, grains, eggs, and dairy products along with moderation in their intake of beef, lamb, and chicken, their GHGE might not rise. For instance, eating out, drinking, and consuming sweets were the main causes of the high carbon footprints found in an urban Japanese population. Meat consumption, on the other hand, contributed significantly less to the carbon footprint—it accounted for just 9% of the difference between low and high dietary carbon footprints. Similar conclusions were drawn from a recent Australian cohort study, in which the majority of the environmental footprint was attributed to "discretionary foods," which included alcohol, confections, sugar-sweetened beverages, and other highly processed foods. Future research in American and European homes should validate this theory, but given the widespread consumption of the Standard Western/American diet, comparable outcomes are probably reasonable. Furthermore, a healthy omnivorous diet rich in whole-food plant and animal sources has the highest carrying capacity for feeding populations in a variety of global regions, according to biophysical simulation of various dietary patterns. Diets high in processed foods, such as the Standard Western Diet, have a lower carrying capacity than vegan and vegetarian diets. On this basis, some argue for the adoption of a plant-based diet; however, a diet consisting solely of plant foods does not effectively use land that could feed more people, nor does it integrate farming and grazing to improve soil fertility, which strengthens both plant and livestock farming systems synergetically. The latter is important because billions of people live on managed livestock grazing, which is an unsuitable use of two-thirds of the planet's land mass for crop production. So, it is essential that no laws that endanger the health and way of life of the world's poorest people be put in place when talking about dietary shifts to plant-based alternatives.

The fact that most crops are grown in monocultures, where life is sacrificed mechanically, chemically, or both, is another crucial factor to take into account. While consuming almond seeds, carrot roots, or plant-based meat substitutes made from soy or peas does not directly result in animal deaths, it does indirectly. There is destruction of other plants' and animals' habitats. One notable example is the loss of over 50% of grassland bird populations in North America over the past 50 years as a result of industrial farming methods, including both plant and animal husbandry. Another striking example can be found in Borneo and the Southern Peninsula of Malaysia, where oil palm plantations have replaced natural forests, resulting in an approximately 85% decline in orangutan populations and an 80% decrease in the total number of bird and butterfly species. The amount of animals killed in agriculture as "collateral damage" is extremely difficult to estimate (i.e., mammals, fish, reptiles, and other amphibious animals); nevertheless, conservative annual estimates in the US place this number in the tens of millions. To put things in perspective, the US is estimated to slaughter

40 million cattle and 120 million pigs annually for human consumption. The question of whether "unintended" deaths from plant agriculture, such as those produced for human consumption and animal feed, are ethically comparable to "intentional" deaths from animal consumption is outside the purview of research. It makes the point that in food systems, to survive, life must consume life. That being said, most people (meat and plant-eaters alike) probably agree that advances in plant farming and livestock production techniques must be made in ways that improve the welfare of livestock and wildlife.

The ecological effects of dietary choices made by humans go beyond simple plant vs. onversations about meat. One-size-fits-all policy recommendations are not feasible due to the global food system's extreme diversity and dependence on distinct environmental and socioeconomic circumstances. According to the most recent IPCC Report, systems combining livestock grazing with mixed plant farming can repair damage caused by years of arable cropping that is dependent on chemical and mechanical inputs. Through this process, we could potentially enhance the quantity of animals grazing in phytochemically rich landscapes that nourish people, plants, animals, and soil. Additionally, we could produce food that is more biochemically rich, making it potentially more nourishing for both humans and the planet (van Vliet et al., 2020).

REFERENCES

1. Alessandrini, R., Brown, M. K., Pombo-Rodrigues, S., Bhageerutty, S., He, F. J., & MacGregor, G. A. (2021). Nutritional quality of plant-based meat products available in the UK: A cross-sectional survey. *Nutrients, 13*(12), 4225.
2. Boukid, F. (2021). Plant-based meat analogues: From niche to mainstream. *European Food Research and Technology, 247*, 297–308.
3. Bryngelsson, S., Moshtaghian, H., Bianchi, M., & Hallström, E. (2022). Nutritional assessment of plant-based meat analogues on the Swedish market. *International Journal of Food Sciences and Nutrition, 73*(6), 889–901.
4. Cole, E., Goeler-Slough, N., Cox, A., & Nolden, A. (2022). Examination of the nutritional composition of alternative beef burgers available in the United States. *International Journal of Food Sciences and Nutrition, 73*(3), 425–432.
5. Covidence systematic review software. (n.d.). Covidence. https://www.covidence.org (accessed on 22 October 2024).
6. Curtain, F., & Grafenauer, S. (2019). Plant-based meat substitutes in the flexitarian age: An audit of products on supermarket shelves. *Nutrients, 11*, 2603.
7. Elkin, E. (2021). Plant-based food sales to increase fivefold by 2030, BI says. *Bloomberg*.
8. Gordon, W., Gantori, S., Gordon, J., Leemann, R., & Boer, R. (2019). The food revolution: The future of food and the challenges we face. *UBS*. https://www.ubs.com/global/en/wealth-management/chief-investment-office/investment-opportunities/sustainable-investing/2019/food-revolution.html (accessed on 22 October 2024).
9. Harnack, L., Mork, S., Valluri, S., Weber, C., Schmitz, K., Stevenson, J., & Pettit, J. (2021). Nutrient composition of a selection of plant-based ground beef alternative products available in the United States. *Journal of the Academy of Nutrition and Dietetics, 121*(12), 2401–2408.e12.
10. Hartmann, C., & Siegrist, M. (2020). Our daily meat: Justification, moral evaluation, and willingness to substitute. *Food Quality and Preference, 80*, 103799.
11. He, J., Evans, N. M., Liu, H., & Shao, S. (2020). A review of research on plant-based meat alternatives: Driving forces, history, manufacturing, and consumer attitudes. *Comprehensive Reviews in Food Science and Food Safety, 19*, 2639–2656.

12. Hu, F. B., Otis, B. O., & McCarthy, G. (2019). Can plant-based meat alternatives be part of a healthy and sustainable diet? *JAMA, 322*, 1547–1548.
13. Ishaq, A., Irfan, S., Sameen, A., & Khalid, N. (2022). Plant-based meat analogs: A review with reference to formulation and gastrointestinal fate. *Current Research in Food Science, 5*, 973–983.
14. Jahn, S., Furchheim, P., & Strässner, A. M. (2021). Plant-based meat alternatives: Motivational adoption barriers and solutions. *Sustainability, 13*, 13271.
15. Lacy-Nichols, J., Hattersley, L., & Scrinis, G. (2021). Nutritional marketing of plant-based meat-analogue products: An exploratory study of front-of-pack and website claims in the USA. *Public Health Nutrition, 24*(15), 4430–4441.
16. Lee, H. J., Yong, H. I., Kim, M., Choi, Y. S., & Jo, C. (2020). Status of meat alternatives and their potential role in the future meat market—A review. *Asian-Australasian Journal of Animal Sciences, 33*(10), 1533–1543.
17. McClements, D. J. (2020). Future foods: Is it possible to design a healthier and more sustainable food supply? *British Nutrition Foundation Nutrition Bulletin, 45*, 341–354.
18. McClements, D. J., & Grossmann, L. (2021). The science of plant-based foods: Constructing next-generation meat, fish, milk, and egg analogs. *Comprehensive Reviews in Food Science and Food Safety, 20*, 4049–4100.
19. Morach, B., Witte, B., Walker, D., von Koeller, E., Grosse-Holz, F., Rogg, J., Brigl, M., Dehnert, N., Obloj, P., & Koktenturk, S. (2021). Food for thought: The protein transformation. *Boston Consulting Group*. https://www.bcg.com/en-au/publications/2021/the-benefits-of-plant-based-meats (accessed on 22 October 2024).
20. Morrison, O. (2023). The alternative protein space is coming to terms with the reality of what is needed to build self-sustaining businesses: The challenge of scaling up plant-based production and manufacturing. *Food Navigator*. https://www.foodnavigator.com/Article/2023/04/28/The-challenge-of-scaling-up-plant-based-production-and-manufacturing (accessed on 22 October 2024).
21. O'Donnell, M., & Murray, S. (2023). A deeper dive into alternative protein investments in 2022: The case for optimism. *Good Food Institute*. https://gfi.org/blog/alternative-protein-investments-update-and-outlook/ (accessed on 22 October 2024).
22. Ogawa, Y., Donlao, N., Thuengtung, S., Tian, J., Cai, Y., Reginio, F. C., Ketnawa, S., Yamamoto, N., & Tamura, M. (2018). Impact of food

structure and cell matrix on digestibility of plant-based food. *Current Opinion in Food Science, 19*, 36–41.

23. Stanley, P. L., Rowntree, J. E., Beede, D. K., Delonge, M. S., & Hamm, M. W. (2018). Impacts of soil carbon sequestration on life cycle greenhouse gas emissions in Midwestern USA beef finishing systems. *Agricultural Systems, 162*, 249–258. https://doi.org/10.1016/j.agsy.2018.02.003.
24. Sun, Z., Scherer, L., Tukker, A., Spawn-Lee, S. A., Bruckner, M., Gibbs, H. K., & Behrens, P. (2022). Dietary change in high-income nations alone can lead to substantial double climate dividend. *Nature Food, 3*, 29–37.
25. Takefuji, Y. (2021). Sustainable protein alternatives. *Trends in Food Science & Technology, 107*, 429–431.
26. Toh, D. W. K., Srv, A., & Henry, C. J. (2022). Unknown impacts of plant-based meat alternatives on long-term health. *Nature Food, 3*, 90–91.
27. Toribio-Mateas, M. A., Bester, A., & Klimenko, N. (2021). Impact of plant-based meat alternatives on the gut microbiota of consumers: A real-world study. *Foods, 10*(9), 2040.
28. Van Vliet, S., Bain, J. R., Muehlbauer, M. J., Provenza, F. D., Kronberg, S. L., Pieper, C. F., & Huffman, K. M. (2021). A metabolomics comparison of plant-based meat and grass-fed meat indicates large nutritional differences despite comparable Nutrition Facts panels. *Scientific Reports, 11*, 13828.
29. van Vliet, S., Kronberg, S. L., & Provenza, F. D. (2020). Plant-based meats, human health, and climate change. *Frontiers in Sustainable Food Systems, 4*, 128. https://doi.org/10.3389/fsufs.2020.00128.

Chapter 8

Environmental and Ethical Considerations

Contents

Introduction 258
8.1. Environmental Impacts of Traditional Meat Production 258
8.2. Sustainability of Plant-Based Meat Alternatives 262
8.3. Ethical Considerations and Animal Welfare 265
References 268

INTRODUCTION

While they come with their own set of considerations, plant-based meat substitutes have advantages over traditional animal-derived meats in terms of the environment and ethics. Because plant-based meats use fewer resources and emit fewer greenhouse gases, they usually have a smaller environmental impact. By far, they use a lot less land and water than livestock farming. For instance, the production of plant-based proteins frequently requires less land and water than the production of conventional meat, which helps to address problems like deforestation and water scarcity.

Furthermore, because these alternatives don't require ruminant methane emissions, they typically result in lower greenhouse gas emissions. Concerns about animal welfare are ethically addressed by plant-based meat substitutes, which offer a way to enjoy meat-like products without involving animal slaughter. This change lessens the demand for factory farming methods, which are frequently condemned for their cruel treatment and subpar living conditions for the animals. Plant-based products adhere to ethical principles that support animal rights and humane treatment by completely avoiding the use of animals.

However, there are environmental aspects of plant-based meats that require consideration. Some plant-based products require a lot of processing during production, and they might contain ingredients that need a lot of agricultural inputs, which can still have an influence on the environment. For instance, using pesticides and monoculture farming are problems that may arise from the large-scale production of soy or pea protein. While plant-based meat substitutes are a positive step toward lowering environmental effects and resolving ethical issues, more work and thought must go into these alternatives. Ensuring the sustainable and ethical production of these products will augment their advantages and foster a more conscientious food system in general.

8.1. ENVIRONMENTAL IMPACTS OF TRADITIONAL MEAT PRODUCTION

The production of meat through traditional means has significant negative effects on the environment, which lead to various ecological problems. The significant amount of greenhouse gas emissions connected to cattle farming is one of the main issues. Methane, a powerful greenhouse gas released by enteric fermentation in ruminants like cows, is largely sourced from this industry. According to the Food and Agriculture Organization (FAO), livestock farming accounts for about 14.5% of greenhouse gas emissions worldwide, of which methane emissions are a major cause of global warming. Conventional meat production uses a lot of natural resources and emits greenhouse gases.

Large amounts of land and water are needed for both animal husbandry and the cultivation of feed crops. For instance, growing one kilogram of beef requires about 15,000 liters of water, not counting the water needed for feed cultivation. Water resources are strained by this high water use, especially in areas where water is scarce.

Another significant effect of conventional meat production is deforestation. Large tracts of forest are cleared to make room for pasture and to grow feed crops; this reduces biodiversity, damages habitats, and raises atmospheric carbon dioxide levels. Deforestation affects wildlife and reduces the capacity of the planet to store carbon.

In addition, conventional meat production produces a lot of waste, such as manure and leftovers from slaughterhouses. If this waste is not properly managed, it can cause water pollution by introducing pathogens and nutrients into rivers and lakes that are harmful to aquatic ecosystems and public health. In general, the negative effects of conventional meat production on the environment draw attention to the need for more sustainable methods and substitutes to lessen the effects of this industry on climate change, resource depletion, and ecosystem degradation.

8.1.1. Production and Composition

There are two methods for creating meat substitutes that have a texture similar to real meat: top-down and bottom-up. Bottom-up structuring involves creating individual fibers first, and then assembling them into larger products. Cultured meat is an illustration of a meat substitute produced through a bottom-up method. Conversely, the top-down method results in fibrousness over a greater length scale by deforming the material to create a fibrous structure.

Food extrusion is an illustration of a top-down method. For a variety of advantages, bottom-up, and top-down processing can be utilized separately or in combination. The nutritional content of various meat substitute products varies, as will be covered later. The ability to precisely control the composition and properties of the final product, enabling optimized nutritional profiles, is a noteworthy benefit of the bottom-up approach. Conversely, meat substitutes developed through top-down methods may be less flexible, but they are more scalable and can make better use of the infrastructure and agricultural resources already in place. Methods that employ the bottom-up approach may be the closest to the structure of meat, per a Wageningen University and Research study titled "Structuring Processes for Meat Analogues." According to a cross-national survey of meat eaters with differing levels of meat alternative consumption, people who ate more meat were more likely to convert to meat alternatives if they more closely resembled real meat. This can be achieved by using bottom-up methods. The study comes to the conclusion that for heavy meat eaters, sensory experience is important when it comes to using plant-based substitutes.

The range of ingredients that can be utilized to make meat alternatives is growing. Plentify, for example, uses high-protein bacteria found in the human microbiome, while Meati Foods grows the mycelium of fungus—in this case, *Neurospora crassa*—to make fish, steaks, or chicken breasts. The majority of meat alternatives on the market are typically made with soy protein isolates, soybean flour, and gluten as their base. With a minimum protein content of 90%, soy protein isolate is an extremely pure form of soy protein. The dehulling, or decortication, of the soybean seeds is the first step in the process of extracting the protein from them. Subsequently, the seeds undergo treatment with solvents like hexane to extract the oil. After that, the oil-free soybean meal is suspended in water and given an alkali treatment to dissolve the protein and leave the carbohydrates behind. After adding acidic materials to the alkaline solution to precipitate the protein, it is dried and cleaned.

The product is largely flavorless after fats and carbohydrates are eliminated. As soy protein includes every essential amino acid required for healthy human growth and development, it is also regarded as a complete protein. Several flavorings can be added to the textured base material to give the product a meaty flavor. Since 1972, a basic vegan chicken flavor recipe that uses the Maillard reaction to create aromas from uncomplicated chemicals has been known. Lipid oxidation and the breakdown of thiamine were later discovered to be significant processes in the explanation of the aroma in cooked meat. To achieve richer and more convincing meat flavors, more complex starting materials are used, such as yeast extract (which is regarded as a natural flavoring in the EU), hydrolyzed vegetable protein, different fermented foods, and spices. These reactions are also replicated during cooking.

8.1.2. Impact

8.1.2.1. Environmental

Better meat substitutes could lessen the environmental impact of meat production, which is a significant concern considering that the demand for meat products worldwide is expected to rise by 15% by 2031. This is in addition to ethical and health-related reasons. In terms of greenhouse gas production, water, and land use, research on meats and meat substitutes indicates that no-meat products can provide significant advantages over the production of beef, and to a lesser extent, pork, and chicken. When compared to other investments, a 2022 report from the Boston Consulting Group indicated that significant greenhouse gas reductions result from improving and increasing the production of meat and dairy substitutes. The Good Food Institute states that attaining sustainability requires increasing the Western diet's efficiency. The way land is used will change as the world's population rises. Three-quarters of Earth's livable land is occupied by animals. Although animal agriculture accounts for only 17% of the world's total food supply, it occupies 77% of all agricultural land. More production opportunities arise from the potential 47–99% reduction in land required for plant-based meat compared to conventional meat. 33% of the water used in agriculture worldwide is used for animal agriculture; under a different approach, this water could be used for drinking water or other growing applications. Meat from plants requires between 72% and 999% less water to produce than meat from animals.

The next biggest cause of wasted water is pollution. In addition to depleting the water supply directly, pesticides used in the production of animal feed and waste runoff into reservoirs can harm the environment and even pose a health risk to humans. The primary source of greenhouse gas emissions in the food sector is animal agriculture. Production of meat substitutes made from plants emits 30–90% less emissions than that of traditional meat production. Much of the land used for animal feed could be used for carbon recycling, soil conservation, and the production of renewable energy, all of which would lessen the overall pollution that humans are causing on the planet. Apart from the damage the industry currently causes to the environment, overuse of antibiotics on animals results in resistant microbes, which could make some of the life-saving medications used in human medicine ineffective. Antibiotic resistance in microbes would be significantly reduced if meat came from plants instead of animals.

It was discovered in Nature Communications that substituting half of the world's consumption of beef, chicken, dairy, and pork products with plant-based alternatives could reduce the amount of land used for agriculture by nearly a third, almost completely stop deforestation, aid in the restoration of biodiversity by rewilding the land, and reduce greenhouse gas emissions from agriculture by 31% by the year 2050. This finding would clear the way for accomplishing both climate and biodiversity goals. of all industrial sectors, the shift to plant-based protein is said to produce the largest reductions in emissions that contribute to climate change per dollar invested.

8.1.2.2. Health

Meat substitutes are higher in carbohydrates, dietary fiber, sodium, iron, and calcium than meat products, but they are lower in zinc, saturated fat, and vitamin B_{12}. The American Heart Association declared in 2021 that there is insufficient data to determine the immediate and long-term health impacts of plant-based meat substitutes. In the same year, the World Health Organization declared that further research is necessary to examine the health effects of meat substitutes because there are substantial knowledge gaps regarding their nutritional makeup.

According to a 2023 review, substituting red and highly processed meat with a range of meat alternatives increased quality-adjusted life years, resulted in significant savings for the health system, and decreased greenhouse gas emissions; substituting meat with vegetarian alternatives that were minimally processed, like legumes, had the biggest impact. According to a 2024 review, foods made from plant sources may be healthier and less harmful to the environment than foods made from animal sources.

8.2. SUSTAINABILITY OF PLANT-BASED MEAT ALTERNATIVES

The sustainability of plant-based meat alternatives represents a significant advancement in the quest for environmentally responsible food systems. These alternatives, which are designed to replicate the taste and texture of animal-derived meats, have gained traction as a more eco-friendly option due to several key factors. Firstly, when compared to conventional animal farming, plant-based meat substitutes typically require fewer natural resources. Less energy, water, and land are usually used in the production of plant-based proteins. A kilogram of beef, for example, is thought to require roughly 15,000 liters of water to produce, whereas a similar amount of plant-based protein is produced using a fraction of that amount. This lower resource intensity is essential for addressing problems that conventional livestock farming exacerbates, such as deforestation and water scarcity. Second, substituting plant-based meat for animal products helps reduce greenhouse gas emissions.

A significant source of methane, a powerful greenhouse gas, is livestock farming. Conversely, the carbon footprint of plant-based products is typically substantially lower. According to studies, eating plant-based burgers instead of beef ones can cut greenhouse gas emissions by up to 90%. To combat climate change and meet global emission reduction targets, this reduction is essential.

Furthermore, substituting plant-based meat for meat often results in reduced environmental pollution. Manure and runoff from traditional meat production pose serious waste management problems because they can contaminate water sources and destroy ecosystems. The production of meat from plants usually produces less waste and is not as likely to cause environmental problems.

It is crucial to remember that the sustainability of plant-based meat substitutes can vary based on many variables, including the location of ingredients and the methods used in production. For instance, the overall environmental benefits of plant-based products may be compromised if they rely on highly processed ingredients or require a significant amount of energy during production. Therefore, to maximize the sustainability of these products, ongoing improvements in production processes and ingredient sourcing are crucial. Although plant-based meat substitutes show promise for creating more environmentally friendly food systems, their full environmental benefits rely on continuous improvements in technology and methods. Plant-based meats provide a feasible way to lessen the ecological impact of our diets by minimizing pollution, lowering greenhouse gas emissions, and reducing the use of resources. These products' sustainability will be further enhanced as the industry develops, making them an increasingly important part of a future where food is produced more sustainably.

It is generally acknowledged that reducing the amount of meat consumed and increasing the amount of plant-based diets is necessary to improve animal welfare, and human health, and reduce the negative effects of the food system on the environment. But the amount of meat we eat as a nation is still rising and is expected to do so for the next 10 years. It is suggested that a significant decrease in average meat consumption levels, beginning in wealthy societies, is vitally necessary to stop the anticipated increase in meat consumption worldwide.

There are several methods to achieve meat reduction: (i) cutting back on meat portions; (ii) substituting plant-based alternatives for some meat-based products (referred to as hybrid meats); or (iii) using the "less but better" theory (less quantity, more quality, i.e., more sustainably farmed meat), (iii) removing the meat from the dish without adding anything in its place, and (iv) substituting meat with another source of protein (which can be anything from plant-based foods like tofu, mushrooms, or legumes to animal-based foods like eggs or cheese)—not to mention alternative protein sources with a minimal, i.e., insects, and seaweed, or a market share that is still non-existent, i.e., the consumption of plant-based meat substitutes (PBMAs), cultured meat, and, last but not least, (v). These tactics suggest that a flexitarian diet should encompass the substitution of other (unprocessed) proteins, both plant- and animal-based, for meat, rather than being limited to the adoption of (processed) meat substitutes. Having said that, the majority of people who currently use these products and the broad and diverse dietary group of flexitarians are unquestionably the primary target group of PBMAs. It is clear that people who follow a flexitarian diet, which is defined as avoiding meat (occasionally, frequently, or often), are looking for and interested in meat substitutes to maintain their lower meat intake. In other words, it makes sense to identify flexitarians as launch clients. From the standpoint of plant-based meat alternatives (PBMAs), the current dominant market strategy is to replicate the flavor, texture, and appearance of traditional meat as closely as possible, e.g., nutritional value (iron, vitamins, etc.), "the bleeding burger," etc.), and even the names of products (using terminology associated with meat); interestingly, "PBMAs" can also be understood as "plant-based meat analogs." Undoubtedly, the food industry's objective of creating plant-based foods that resemble meat helps many flexitarians, vegetarians, and vegans who may be put off by the idea of PBMAs being associated with meat.

It seems safe to say that PBMAs facilitate the need of many modern food consumers in various high-income countries to be supplied with tasty, affordable, and accessible alternative protein products to satisfy their cravings to eat beyond meat, even though food consumers' adoption and acceptance of PBMAs are not self-evident, as will be shown in the remainder of research. Numerous factors attest to the fact that the field of PBMAs is currently thriving and deserving of more investigation and critical evaluation. These include the notable achievements of recent decades in enhancing the product qualities of plant-based meat alternatives (PBMAs) and the widespread availability of PBMAs in food service establishments such as McDonald's, Burger King, and KFC, which have introduced plant-based substitutes for beef burgers and chicken nuggets. Additional factors that support the critical role of PBMAs include the sizeable investments made in the market, the notable growth rates it is seeing in front-running nations (like Germany, the UK, and the Netherlands), and the projected growth rates in global sales shortly.

8.2.1. Plant-Based Meat Alternatives

A meat substitute, also known as plant-based meat, mock meat, or fake meat in a derogatory sense, is a food item that is consumed in place of meat that is made with vegetarian or vegan ingredients. Meat substitutes usually mimic the mouth feel, flavor, appearance, and chemical properties of particular meat varieties. Replacements based on plants and fungi are commonly made with soy (e.g., textured vegetable protein, tofu, and tempeh), but they can also be made with mycoprotein, wheat gluten, or pea protein, like in Quorn, Beyond Burgers, and Beyond Burgers. Other methods of producing alternative protein foods include precision fermentation, in which single-celled organisms, like yeast, use a carbon source to produce specific proteins; or laboratory-grown, utilizing techniques based on tissue engineering. The contents of the meat substitute consist of 50–80% water, 10–25% textured vegetable proteins, 4–20% non-textured proteins, 0–15% fat and oil, 3–10% flavors/spices, 1–5% binding agents, and 0–5% coloring agents.

Meatless tissue engineering is the process of growing stem cells on synthetic or natural scaffolds to produce products that resemble meat. Many materials, such as synthetic polymers, animal-based proteins, plant-derived biomaterials, and self-assembling polypeptides, can be used to create scaffolds. These three-dimensional scaffold-based techniques offer a unique structural setting for the development of cells. On the other hand, scaffold-free techniques encourage cell aggregation, enabling cells to arrange themselves into structures that resemble tissues. Meat substitutes are generally used by vegetarians, vegans, and those adhering to dietary restrictions based on religion or culture as a source of protein. However, the need for sustainable diets around the world has also made them more popular among flexitarians and non-vegetarians who want to lessen the negative environmental effects of animal husbandry.

8.3. ETHICAL CONSIDERATIONS AND ANIMAL WELFARE

The popularity of plant-based meat substitutes signifies a profound change in the way we think about food choices, with important ramifications for nutritional profiles and ethical issues. Examining these products' effects on the environment, their ethical implications (especially concerning animal welfare), and their nutritional value about conventional meats is crucial as they become more widely available. Animal welfare concerns and the ethical issues surrounding conventional meat production are closely related. Conventional livestock farming frequently uses methods that have the potential to cause serious suffering to animals. Animals raised in intensive farming systems may be subjected to cruel methods of slaughter, small quarters, and other hardships. Concerns about the treatment of animals, their standard of living, and the ethics of killing them for food are raised by these practices. In contrast, since they do not require the exploitation of animals, plant-based meat substitutes directly address these moral dilemmas.

These products support ethical values that put animal welfare first by offering a source of protein and other nutrients without depending on animal agriculture. In addition to avoiding the psychological and physical strains of industrial farming, plant-based meat substitutes do not require the death of animals, allaying worries about animal abuse and suffering.

Despite these benefits, ethical considerations do not end with the elimination of animal-based ingredients. The production of plant-based meats must also be scrutinized for other ethical concerns, including labor practices and resource use. Ensuring that these products are manufactured under fair labor conditions and sourced sustainably is crucial for addressing the broader spectrum of ethical issues associated with food production.

- **Nutritional Comparison:** When comparing the nutritional profiles of plant-based meat alternatives and traditional meats, several factors come into play, including protein content, fat composition, vitamins, and minerals.
- **Protein Content:** One important aspect of a balanced diet is protein, which traditional meats are well known for having a high protein content. A 100-gram portion of cooked chicken breast, for instance, has about 31 grams of protein. Plant-based meat substitutes, like those derived from soy or pea protein, are also high in protein. About 20 grams of protein can be found in a typical serving of a plant-based burger, though this can vary based on the product and formulation. Although plant-based alternatives can have protein contents that are on par with or higher than those of conventional meats, they frequently do so by adding additives and isolated proteins.
- **Fat Composition:** When comparing the nutritional value of traditional meats and plant-based meats, it's crucial to consider the type and content of fat. Saturated fats, which are connected to conditions like heart disease, can be found in large amounts in traditional meats, especially red and processed meats. A 100-gram portion of ground beef, for example, may have about 25 grams of fat, of which saturated fat makes up a sizable amount.
- Alternatives to meat made from plants typically have a better fat composition. Vegetable oils that are higher in unsaturated fats and lower in saturated fats are used in many products. Some plant-based burgers, for instance, use sunflower or canola oil, which has lower saturated fat content than animal fats. Nonetheless, it's important to read the nutritional labels because some plant-based meats are processed and might have extra oils added to improve texture and flavor.
- **Vitamins and Minerals:** Vitamins and minerals are crucial for overall health and can differ between plant-based and traditional meats. Traditional meats are rich in essential nutrients like iron, zinc, and B vitamins (e.g., B12). Iron from animal sources is home iron, which is more readily absorbed by the body compared to non-heme iron from plant sources.
- Plant-based meat alternatives often fortify their products to address these nutritional gaps. For example, many plant-based meats are fortified with vitamins B12 and iron to match the nutritional profile of traditional meats. However, the bioavailability of these nutrients can vary. Non-heme iron found in plant-based products may require consumption with vitamin C-rich foods to enhance absorption.

- **Health Considerations:** Beyond their basic nutritional content, traditional meats and plant-based alternatives have implications for health. Traditional meats have been linked to higher risks of chronic illnesses like diabetes, heart disease, and some types of cancer, especially red and processed meats. High concentrations of cholesterol, saturated fat, and preservatives found in processed meats are frequently associated with these risks. In many ways, plant-based meat substitutes have a healthier profile.

They typically have reduced cholesterol and saturated fat content, which is good for cardiovascular health. Furthermore, a lot of plant-based alternatives are high in fiber, which is lacking in animal products and promotes healthy digestive systems. It's crucial to remember that not all plant-based meats are made equal. Highly processed foods may have artificial additives, added sugars, and sodium that are harmful to general health.

The trend toward plant-based meat substitutes raises important nutritional and ethical issues. Because they lessen animal suffering and exploitation, these products provide a compassionate option from an ethical point of view. While plant-based meats offer similar protein contents and generally healthier fat profiles, there are certain nutritional considerations to take into account, such as processing levels and nutrient fortification. Ongoing research and development will be essential in filling in these nutritional gaps and guaranteeing that ethical standards are maintained throughout the production process as the plant-based meat market develops. By weighing the ethical ramifications and nutritional value of the products they select, consumers can make well-informed decisions that will ultimately contribute to a more humane and sustainable food system (He et al., 2020).

REFERENCES

1. Afshin, A., Sur, P. J., Fay, K. A., Cornaby, L., Ferrara, G., Salama, J. S., Mullany, E. C., Abate, K. H., Abbafati, C., Abebe, Z., et al. (2019). Health effects of dietary risks in 195 countries, 1990–2017: A systematic analysis for the Global Burden of Disease Study 2017. *The Lancet, 393*(10184), 1958–1972.
2. Bager, S. L., Persson, U. M., & dos Reis, T. N. P. (2021). Eighty-six EU policy options for reducing imported deforestation. *One Earth, 4*(3), 289–306.
3. Banovic, M., & Sveinsdóttir, K. (2021). Importance of being analogue: Female attitudes towards meat analogue containing rapeseed protein. *Food Control, 123*, 107833.
4. Boukid, F. (2021). Plant-based meat analogues: From niche to mainstream. *European Food Research and Technology, 247*(2), 297–308.
5. Boukid, F., & Castellari, M. (2021). Veggie burgers in the EU market: A nutritional challenge? *European Food Research and Technology, 247*(10), 2445–2453.
6. Boukid, F., Sogari, G., & Rosell, C. M. (2022). Edible insects as foods: Mapping scientific publications and product launches in the global market (1996–2021). *Journal of Insects as Food and Feed*, 1–16.
7. Bryant, C. J. (2022). Plant-based animal product alternatives are healthier and more environmentally sustainable than animal products. *Future Foods, 6*, 100174.
8. Bryngelsson, S., Moshtaghian, H., Bianchi, M., & Hallström, E. (2022). Nutritional assessment of plant-based meat analogues on the Swedish market. *International Journal of Food Sciences and Nutrition, 73*(6), 889–901.
9. Caparros Megido, R., Gierts, C., Blecker, C., Brostaux, Y., Haubruge, E., Alabi, T., & Francis, F. (2016). Consumer acceptance of insect-based alternative meat products in Western countries. *Food Quality and Preference, 52*, 237–243.
10. Caputo, V., Sogari, G., & Van Loo, E. J. (2022). Do plant-based and blend meat alternatives taste like meat? A combined sensory and choice experiment study. *Applied Economic Perspectives and Policy*, 1–20.
11. Cole, E., Goeler-Slough, N., Cox, A., & Nolden, A. (2022). Examination of the nutritional composition of alternative beef burgers available in the United States. *International Journal of Food Sciences and Nutrition, 73*(3), 425–432.

12. Crimarco, A., Landry, M. J., Carter, M. M., & Gardner, C. D. (2022). Assessing the effects of alternative plant-based meats vs. animal meats on biomarkers of inflammation: A secondary analysis of the SWAP-MEAT randomized crossover trial. *Journal of Nutritional Science, 11*, e82.
13. Cutroneo, S., Angelino, D., Tedeschi, T., Pellegrini, N., Martini, D., SINU Young Working Group, Dall'Asta, M., Russo, M. D., Nucci, D., Moccia, S., et al. (2022). Nutritional quality of meat analogues: Results from the food labelling of Italian products (FLIP) project. *Frontiers in Nutrition, 9*, 852831.
14. Dagevos, H. (2016). Exploring flexitarianism: Meat reduction in a meat-centred food culture. In T. Raphaely & D. Marinova (Eds.), *Impact of Meat Consumption on Health and Environmental Sustainability* (pp. 233–243). IGI Global.
15. Dagevos, H. (2021). Finding flexitarians: Current studies on meat eaters and meat reducers. *Trends in Food Science & Technology, 114*, 530–539.
16. De Angelis, D., Kaleda, A., Pasqualone, A., Vaikma, H., Tamm, M., Tammik, M. L., Squeo, G., & Summo, C. (2020). Physicochemical and sensorial evaluation of meat analogues produced from dry-fractionated pea and oat proteins. *Foods, 9*(12), 1754.
17. De Marchi, M., Costa, A., Pozza, M., Goi, A., & Manuelian, C. L. (2021). Detailed characterization of plant-based burgers. *Scientific Reports, 11*(1), 2049.
18. Demartini, E., Vecchiato, D., Finos, L., Mattavelli, S., & Gaviglio, A. (2022). Would you buy vegan meatballs? The policy issues around vegan and meat-sounding labelling of plant-based meat alternatives. *Food Policy, 111*, 102310.
19. Detzel, A., Krüger, M., Busch, M., Blanco-Gutiérrez, I., Varela, C., Manners, R., Bez, J., & Zannini, E. (2022). Life cycle assessment of animal-based foods and plant-based protein-rich alternatives: An environmental perspective. *Journal of the Science of Food and Agriculture, 102*(11), 5098–5110.
20. Eckl, M. R., Biesbroek, S., van't Veer, P., & Geleijnse, J. M. (2021). Replacement of meat with non-meat protein sources: A review of the drivers and inhibitors in developed countries. *Nutrients, 13*(10), 3602.
21. Elzerman, J. E., Hoek, A. C., van Boekel, M. A. J. S., & Luning, P. A. (2011). Consumer acceptance and appropriateness of meat substitutes in a meal context. *Food Quality and Preference, 22*(3), 233–240.
22. European Union. (2020). *Farm to Fork Strategy: For a Fair, Healthy and Environmentally-Friendly Food System*. European Union: Brussels, Belgium.

23. Fiorentini, M., Kinchla, A. J., & Nolden, A. A. (2020). Role of sensory evaluation in consumer acceptance of plant-based meat analogs and meat extenders: A scoping review. *Foods, 9*(9), 1334.
24. Froldi, F., Lamastra, L., Trevisan, M., Mambretti, D., & Moschini, M. (2022). Environmental impacts of cow's milk in Northern Italy: Effects of farming performance. *Journal of Cleaner Production, 363*, 132600.
25. Giacalone, D., Clausen, M. P., & Jaeger, S. R. (2022). Understanding barriers to consumption of plant-based foods and beverages: Insights from sensory and consumer science. *Current Opinion in Food Science, 48*, 100919.
26. Godfray, H. C. J., Aveyard, P., Garnett, T., Hall, J. W., Key, T. J., Lorimer, J., Pierrehumbert, R. T., Scarborough, P., Springmann, M., & Jebb, S. A. (2018). Meat consumption, health, and the environment. *Science, 361*(6399), eaam5324.
27. Goldstein, B., Moses, R., Sammons, N., & Birkved, M. (2017). Potential to curb the environmental burdens of American beef consumption using a novel plant-based beef substitute. *PLoS ONE, 12*(12), e0189029.
28. Gorissen, S. H. M., Crombag, J. J. R., Senden, J. M. G., Waterval, W. A. H., Bierau, J., Verdijk, L. B., & van Loon, L. J. C. (2018). Protein content and amino acid composition of commercially available plant-based protein isolates. *Amino Acids, 50*(12), 1685–1695.
29. Grasso, S., Rondoni, A., Bari, R., Smith, R., & Mansilla, N. (2022). Effect of information on consumers' sensory evaluation of beef, plant-based and hybrid beef burgers. *Food Quality and Preference, 96*, 104417.
30. Grossmann, L., & Weiss, J. (2021). Alternative protein sources as technofunctional food ingredients. *Annual Review of Food Science and Technology, 12*, 93–117.
31. Grosso, G., La Vignera, S., Condorelli, R. A., Godos, J., Marventano, S., Tieri, M., Ghelfi, F., Titta, L., Lafranconi, A., Gambera, A., et al. (2022). Total, red and processed meat consumption and human health: An umbrella review of observational studies. *International Journal of Food Sciences and Nutrition, 73*(6), 726–737.
32. Hartmann, C., Furtwaengler, P., & Siegrist, M. (2022). Consumers' evaluation of the environmental friendliness, healthiness and naturalness of meat, meat substitutes, and other protein-rich foods. *Food Quality and Preference, 97*, 104486.
33. Hashempour-Baltork, F., Khosravi-Darani, K., Hosseini, H., Farshi, P., & Reihani, S. F. S. (2020). Mycoproteins as safe meat substitutes. *Journal of Cleaner Production, 253*, 119958.

34. He, J., Evans, N. M., Liu, H., & Shao, S. (2020). A review of research on plant-based meat alternatives: Driving forces, history, manufacturing, and consumer attitudes. *Comprehensive Reviews in Food Science and Food Safety, 19*(5), 2639–2656.
35. Heller, M. C., & Keoleian, G. A. (2018). *Beyond Meat's Beyond Burger Life Cycle Assessment: A Detailed Comparison Between a Plant-Based and An Animal-Based Protein Source* (Report No. CSS18-10). University of Michigan: Ann Arbor, MI, USA.
36. Hunt, J. R. (2003). Bioavailability of iron, zinc, and other trace minerals from vegetarian diets. *American Journal of Clinical Nutrition, 78*(3), 633S–639S.
37. ISO. (2006a). *ISO 14040: Environmental Management—Life Cycle Assessment—Principles and Framework.* International Organization for Standardization, Geneva, Switzerland.
38. ISO. (2006b). *ISO 14044: Environmental Management—Life Cycle Assessment—Requirements and Guidelines.* International Organization for Standardization, Geneva, Switzerland.
39. Kerslake, E., Kemper, J. A., & Conroy, D. (2022). What's your beef with meat substitutes? Exploring barriers and facilitators for meat substitutes in omnivores, vegetarians, and vegans. *Appetite, 170*, 105864.
40. Kumar, M., Tomar, M., Punia, S., Dhakane-Lad, J., Dhumal, S., Changan, S., Senapathy, M., Berwal, M. K., Sampathrajan, V., Sayed, A. A. S., et al. (2022). Plant-based proteins and their multifaceted industrial applications. *LWT, 154*, 112620.
41. Lawrence, M. A., Dickie, S., & Woods, J. L. (2018). Do nutrient-based front-of-pack labelling schemes support or undermine food-based dietary guideline recommendations? Lessons from the Australian Health Star Rating System. *Nutrients, 10*(1), 32.
42. Lemken, D., Spiller, A., & Schulze-Ehlers, B. (2019). More room for legume—Consumer acceptance of meat substitution with classic, processed and meat-resembling legume products. *Appetite, 143*, 104412.
43. Lusk, J. L., Blaustein-Rejto, D., Shah, S., & Tonsor, G. T. (2022). Impact of plant-based meat alternatives on cattle inventories and greenhouse gas emissions. *Environmental Research Letters, 17*(2), 24035.
44. McLaren, S., Berardy, A., Henderson, A., Holden, N., Huppertz, T., Jolliet, O., De Camillis, C., Renouf, M., Rugani, B., Saarinen, M., et al. (2021). Integration of environment and nutrition in life cycle assessment of food items: Opportunities and challenges. Food and Agriculture Organization of the United Nations FAO: Rome, Italy.

45. Mejia, M., Fresán, U., Harwatt, H., Oda, K., Uriegas-Mejia, G., & Sabaté, J. (2020). Life cycle assessment of the production of a large variety of meat analogs by three diverse factories. *Journal of Hunger and Environmental Nutrition, 15*(5), 699–711.

46. Michel, F., Hartmann, C., & Siegrist, M. (2021). Consumers' associations, perceptions and acceptance of meat and plant-based meat alternatives. *Food Quality and Preference, 87*, 104063.

47. Neuhofer, Z. T., & Lusk, J. L. (2022). Most plant-based meat alternative buyers also buy meat: An analysis of household demographics, habit formation, and buying behavior among meat alternative buyers. *Scientific Reports, 12*(1), 13062.

48. Neville, M., Tarrega, A., Hewson, L., & Foster, T. (2017). Consumer-oriented development of hybrid beef burger and sausage analogues. *Food Science and Nutrition, 5*(4), 852–864.

49. Nezlek, J. B., & Forestell, C. A. (2022). Meat substitutes: Current status, potential benefits, and remaining challenges. *Current Opinion in Food Science, 47*, 100890.

50. Nishinari, K., Fang, Y., Guo, S., & Phillips, G. O. (2014). Soy proteins: A review on composition, aggregation and emulsification. *Food Hydrocolloids, 39*, 301–318.

51. OECD and FAO. (2020). *OECD-FAO Agricultural Outlook 2020–2029*. OECD: Paris, France.

52. Onwezen, M. C., Bouwman, E. P., Reinders, M. J., & Dagevos, H. (2021). A systematic review on consumer acceptance of alternative proteins: Pulses, algae, insects, plant-based meat alternatives, and cultured meat. *Appetite, 159*, 105058.

53. Onwezen, M. C., Verain, M. C. D., & Dagevos, H. (2022). Social norms support the protein transition: The relevance of social norms to explain increased acceptance of alternative protein burgers over 5 years. *Foods, 11*(2).

54. Oussalah, A., Levy, J., Berthezène, C., Alpers, D. H., & Guéant, J. L. (2020). Health outcomes associated with vegetarian diets: An umbrella review of systematic reviews and meta-analyses. *Clinical Nutrition, 39*(10), 3283–3307.

55. Patinho, I., Saldaña, E., Selani, M. M., Teixeira, A. C. B., Menegali, B. S., Merlo, T. C., Rios-Mera, J. D., Dargelio, M. D. B., Rodrigues, H., & Contreras-Castillo, C. J. (2021). Original burger (traditional) or burger with mushroom addition? A social representation approach to novel foods. *Food Research International, 147*, 110551.

56. Pereira, P. M. d. C. C., & Vicente, A. F. d. R. B. (2013). Meat nutritional composition and nutritive role in the human diet. *Meat Science, 93*(3), 586–592.
57. Petersen, T., Hartmann, M., & Hirsch, S. (2021). Which meat (substitute) to buy? Is front of package information reliable to identify the healthier and more natural choice? *Food Quality and Preference, 94*, 104298.
58. Pointke, M., & Pawelzik, E. (2022). Plant-based alternative products: Are they healthy alternatives? Micro- and macronutrients and nutritional scoring. *Nutrients, 14*(3), 601.
59. Pulvento, C., Riccardi, M., Lavini, A., D'Andria, R., & Ragab, R. (2015). Parameterization and field validation of Saltmed model for grain amaranth tested in South Italy. *Irrigation and Drainage, 64*(1), 59–68.
60. Rotz, C. A., Montes, F., & Chianese, D. S. (2010). The carbon footprint of dairy production systems through partial life cycle assessment. *Journal of Dairy Science, 93*(3), 1266–1282.
61. Saget, S., Costa, M., Barilli, E., Wilton de Vasconcelos, M., Santos, C. S., Styles, D., & Williams, M. (2020). Substituting wheat with chickpea flour in pasta production delivers more nutrition at a lower environmental cost. *Sustainable Production and Consumption, 24*, 26–38.
62. Schutyser, M. A. I., Pelgrom, P. J. M., van der Goot, A. J., & Boom, R. M. (2015). Dry fractionation for sustainable production of functional legume protein concentrates. *Trends in Food Science & Technology, 45*, 327–335.
63. Sha, L., & Xiong, Y. L. (2020). Plant protein-based alternatives of reconstructed meat: Science, technology, and challenges. *Trends in Food Science & Technology, 102*, 51–61.
64. Smetana, S., Mathys, A., Knoch, A., & Heinz, V. (2015). Meat alternatives: Life cycle assessment of most known meat substitutes. *International Journal of Life Cycle Assessment, 20*, 1254–1267.
65. Smetana, S., Profeta, A., Voigt, R., Kircher, C., & Heinz, V. (2021). Meat substitution in burgers: Nutritional scoring, sensorial testing, and life cycle assessment. *Futur. Foods, 4*, 100042.
66. Sogari, G., Li, J., Wang, Q., Lefebvre, M., Gómez, M. I., & Mora, C. (2021). Factors influencing the intention to purchase meat-mushroom blended burgers among college students. *Food Quality and Preference, 90*, 104169.
67. Stagnari, F., Maggio, A., Galieni, A., & Pisante, M. (2017). Multiple benefits of legumes for agriculture sustainability: An overview. *Chemical and Biological Technologies in Agriculture, 4*, 2.

68. Steinfeld, H., Gerber, P., Wassenaar, T., Castel, V., Rosales, M., & De Haan, C. (2006). *Livestock's Long Shadow: Environmental Issues and Options*. FAO: Rome, Italy.

69. The Smart Protein Project. (2022). *Plant-Based Foods in Europe: How Big is the Market? The Smart Protein Plant-Based Food Sector Report*. The Smart Protein Project: New Delhi, India.

70. Tonsor, G. T., Lusk, J. L., & Schroeder, T. C. (2022). Market potential of new plant-based protein alternatives: Insights from four US consumer experiments. *Applied Economic Perspectives and Policy, 1–18*.

71. Tso, R., & Forde, C. G. (2021). Unintended consequences: Nutritional impact and potential pitfalls of switching from animal- to plant-based foods. *Nutrients, 13*(8), 2527.

72. Tucker, C. A. (2014). The significance of sensory appeal for reduced meat consumption. *Appetite, 81*, 168–179.

73. Tukker, A., Goldbohm, R. A., de Koning, A., Verheijden, M., Kleijn, R., Wolf, O., Pérez-Domínguez, I., & Rueda-Cantuche, J. M. (2011). Environmental impacts of changes to healthier diets in Europe. *Ecological Economics, 70*, 1776–1788.

74. Van Loo, E. J., Caputo, V., & Lusk, J. L. (2020). Consumer preferences for farm-raised meat, lab-grown meat, and plant-based meat alternatives: Does information or brand matter? *Food Policy, 95*, 101931.

75. Verain, M., Dagevos, H., & Antonides, G. (2022). Flexitarianism: A range of sustainable food styles. In L. A. Reisch & J. Thøgersen (Eds.), *Handbook of Research on Sustainable Consumption* (pp. 209–223). Edward Elgar Publishing: Cheltenham, UK.

76. Wittek, P., Ellwanger, F., Karbstein, H. P., & Emin, M. A. (2021). Morphology development and flow characteristics during high moisture extrusion of a plant-based meat analogue. *Foods, 10*(8), 1753.

77. World Health Organization, & Food and Agriculture Organization of the United Nations. (2019). *Sustainable Healthy Diets: Guiding Principles*. World Health Organization: Geneva, Switzerland.

78. Zahari, I., Östbring, K., Purhagen, J. K., & Rayner, M. (2022). Plant-based meat analogues from alternative protein: A systematic literature review. *Foods, 11*(11), 2870.

79. Zhang, T., Dou, W., Zhang, X., Zhao, Y., Zhang, Y., Jiang, L., & Sui, X. (2021). The development history and recent updates on soy protein-based meat alternatives. *Trends in Food Science & Technology, 109*, 702–710.

Chapter 9

Accelerate Consumer Adoption of Plant-Based Meat

Contents

Introduction 276
9.1. Understand the Consumer 276
9.2. Product Development 283
9.3. Choice Architecture 289
9.4. Targeted Messaging 291
References 298

INTRODUCTION

Although it has been commercially available for a while, plant-based meat has reached a turning point. With proper marketing, plant-based meat substitutes can fulfill consumers' expectations for a product and potentially satisfy the growing demand for a just, sustainable, and healthy source of protein. Though currently purchased by "early adopters," plant-based meat has the potential to swiftly spread to a wider range of people. Plant-based meat will become more widely accepted by consumers if it is developed and marketed using an evidence-based strategy.

9.1. UNDERSTAND THE CONSUMER

To make a substantial shift in meat consumption in favor of plant-based proteins that are healthier, more sustainable, and socially conscious, we must first comprehend the underlying reasons why people choose certain foods over others, and protein in particular. To give consumers more options to select a product that is not only tasty, economical, and convenient but also healthier, more sustainable, and just, plant-based product development should aim to satisfy these motivations and remove the obstacles to the consumption of plant-based meat.

9.1.1. Motivations

Food choices are influenced by many factors, including biological ones (hunger, appetite, taste preferences), social, economic, and physical ones (culture, family, peers, habits), psychological, and physical ones (cost, income, availability, access, education, skills, time).

9.1.2. Foundational Drivers

The most important elements are referred to as "foundational drivers," or those that satisfy urgent needs and desires. Taste, cost, and convenience make up these three main motivators. Taste has been found to be the most significant factor in influencing consumer decisions about what to eat and how much protein to consume.

9.1.3. Evolving Drivers

Before a consumer can make healthier food choices that are in line with higher, more aspirational values, like animal welfare and environmental impact, foundational food motivations must be satisfied. These are referred to as "evolving drivers," or those that have no direct bearing on the needs and desires of customers right now. Even though evolving drivers are becoming more significant, they still have less of an impact on consumers' decisions to buy than foundational drivers. Although a product's health profile indicates how it will benefit the user personally, most users are not likely to give health benefits much thought until they find the product to be appetizing, reasonably priced, and easily accessible (Figure 9.1).

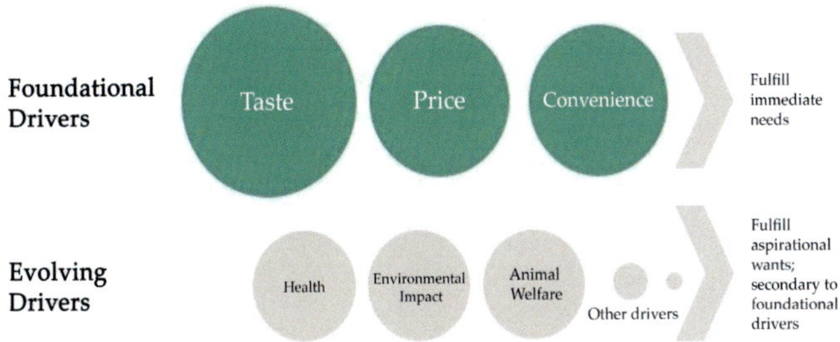

Figure 9.1. *Foundational drivers vs. evolving drivers.*

Source: Szejda, K., and Parry, J. (2020). Strategies to accelerate consumer adoption of plant-based meat: Recommendations from a comprehensive literature review. Research Report. Washington, DC: The Good Food Institute. Available at go.gfi.org/plant-based-meat-consumer-adoption-recommendations.

9.1.4. Point of Purchase Motivations to Select Plant-Based Meat

The reasons for choosing plant-based meat substitutes are largely consistent with the reasons for making overall dietary choices. The main factors that influence a consumer's decision to buy a plant-based product are all related to how the product will help them personally. Taste is the factor that matters the most for all diet types and age groups. In this category, familiarity, and tradition also play a significant role. Customers show a preference for familiar-appearing plant-based meat alternatives over unfamiliar ones. This is particularly true for consumers of traditional meat, who have the greatest desire for plant-based meat that has the same sensory qualities as traditional meat.

Though they are still strong predictors of purchase intent in this category, taste, tradition, and familiarity have a greater impact than health and nutrition considerations. Protein and fiber content seem to be the most crucial factors for consumers when it comes to their health. Freshness influences consumer decisions more favorably than any health factor and is a major factor in their purchasing decisions. When consumers are deciding whether to buy plant-based meat, the products' altruistic benefits are ranked lower on the hierarchy of benefits. However, if consumers were better informed about the extent to which meat production is linked to environmental damage or animal suffering, these benefits might become more significant. Taste, familiarity, and health are more likely to drive a purchase than these factors, even though millennials are the demographic group most likely to find these factors appealing. The majority of consumers can be persuaded to prefer plant-based meat substitutes by emphasizing personal qualities like flavor, freshness, and familiarity over perceived societal benefits (Figure 9.2).

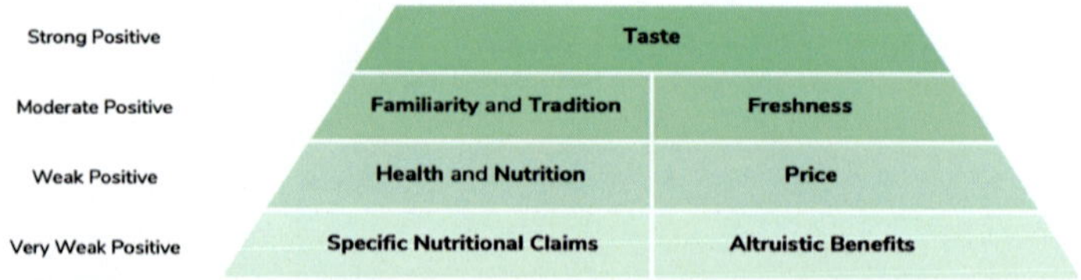

Findings from an implicit purchasing study of the plant-based category. Courtesy of GFI.

Figure 9.2. *Findings from an implicit purchasing study of the plant-based category.*

Source: Szejda, K., and Parry, J. (2020). Strategies to accelerate consumer adoption of plant-based meat: Recommendations from a comprehensive literature review. Research Report. Washington, DC: The Good Food Institute. Available at go.gfi.org/plant-based-meat-consumer-adoption-recommendations.

9.1.5. Barriers

Recognizing the obstacles to dietary change is even more crucial than knowing the main drivers behind decisions about what foods and proteins to buy. Obstacles have the power to negate incentives and frequently have a greater impact on consumer purchasing decisions than drivers. The main consumer group willing to try plant-based meat is meat reducers. Here, we'll concentrate on the obstacles to meat reduction and how plant-based meat can get past them (Figure 9.3).

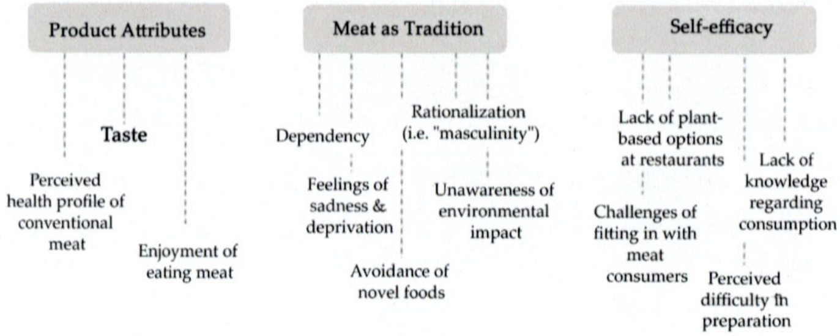

Figure 9.3. *The barriers to meat reduction.*

Source: Szejda, K., and Parry, J. (2020). Strategies to accelerate consumer adoption of plant-based meat: Recommendations from a comprehensive literature review. Research Report. Washington, DC: The Good Food Institute. Available at go.gfi.org/plant-based-meat-consumer-adoption-recommendations.

9.1.6. Product Attributes

The perception of conventional meat's health profile is the biggest obstacle. Most consumers, especially those who consume traditional omnivorous diets, still view conventional meat as nutritious and healthful. The conviction that conventional meat is safe and does not contribute to illness is a strong indicator of resistance to cutting back on consumption. Hedonism, or the pleasure people get from eating meat, is a significant impediment to reduction. This enjoyment stems from several things, including taste, a sense of accomplishment, and satisfaction. Taste is another obstacle for those following flexitarian diets, and it's a common excuse given by former vegetarians for their return to meat consumption.

9.1.7. Meat as Tradition

When traditional meat eaters are asked to consider giving up meat, they display feelings of deprivation, dependency, and sadness. Strong attachments to conventional meat in consumers may make them more resistant to dietary changes and provide excuses for their actions. These ideas are frequently connected to a more "conventional" conception of masculinity. Eating conventional meat is more likely to be justified by those who consider themselves to be more manly.

A large number of consumers do not know enough about how much the production of meat affects the environment. Customers' reluctance to cut back on their consumption of conventional meat may be partially explained by their lack of awareness of the environmental impact of such meat. Even with greater awareness, though, at the point of purchase, more direct and beneficial factors (like convenience, familiarity, freshness, and taste) will most likely be much more important.

Customers may have a strong attachment to traditional meat as well as food neophobia. This particular characteristic behavior is a general aversion to trying new foods. Tofu and tempeh, two popular plant-based meat substitutes, are still unfamiliar to some people, and their resistance to them serves as a major obstacle to dietary change.

9.1.8. Self-Efficacy

Commonly mentioned obstacles to dietary change include a lack of plant-based options at restaurants, the perception of difficulty in preparing plant-based foods, and a general lack of knowledge regarding the consumption of plant-based foods. Furthermore, adopting a plant-based diet may make it harder to modify behavior because it can be challenging to blend in with traditional meat eaters in social situations.

9.1.9. Consumer Segmentation

Individuals and groups have different reasons for eating and different obstacles. Creating a clear consumer segmentation strategy and comprehending the complexities of meat consumption options are essential to creating an evidence-based plan that will hasten the adoption of plant-based meat. Using this approach will give product developers and marketers the ability to choose messaging, branding, labeling, and formulation that are specifically targeted.

Starting with the customers who are most open to changing their diet is the best course of action. As a result, the adoption process will quicken and more people will be able to eat plant-based meat. Thoroughly examined theoretical models offer useful guidance on consumer targeting tactics and long-term adoption acceleration. Marketers can use these to pinpoint the groups that are most receptive to their messages, and then they can methodically target the harder-to-influence groups.

9.1.10. Diffusion of Innovations

The Diffusion of Innovations Theory (DOI), which was first developed in 1962, describes how new concepts, goods, or behaviors gradually gain traction among a population. According to the DOI, some members of a social system are initially more likely than others to accept or consider a given change. Over time, the new innovation spreads throughout society (Figure 9.4).

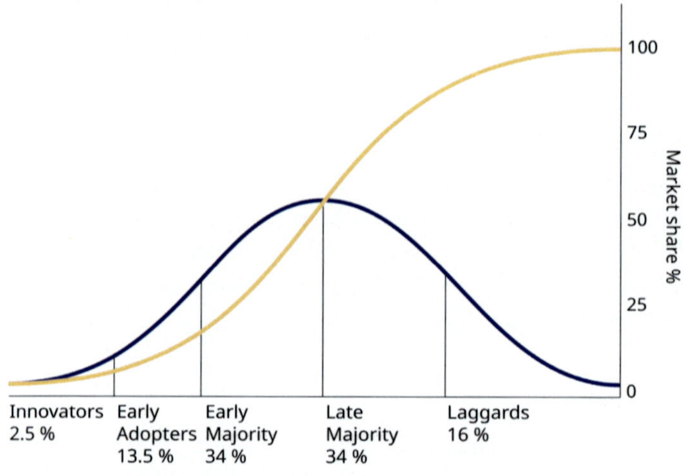

Figure 9.4. *Adopter categorization on the basis of innovativeness.*

Source: https://upload.wikimedia.org/wikipedia/commons/thumb/1/11/Diffusion_of_ideas.svg/640px-Diffusion_of_ideas.svg.png.

The (blue) bell curve representing a population's distribution among adopter categories is displayed in Figure 9.5. Each category's population is typical and has been seen in a variety of (successful) innovation contexts. The (yellow) S curve illustrates the rising market share that results from successive customer groups adopting the innovation and eventually reaching market saturation.

DOI segments the population into five groups (Table 9.1):

Table 9.1. *Diffusion of Innovations Theory (DOI) Segments*

Segment	**Description**	**Estimated Population Percentage**	**Successive Market Share**
Innovators	Innovators are the first to try a new behavior, product, or idea (and may even be its creators, as the term implies).	2.5%	2.5%
Early Adopters	Early adopters are comfortable with innovations and cognizant that change is often inevitable.	13.5%	16%
Early Majority	The early majority must see evidence of the innovation's worth before their adoption of it.	34%	50%
Late Majority	The late majority is skeptical and more reluctant to embrace change, only adopting an innovation once it becomes the norm in their society.	34%	84%
Laggards	Laggards are bound by tradition and suspicion and fervently dislike change.	16%	100%

Source: Szejda, K., and Parry, J. (2020). Strategies to accelerate consumer adoption of plant-based meat: Recommendations from a comprehensive literature review. Research Report. Washington, DC: The Good Food Institute. Available at go.gfi.org/plant-based-meat-consumer-adoption-recommendations.

Change is resisted by both the laggard and late majority segments. It makes sense to hold off on using marketing techniques to specifically target these groups until eating plant-based meat becomes the norm. When the benefits of plant-based meat products (taste, ease of preparation) are readily apparent and the products gain more social acceptance, the late majority will be more inclined to consider adoption. In the same way, laggards won't start altering their diets until they start seeing more plant-based meat in supermarkets and hearing about it more often in social settings. Until then, targeting this group with marketing efforts might not be the best use of your resources.

9.1.11. Target Meat Reducers

Targeting consumers based on their current meat consumption habits is the most practical approach, even though creating detailed profiles of early adopter and early majority segments is the ideal marketing strategy. Customers' willingness to eat plant-based meat can be accurately predicted by segmenting the market based on their current dietary choices. Meat avoiders, meat reducers, and traditional meat consumers are the three main consumer categories that arise from this segmentation strategy. The majority of consumers prioritize fundamental and individualized factors such as taste, cost, convenience, and health over value-based factors like sustainability and animal welfare when making food choices. However, there are differences in the reasons behind these three consumer group decisions to consume less meat.

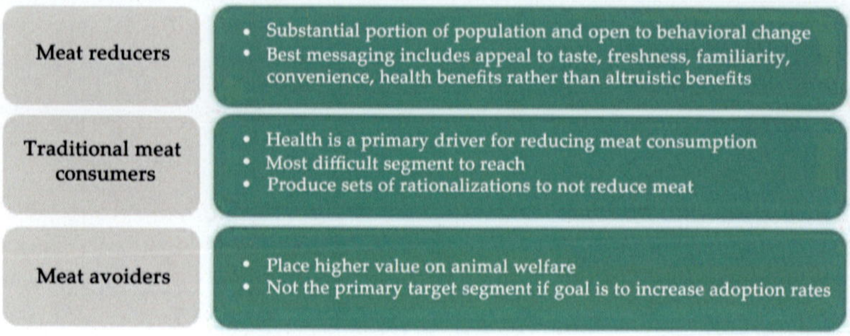

Figure 9.5. *Reducing meat consumption.*

Source: Szejda, K., and Parry, J. (2020). Strategies to accelerate consumer adoption of plant-based meat: Recommendations from a comprehensive literature review. Research Report. Washington, DC: The Good Food Institute. Available at go.gfi.org/plant-based-meat-consumer-adoption-recommendations.

All three consumer groups have reduced their meat consumption primarily due to health concerns. Vegetarians and vegans place a high priority on animal welfare, while traditional meat eaters and flexitarians seem more concerned with their health than the environment or the welfare of other animals.

The primary target segment should not be current plant-based meat consumers, particularly vegetarians and vegans, if the goal is to increase the rate of adoption of plant-based products. It is far more economical to focus on the meat reducer market. This group makes up a sizable section of the populace and is already willing to modify its behavior. The hardest market to penetrate is traditional meat eaters. When asked to reduce their intake, these customers frequently come up with a litany of excuses, such as the notion that eating traditional meat is vital for good health and the impossibility of adopting an alternative diet.

Instead of emphasizing environmental effect and animal welfare, messaging aimed at meat reducers should emphasize excellent taste, freshness, familiarity, convenience, and health benefits. Additionally, appealing to taste and familiarity won't likely increase traditional meat consumers' resistance to the products, which could happen when emphasizing altruistic messaging (Szejda and Parry, 2020).

9.2. PRODUCT DEVELOPMENT

Reduced consumption of conventional meat can be achieved with relative ease by switching to high-fidelity plant-based meat. But in terms of a few crucial product characteristics, plant-based meat as a category still lags behind conventional meat. To create products that consumers want and need, marketers, and product developers need to identify the qualities that consumers look for in products and create products with those features. The benefits of plant-based meat substitutes that are currently available are discussed below, along with some of their drawbacks. The main goal in developing plant-based meat substitutes should be to benefit the consumer directly by offering cheaper, more convenient, higher-quality nutritional value, better taste and other sensory aspects, and so on. The merchandise ought to be thrilling while maintaining a strong sense of familiarity. This familiarity is attained by sharing characteristics with meat derived from animals and by being able to "fit" into enticing recipes and traditional meal settings.

9.2.1. Sensory Aspects and Likeness to Meat

Consumer appeal will rise with the development of products that have more characteristics of meat, such as taste, texture, juiciness, smell, and appearance. One of the most obvious obstacles to eating plant-based meat substitutes is its low sensory appeal. Even for new high-fidelity products, most consumers do not believe that the sensory qualities of plant-based meat currently match those of conventional meat. Omnivores have the greatest desire for plant-based meat to replicate the sensory qualities of conventional meat. Vegetarians and vegans have differing levels of desire. However, regardless of the type of product, highly desirable sensory qualities are essential.

9.2.2. Price Adjustments

The majority of people as well as those who eat the most conventional meat can be reached by focusing on price-driven consumers. Strongly attached meat consumers are more likely to accept products from plant-based meat producers when they are able to lower the price of their offerings. With the help of coupons, you can persuade people to try the products and eventually turn into loyal customers. One of the key deterrents for omnivores to switch to plant-based meat is price, which is also one of the factors that influences consumer decisions the most. Another stated reason why consumers cut back on their meat consumption is the cost of conventional meat. Developing plant-based meat substitutes that are priced similarly to or less than their traditional counterparts is probably a good way to hasten adoption across all consumer segments.

9.2.3. Convenience

In many ways, people believe that conventional meat is more convenient than plant-based meat. Some plant-based meals are poorly understood, and making them may require more time than making their conventional counterparts. Convenient products are the ones that consumers are most likely to purchase, so efforts should be made to create products that are easy to prepare and conform to social norms and consumer habits. Moreover, products should be simple to find in the grocery store for the intended customer. This creates another barrier.

9.2.4. Meal Context

One social and cultural factor that affects consumer purchasing decisions is meal-context appropriateness. Product designers ought to design goods that complement the presently fashionable foods. Creating goods that complement meals rather than serving as stand-alone components is a viable strategy to boost consumer acceptance of plant-based meat substitutes. Regular plant-based meat eaters will want to introduce more meat-like plant-based meat products to their social networks because they blend in well with everyday meal situations.

The appropriateness of a product in the context of a meal can be communicated through naming strategies. Giving a product's description as much detail as you can (e.g., the product label should say "spaghetti with plant-based ground beef" rather than "spaghetti with plant-based meat,") emphasizing provenance (e.g., "Cumberland-spiced veggie sausages"), or employing language that highlights the mouthwatering flavor of meat could be useful techniques.

9.2.5. Familiarity and Novelty

Plant-based meat is in a unique position to satisfy the craving for comfort food as well as the urge to try novel and interesting dishes. When creating and promoting new products, it is important to take these two opposing forces into account. Customers prefer familiar products with well-known flavors, looks, and preparations over ones that are completely new. A dish's familiarity has a big impact on how much one wants to eat it. Plant-based meat substitutes might be less popular because they are not yet commonly regarded as familiar. It is recommended that products be developed that are similar to their conventional versions in terms of sensory aspects, packaging, preparation, and marketing.

Customers still indicate a relatively strong desire for plant-based products to be exciting and interesting, even though novelty isn't as strong a driver of purchase intent as familiarity is. This is particularly true for people who don't have a lot of food phobia. As long as the products retain their enticing sensory qualities, creating, and promoting plant-based meat substitutes as novel and intriguing takes on well-known products may prove to be a successful marketing tactic.

9.2.6. Health Benefits

Though not as much as factors like taste, familiarity, and freshness, health, and nutrition play a role in consumers' decisions at the point of purchase. Nowadays, a lot of consumers believe that plant-based meat is low in saturated fat, high in protein, and great for helping people control their weight. These qualities ought to be retained in plant-based meat. When compared to traditional meat, consumers believe it to have a lower overall nutritional value. Along with other aspects of health, more attention needs to be paid to improving the products. For instance, zinc, and vitamin B12, which are present in traditional beef, could be added to a plant-based beef patty to enhance its nutritional value. One of the main obstacles to consumption will be lessened if these health issues are resolved.

9.2.7. Altruistic Benefits

Taste, familiarity, freshness, health, and convenience are stronger motivators than ethical concerns, even though plant-based meat has obvious altruistic advantages over conventional meat. Only when products satisfy fundamental requirements for food choice should ethics-focused messaging be emphasized as a "bonus" benefit, as it is insufficient for targeting non-vegetarian groups.

9.2.8. PBM Pleasure Drivers and Factors Involved in Product Success

9.2.8.1. Nutritional and Sensory Qualities

Enhancing sensory aspects is crucial to guaranteeing product acceptability since they have a big influence on consumers' purchase decisions. This is especially true for meat substitutes, since the success of these products depends on their sensory appeal. It is thought that the perceived sensory differences from those of meat are the main obstacle to the acceptance of PBM. According to reports, flavor, sensory, and nutritional qualities are important for PBM's effectiveness. Furthermore, it has been stated that the primary obstacles to PBM acceptance are unfamiliarity and low sensory appeal. It is imperative to acknowledge that PBM ought to possess nutritional attributes that are either equivalent to or similar to those of products derived from animals. Kurek along with others stated that a product is considered high quality if it has an amino acid score that is comparable to a meat protein digestibility-corrected amino acid score, is low in fat, and contains more than 30% protein. In fact, if they take the time to taste a product they are unfamiliar with and decide they do not like it, they might have a negative opinion of the whole PBM product line, which would greatly lower the likelihood that they will become a regular customer. Wang et al. described that the flavor profile of the final product is greatly influenced by the protein source. Comprehending the mechanisms underlying the development of flavors and off-flavors in PBM products can facilitate the development of effective strategies for mitigating off-flavors and enhancing desired flavors. In an examination, Caputo et al. discovered that taste and product names/labeling play a significant role in PBM acceptance. The transition from meat consumption to PBM substitutes is currently hindered by the requirement that PBM and meat be similar, especially in terms of texture, taste, and flavor. The primary factors influencing consumer acceptability are texture and flavor. The composition and moisture content of the raw material, and the extrusion conditions (residence time, extruder temperature, compression, and pressure, vapor loss during expansion, and diffusivity of the volatiles in the mass) are some of the factors that affect the retention or loss of volatile compounds during the extrusion process. Even health-beneficial products might be rejected by consumers if their sensory qualities fall short of their expectations. Hobbs claims that when purchasing a meat substitute, consumer attitudes and the acceptance or rejection of products may be influenced by product-related beliefs, such as sensory qualities and product details, as opposed to person-related beliefs, which include taste and environmental or health concerns. Padilha de Oliveira et al. discovered in a survey that PBM is perceived as less appetizing; given that

PBM's flavor continues to impact dietary decisions, these views might serve as obstacles to PBM being substituted for animal meat. Questions concerning sensory attributes, nutrition, clean labeling, and consumer confidence are raised by the use of various additives to produce a texture, juiciness, tongue feel, and flavor similar to that of meat. To satisfy consumer expectations regarding health benefits, PBMS should also be manufactured with natural ingredients and be as clean as possible. A person's familiarity with food products helps to lower uncertainty about them and improves the fit between their expectations and sensory qualities.

9.2.8.2. Consumer Attitudes Towards the Consumption and the Price of PBM

Eating habits are often influenced by a number of interrelated situations that occur in daily life. One of these things is culinary knowledge. When dining at home, people frequently select meals that require less processing and are simpler to prepare, which is a trend in modern living observed in many different nations. Vaskelainen et al. described that consumers want quick- and easy-to-cook food that can be fitted to everybody's taste and is suitable for all palates Fast food items, like pizza and burgers, are linked to status-related factors that could influence people's preference for more traditional choices when attempting to differentiate themselves from their social groups. Traditional cuisines are believed to be associated with lower social status, whereas modern foods are seen to be associated with high social status. Siddiqui Anusha et al. revealed that, taking into account all the psychological aspects involved in a new product's success, beliefs, and attitudes related to PBM still need to be influenced. The introduction of new goods is also necessary to provide consumers with more options than just traditional meat products like sausages, nuggets, burgers, and minced meat. To make PBMSs more appealing to the general public, their price must also be lowered. The best course of action would be to rank PBM products below meat because some customers liken them to meat. A discount plan is advised before reaching this price point because it would enable customers who find alternatives too costly to test to possibly have a positive first impression and then purchase them again, even though they increase in price.

9.2.8.3. Awareness of New Technologies and Novel Foods

A common reaction to the use of new technologies or to food that defies consumers' food/gastronomic cultures is food neophobia. Some think that the failure of food is caused by new technologies that are unethical and break social norms, creating moral dilemmas. Since they are linked to the acceptance of products, uncertainties, risks to one's health and the environment, and worries about emerging technologies should all be addressed. Michel et al. expressed doubts about the knowledge of individuals who are driven to consume a more ecologically friendly diet and whether well-meaning individuals could make bad decisions as a result of ignorance. Nonetheless, people may find it easier to select sustainable foods if they have more knowledge and information, and they might also need an information interpretation guide. Because PBM is a sustainable protein source and does not present any ethical issues, it is growing in popularity among consumers as an alternative to animal meat. It is advisable to take into account less costly and energy-intensive processing techniques to satisfy the growing market demand.

9.2.8.4. Food Literacy and Cooking Ability

Food literacy is the capacity to organize, select, cook, and consume healthful foods to meet needs and monitor intake. It is a contemporary concept that provides a framework for understanding the relationship between dietary consumption and food-related knowledge and skills. Nutritionists have always been curious about how individuals prepare their own food. Cooking, nutritional awareness, food involvement, food safety, and food skills have all been taken into account. People use the idea of food choice value as a guide when deciding what to buy or eat. Food choices made in daily life are frequently habitual and impacted by a multitude of interrelated factors. To encourage PBM consumption, food literacy is required. Furthermore, the ability to cook is essential for the adoption of any food ingredient since knowing how to use them and which recipes go well with them can boost demand and acceptance. Furthermore, most customers are not always familiar with how to use new products that are introduced to the market. As a result, recipes or information on simple ways to incorporate the products into daily routines should be given to customers. Given that culinary proficiency is believed to be acquired through experience, a consumer's capacity for sensory discrimination may differ based on their degree of culinary proficiency. A lower level of interest in food may be the cause of consumers with less or no cooking skills paying less attention to a food's intrinsic qualities than consumers with better skills. When creating new products, it could be beneficial to use recipes from traditional food-cooking cultures. Thus, taking into account the cooking process could aid in enhancing flavor and taste. People's willingness to try new foods isn't always increased when they are served with familiar ingredients. Recipes for cuisine that are publicized in the media also showcase the uniqueness and legacy of the goods. Prominent chefs should have promoted the use of PBM as a meat substitute in both traditional and contemporary recipes (think Asian cuisine, for example).

9.2.8.5. Influence of Distribution Channels and Promotion

When a new product enters the market, a whole new set of relationships must be built around it because consumers, retailers, and lawmakers are probably not familiar with it. As a result, it's critical to present the products correctly so that customers can find them in supermarkets and decide what to buy. To attract as many consumers as possible, PBM products must continue to be made more obvious and prominent in supermarkets. For customers to find them and compare them with meat, it would be ideal if they were placed close to the meat. Additionally, it is advised that PBMSs be given out in restaurants to entice patrons to try them and perhaps adopt them later. Restaurants have the benefit of being able to display PBMs and provide customers with ideas on how to incorporate these goods into their everyday meals. Through the reduction of physical distances, the ability for businesses to expand their markets, and the recognition that the establishment of direct relationships between producers and end users can result in long-term advantages for PBM acceptance, social media and the internet can increase potential benefits. Posts on social media with rich colors and symmetry in both vertical and horizontal directions receive more likes. Promotion is a crucial element in creating an impression of a company's commercial events. One of the most successful strategies is celebrity endorsement advertising, which combines social culture and marketing and is well-liked in the international market. Vaskelainen et al. claimed that to facilitate the shift toward PBM, hype—a rise in public awareness and expectations about the innovation potential, such as media hype—can swiftly spur demand for sustainable products and cause a bandwagon effect that pushes numerous similar niche products into the mainstream.

9.3. CHOICE ARCHITECTURE

Choice architecture is the design of how options are presented to consumers in environments where they are making purchases, like restaurants or grocery stores. Interventions in choice architecture "nudge" or encourage customers to make the desired decision. Product labels, product placement at a grocery store, "default" menu items, placing food items first or last on a menu, and combinations of these tactics are examples of nudge tactics. By using unconscious processes to influence consumer decision-making, such strategies have a great deal of potential to increase consumer adoption of plant-based meat.

9.3.1. Product Labeling

In the market, different labels are used to say that a product doesn't contain any ingredients derived from animals (e.g., "plant-based," "vegan," "meatless," and "meat-free."). According to research, the terms "plant-based" and "plant protein" are more enticing than alternative descriptions. Customers are most likely to make a purchase when they see these two terms. Restrictive terms, like "vegetarian," "meat-free," and "vegan," are less effective because they emphasize a loss and have unfavorable connotations for a large number of omnivores.

Consumers prioritize cost, taste, and convenience when choosing foods, so sustainability and health are not the main factors influencing their choices, but many consumers still find these features appealing. Studies show that descriptive labels with a health focus positively influence consumers' choice of plant-based meat substitutes to a moderate extent. When describing a meal or product's health profile, the most effective language highlights health benefits like high protein or fiber content. Health labels that are too restrictive, like "dairy-free," are met with less positivity by consumers."

9.3.2. Product Location

A key consideration when buying food is convenience. Studies indicate that putting plant-based meat in a convenient location makes it more likely to be bought and eaten. Products can be encouraged to be purchased by positioning them in the mainstream area of their respective categories, such as in a nearby segregated set. Accessible product placement makes sure traditional meat eaters don't have to work as hard mentally or physically to get plant-based products.

More significant than the number of locations a food item is placed in is the prominence of that location. Putting products in places where they will be seen and frequented effectively encourages people to choose them. Supermarket and buffet environments have both shown the value of prominent placement.

9.3.3. Menu Design

Studies on menu design have demonstrated that rather than having plant-based dishes in a separate vegetarian section, omnivores are more likely to choose plant-based dishes that are integrated into the entire menu. Customers can be persuaded to select the plant-based options on a menu by changing the context in which plant-based dishes are presented. Menu structure interventions encourage the choice of plant-based meat by either making the plant-based options the easier option or making the conventional meat option require more thought and effort.

A salad on the menu that can be made with different protein components will come with a plant-based protein component unless the customer requests otherwise. Default options are implemented if a person does not take action.

Giving consumers default plant-based meal options encourages them to select them over traditional meat options. People tend to avoid the mental and occasionally physical effort involved in decision-making, so altering the default meal components influences food selection. Additionally, the consumer is implicitly advised that the default option is the best one in a particular situation.

The descriptions of menu items affect the choice of food. Studies reveal that words that convey the feeling and satisfaction of eating are most effective (e.g., "comforting," "melt in the mouth." Slurred words (e.g., More so than simple descriptive language or language emphasizing health benefits, the phrase "sweet sizzlin' green beans and crispy shallots") have also proven to be particularly effective at increasing the volume of vegetables consumed.

The way that descriptive language is framed affects how people decide what to buy. Emphasizing the origins of ingredients geographically (e.g., "Sausages spiced with Cumberland vegetables") or by employing favorable perceptions of plant-related terms (e.g., Using terms like "field-grown" or "garden") encourages customers to order plant-based foods. Comparably, appealing, descriptive terms like "fresh" and "seasonal" expand the range of plant-based options available to people who don't often eat vegetarian meals.

Gains-focused framing of health benefits is more effective in steering consumer choice than loss-focused framing. For example, implicit testing of plant-based food selection showed that emphasizing high nutritional content (e.g., "high-protein," "high-fiber") was more effective than emphasizing low negative nutritional content (e.g., "low-fat," "low-calorie," "low-sugar"). In cafeteria settings, framing a vegetable option as healthy positive (e.g., "healthy, energy-boosting green beans and shallots") resulted in a greater selection of that option than framing it as healthy restrictive (e.g., "light "n" low-carb green beans and shallots").

9.4. TARGETED MESSAGING

When done correctly, direct-to-consumer messaging can play a vital role in encouraging consumers to embrace plant-based meat substitutes. Message designers need to think about the message's source, channel, target audience, and framing in addition to the message's content when creating a marketing plan. The suggested message content areas are summarized here, along with ideas for incorporating evidence-based messaging principles into message design.

9.4.1. Message Content

Overall, the most persuasive messaging claims will emphasize the fundamental factors that influence food choice, particularly taste and other sensory aspects.

Other factors to consider are meal context, familiarity, tradition, ease of preparation, and protein content. Although these drivers offer a starting point, developing an efficient messaging strategy for every brand and product requires a well-defined approach to consumer segmentation. Consumers are often drawn to the product category by health, which is a common driver of meat reduction. However, taste appeals should take center stage in point-of-purchase communications, with health appeals coming in second. Another interesting area for message content is social norms messaging. Emphasizing the growing trend of plant-based meat consumption is probably going to encourage purchases.

9.4.1.1. Taste and Sensory Appeal

In the plant-based category, the most frequently mentioned factors influencing product choice among all groups are taste and sensory appeal. Many consumers find that most plant-based alternatives to traditional meat products lack sufficient flavor, juiciness, and texture, and they express sensory dissatisfaction with them. Since traditional meat consumers and meat reducers are the ones who want plant-based meat products to have the same sensory qualities as conventional meat the most, emphasizing how they are similar to conventional meat products may be a good way to win them over.

9.4.1.2. Familiarity and Novelty

It is helpful to highlight the familiarity as well as the novelty of plant-based meat substitutes. By promoting plant-based meat as a practical and simple alternative to traditional meat in one's favorite dishes, consumers can continue to make dietary choices that support their habits without having to make a major effort to alter them. Product adoption among more daring, non-neophobic consumers may rise if it is marketed as exciting and innovative. It is important to maintain the product within the familiar, though.

Convenience—that is, the simplicity of preparation and product accessibility—is a primary factor in determining food preferences and a driving force behind the consumption of plant-based meat. According to earlier consumer research, people believe that conventional meat is more convenient than plant-based meat because it is easier to prepare and more readily available in stores. A wealth of behavioral dietary research suggests that individuals won't alter their eating patterns unless they believe they can do so. Advertising that highlights the ease of preparation and convenience of plant-based meat is likely to reach as many people as possible. A successful marketing approach might include preparation guidelines, emphasizing how simple and quick it is to prepare plant-based meat compared to conventional meat and how well-suited it is for a wider range of meal situations.

One of the most noticeable aspects of plant-based meat's perceived health benefits is its protein content, which is frequently mentioned in plant-based food research as either a desired or a cause for concern. High protein content should be the main feature emphasized in health-focused marketing messages since it is both a highly desired core attribute of plant-based meat and a salient perceived barrier, and it will appeal to most consumers. It is also probably more effective to emphasize the presence of a positive nutritional component, like protein, rather than the absence of a negative nutritional component, like saturated fat.

9.4.1.3. Altruistic Benefits

Research indicates that most consumers recognize and value the humanitarian advantages of plant-based meat e.g., taste, price, convenience, and familiarity—are the main factors that drive the majority of consumers. Other considerations include the environment and animal welfare. Furthermore, people who already abstain from meat are the target audience for messaging about the altruistic benefits. Only when consumers can clearly see that the product also satisfies their fundamental needs related to food choice will this messaging strategy have an impact on their purchasing behavior.

9.4.1.4. Social Norms

Social pressure—real, perceived, or inferred—has a significant impact on people's attitudes, emotions, and actions. Social norms, or the normative influence of others, might even be a better indicator of food choice than personal preference. Norm messaging drives consumers to act in a desired way because they want to fit into a preconceived notion of what is expected of them. As a result, the product is more widely accepted by consumers and is seen as more mainstream.

In behavioral interventions generally and in the context of meat reduction specifically, social norms messaging has proven effective. Adoption of plant-based meat is likely to be successful in the new context if social norms messaging techniques are applied well. In particular, dynamic descriptive norms work well. These are declarations that indicate anticipated expansion or the rising acceptance of the norm over time e.g., "Three out of 10 people have altered their behavior in the past four years and started buying plant-based meat on a weekly basis." Dynamic descriptive norms highlight positive changes in consumer behavior and display the rate of change. It works best to frame these messages as desirable behaviors rather than conspicuous undesirable behaviors.

9.4.2. Elements of Effective Message Design

A message's ability to persuade a target audience to consume more plant-based meat may be aided or hindered by several factors outside of the message's content. These extra crucial factors are message source, channel selection, and targeting.

9.4.2.1. Targeted Messages

Consumer segments differ in their receptivity to a message. The meat reducer (flexitarian) segment is the perfect target market for messaging encouraging the adoption of plant-based meat substitutes. Targeting a particular subset of meat reducers with a unique profile of incentives, obstacles, demographics, cultures, and purchasing habits should be another area of emphasis for message design. A different tactic would be to go after customers who have sway over other members of the target group, like those who make purchasing decisions for the family.

9.4.2.2. Message Source

The sender—a person or thing—has a significant impact on the attitudes of the target audience. This is particularly true when a customer makes a decision passively rather than carefully, like picking food quickly at the grocery store. Customers need to have faith in the message source, and this faith may be founded on similarity, popularity, or experience, among other things. Experts such as scientists, fitness instructors, registered dietitians, medical professionals, and health coaches are the most reliable sources of information about food choices. But social influence is also significant, and celebrity endorsements ought to be investigated as a means of influencing and broadening societal norms regarding plant-based meat.

9.4.2.3. Channels

Choosing a communication channel, or the means of delivering a message, is another crucial factor to take into account when trying to sway customers. Message designers need to know which communication media their target audience uses or is exposed to the most to target them with messages. Additionally, when it comes to information about food, health, and diet, people tend to view different communication methods as more or less credible. Oral face-to-face and video communication generally have the biggest impacts on the desired results.

9.4.2.4. Message Framing

According to framing theory, an audience's interpretation and receptivity to information are influenced by the context in which it is presented. As an illustration, when a message regarding plant-based meat is presented in terms of flavor (e.g., luxurious) or a health advantage (e.g., high protein), readers of the message will view plant-based meat in a different way than if it is presented from the perspective of the environment or animal welfare. Customers will be more open to claims that support plant-based meat if they have already been considering and analyzing plant-based meat. An important predictor of the intention to purchase plant-based meat is prior familiarity. This familiarity makes sure that customers can access and form their opinions and feelings regarding the product. Future marketing communications will therefore be more successful since customers will already be considering the product. A message that is presented as a story will be more likely to be remembered, considered, and liked by the target market. Stories ought to illustrate a deeper point. A story about a family enjoying a holiday meal that includes plant-based meat, for instance, will naturally convey the message that plant-based meat is comfortable, customary, and comparable to conventional meat.

9.4.3. Antecedents of Behavior Change

It is more likely that desired behavioral outcomes will be attained when tested behavioral and social influence theories are applied to message design and distribution. The most thorough knowledge of the many factors influencing behavior is offered by the Integrative Model. Persuading a customer to buy plant-based meat requires far more skill than just highlighting the advantages of the product's affordability or health benefits. A consumer's path to changing their behavior is influenced by a variety of factors (Figure 9.6).

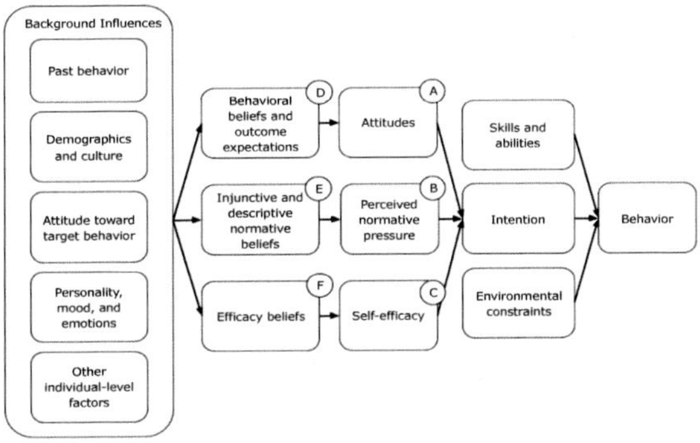

Figure 9.6. *Integrative model of behavior change.*

Source: Szejda, K., and Parry, J. (2020). Strategies to accelerate consumer adoption of plant-based meat: Recommendations from a comprehensive literature review. Research Report. Washington, DC: The Good Food Institute. Available at go.gfi.org/plant-based-meat-consumer-adoption-recommendations.

In the current context, the model posits that the intention to purchase plant-based meat is influenced by the following:

a. Attitudes toward the behavior.

b. Perceptions of normative pressure about the behavior.

c. Perceptions of the ability to execute the behavior.

In other words, to affect behavior, communicators should try to affect the intention of the individual to engage in that behavior. To affect a customer's decision to participate in a behavior (i.e., eating plant-based meat), three elements need to be at the center of an efficient messaging influence strategy: self-efficacy, normative beliefs, and attitudes. This model can serve as a guide for marketers. They can choose which of the main constructs influencing intention they want to have an impact on. For instance, if the objective of a marketing campaign is to raise awareness and adoption of plant-based meat, then messaging needs to take into account shifting normative beliefs and attitudes toward plant-based meat.

9.4.3.1. Attitudes

For maximum impact, message designers ought to investigate the preexisting opinions and attitudes of a target audience regarding a product, utilizing this information to inform message framing and content selection. Despite being only one of many variables that eventually lead to behavior change, attitudes are among the most reliable indicators of intentions. Beliefs about whether a particular behavior will produce the intended result will impact consumer attitudes. To change people's opinions about plant-based meat, messaging should demonstrate or attest to the achievement of the intended result (e.g., better taste, and health) as a consequence of eating meat substitutes made from plants.

9.4.3.2. Normative Beliefs

A person's preference for food may change if they base their beliefs on whether the majority, especially those in positions of power, engage in a particular behavior. Using easily comprehensible data to demonstrate the number of people consuming plant-based meat is one strategy, as is persuading customers that others think this is a healthier option. Emphasizing the adoption rate in recent years is also a good tactic because it demonstrates the growing normalcy of plant-based meat consumption.

9.4.3.3. Self-Efficacy

The acceptance of plant-based meat has a strong correlation with self-efficacy. When it comes to cutting back on meat, people frequently cite ignorance or insufficient experience with alternative eating methods as a barrier. Research looking into the consumption of plant-based meat also reveals this finding: participants claim not to know how to cook plant-based meat. Plant-based meat marketing initiatives will guarantee that consumers feel confident in their ability to modify their behavior by making food preparation as familiar as possible or by providing on-package and online preparation instructions with well-known recipes that are simple to access and follow (Szejda and Parry, 2020).

REFERENCES

1. Aday, S., & Aday, M. S. (2020). Impact of COVID-19 on the food supply chain. *Food Quality and Safety, 4*, 167–180.
2. Ahmad, M., Qureshi, S., Akbar, M. H., Siddiqui, S. A., Gani, A., Mushtaq, M., Hassan, I., & Dhull, S. B. (2022). Plant-based meat alternatives: Compositional analysis, current development and challenges. *Applied Food Research, 2*, 100154.
3. Alae-Carew, C., Green, R., Stewart, C., Cook, B., Dangour, A. D., & Scheelbeek, P. F. D. (2022). The role of plant-based alternative foods in sustainable and healthy food systems: Consumption trends in the UK. *Science of the Total Environment, 807*, 151041.
4. Alcorta, A., Porta, A., Tárrega, A., Alvarez, M. D., & Pilar Vaquero, M. (2021). Foods for plant-based diets: Challenges and innovations. *Foods, 10*, 293.
5. Anusha Siddiqui, S., Bahmid, N. A., Mahmud, C. M. M., Boukid, F., Lamri, M., & Gagaoua, M. (2022). Consumer acceptability of plant-, seaweed-, and insect-based foods as alternatives to meat: A critical compilation of a decade of research. *Critical Reviews in Food Science and Nutrition*, 1–22.
6. Attwood, S., & Hajat, C. (2020). How will the COVID-19 pandemic shape the future of meat consumption? *Public Health Nutrition, 23*, 3116–3120.
7. Banovic, M., & Sveinsdóttir, K. (2021). Importance of being analogue: Female attitudes towards meat analogue containing rapeseed protein. *Food Control, 123*, 107833.
8. Birke Rune, C. J., Song, Q., Clausen, M. P., & Giacalone, D. (2022). Consumer perception of plant-based burger recipes studied by projective mapping. *Future Foods, 6*, 100168.
9. Bogueva, D., Marinova, D., & Bryant, C. (2022). Meat me halfway: Sydney meat-loving men's restaurant experience with alternative plant-based proteins. *Sustainability, 14*, 1290.
10. Bryant, C., & Sanctorum, H. (2021). Alternative proteins, evolving attitudes: Comparing consumer attitudes to plant-based and cultured meat in Belgium in two consecutive years. *Appetite, 161*, 105161.
11. Bryngelsson, S., Moshtaghian, H., Bianchi, M., & Hallström, E. (2022). Nutritional assessment of plant-based meat analogues on the Swedish market. *International Journal of Food Science & Nutrition, 73*, 889–901.
12. Caputo, V., Sogari, G., & Van Loo, E. J. (2022). Do plant-based and blend meat alternatives taste like meat? A combined sensory and choice experiment study. *Applied Economic Perspectives and Policy, 45*(2).

13. Carlsson, F., Kataria, M., & Lampi, E. (2022). How much does it take? Willingness to switch to meat substitutes. *Ecological Economics, 193*, 107329.
14. Chen, Y., & Zhang, Z. (2022). Exploring public perceptions on alternative meat in China from social media data using transfer learning method. *Food Quality and Preference, 98*, 104530.
15. Choudhury, D., Singh, S., Seah, J. S. H., Yeo, D. C. L., & Tan, L. P. (2020). Commercialization of plant-based meat alternatives. *Trends in Plant Science, 25*, 1055–1058.
16. Cordelle, S., Redl, A., & Schlich, P. (2022). Sensory acceptability of new plant protein meat substitutes. *Food Quality and Preference, 98*, 104508.
17. Cutroneo, S., Angelino, D., Tedeschi, T., Pellegrini, N., & Martini, D. (2022). Nutritional quality of meat analogues: Results from the Food Labelling of Italian Products (FLIP) project. *Frontiers in Nutrition, 9*, 676.
18. Davis, S. G., Harr, K. M., Farmer, K. J., Beyer, E. S., Bigger, S. B., Chao, M. D., Tarpoff, A. J., Thomson, D. U., Vipham, J. L., Zumbaugh, M. D., et al. (2021). Quality of plant-based ground beef alternatives in comparison with ground beef of various fat levels. *Meat Muscle Biology, 5*, 12989.
19. de Koning, W., Dean, D., Vriesekoop, F., Aguiar, L. K., Anderson, M., Mongondry, P., Oppong-Gyamfi, M., Urbano, B., Luciano, C. A. G., Jiang, B., et al. (2020). Drivers and inhibitors in the acceptance of meat alternatives: The case of plant and insect-based proteins. *Foods, 9*, 1292.
20. de Oliveira Padilha, L. G., Malek, L., & Umberger, W. J. (2022). Consumers' attitudes towards lab-grown meat, conventionally raised meat and plant-based protein alternatives. *Food Quality and Preference, 99*, 104573.
21. Dreher, J., König, M., Herrmann, K., Terjung, N., Gibis, M., & Weiss, J. (2021). Varying the amount of solid fat in animal fat mimetics for plant-based salami analogues influences texture, appearance and sensory characteristics. *LWT, 143*, 111140.
22. Elzerman, J. E., van Dijk, P. E. M., & Luning, P. A. (2022). Substituting meat and the role of a situational context: Exploring associations and motives of Dutch meat substitute-users. *British Food Journal, 124*, 93–108.
23. Escribano, A. J., Peña, M. B., Díaz-Caro, C., Elghannam, A., Crespo-Cebada, E., & Mesías, F. J. (2021). Stated preferences for plant-based and cultured meat: A choice experiment study of Spanish consumers. *Sustainability, 13*, 8235.

24. Estell, M., Hughes, J., & Grafenauer, S. (2021). Plant protein and plant-based meat alternatives: Consumer and nutrition professional attitudes and perceptions. *Sustainability, 13*, 1478.
25. Font-i-Furnols, M., & Guerrero, L. (2022). Spanish perspective on meat consumption and consumer attitudes. *Meat Science, 191*, 108874.
26. Grasso, S., Rondoni, A., Bari, R., Smith, R., & Mansilla, N. (2022). Effect of information on consumers' sensory evaluation of beef, plant-based and hybrid beef burgers. *Food Quality and Preference, 96*, 104417.
27. Grundy, E. A., Slattery, P., Saeri, A. K., Watkins, K., Houlden, T., Farr, N., Askin, H., Lee, J., Mintoft-Jones, A., Cyna, S., et al. (2022). Interventions that influence animal-product consumption: A meta-review. *Future Foods, 5*, 100111.
28. Guo, Z., Teng, F., Huang, Z., Lv, B., Lv, X., Babich, O., Yu, W., Li, Y., Wang, Z., & Jiang, L. (2020). Effects of material characteristics on the structural characteristics and flavor substances retention of meat analogs. *Food Hydrocolloids, 105*, 105752.
29. Hambardzumyan, G., & Gevorgyan, S. (2022). The impact of COVID-19 on the small and medium dairy farms and comparative analysis of customers' behavior in Armenia. *Futur. Foods, 5*, 100110.
30. Hartmann, C., Furtwaengler, P., & Siegrist, M. (2022). Consumers' evaluation of the environmental friendliness, healthiness and naturalness of meat, meat substitutes, and other protein-rich foods. *Food Quality and Preference, 97*, 104486.
31. Herz, E., Herz, L., Dreher, J., Gibis, M., Ray, J., Pibarot, P., & Weiss, J. (2021). Influencing factors on the ability to assemble a complex meat analogue using a soy-protein-binder. *Innovative Food Science & Emerging Technologies, 73*, 102806.
32. Hobbs, J. E. (2021). The COVID-19 pandemic and meat supply chains. *Meat Science, 181*, 108459.
33. Hwang, J., You, J., Moon, J., & Jeong, J. (2020). Factors affecting consumers' alternative meats buying intentions: Plant-based meat alternative and cultured meat. *Sustainability, 12*, 5662.
34. Jung, A. H., Hwang, J. H., Jun, S., & Park, S. H. (2022). Application of ohmic cooking to produce a soy protein-based meat analogue. *LWT, 160*, 113271.
35. Kaczmarska, K., Taylor, M., Piyasiri, U., & Frank, D. (2021). Flavor and metabolite profiles of meat, meat substitutes, and traditional plant-based high-protein food products available in Australia. *Foods, 10*, 801.

36. Kerslake, E., Kemper, J. A., & Conroy, D. (2022). What's your beef with meat substitutes? Exploring barriers and facilitators for meat substitutes in omnivores, vegetarians, and vegans. *Appetite, 170*, 105864.
37. Knaapila, A., Michel, F., Jouppila, K., Sontag-Strohm, T., & Piironen, V. (2022). Millennials' consumption of and attitudes toward meat and plant-based meat alternatives by consumer segment in Finland. *Foods, 11*, 456.
38. Kołodziejczak, K., Onopiuk, A., Szpicer, A., & Poltorak, A. (2022). Meat analogues in the perspective of recent scientific research: A review. *Foods, 11*, 105.
39. Kurek, M. A., Onopiuk, A., Pogorzelska-Nowicka, E., Szpicer, A., Zalewska, M., & Półtorak, A. (2022). Novel protein sources for applications in meat-alternative products—Insight and challenges. *Foods, 11*, 957.
40. Kwasny, T., Dobernig, K., & Riefler, P. (2022). Towards reduced meat consumption: A systematic literature review of intervention effectiveness, 2001–2019. *Appetite, 168*, 105739.
41. Kyriakopoulou, K., Keppler, J. K., & van der Goot, A. J. (2021). Functionality of ingredients and additives in plant-based meat analogues. *Foods, 10*, 600.
42. Lacy-Nichols, J., Hattersley, L., & Scrinis, G. (2021). Nutritional marketing of plant-based meat-analogue products: An exploratory study of front-of-pack and website claims in the USA. *Public Health Nutrition, 24*, 4430–4441.
43. Lee, J. S., Oh, H., Choi, I., Yoon, C. S., & Han, J. (2022). Physico-chemical characteristics of rice protein-based novel textured vegetable proteins as meat analogues produced by low-moisture extrusion cooking technology. *LWT, 157*, R713–R715.
44. Lin, Q., Pan, L., Deng, N., Sang, M., Cai, K., Chen, C., Han, J., & Ye, A. (2022). Protein digestibility of textured-wheat-protein (TWP) -based meat analogues: (I) Effects of fibrous structure. *Food Hydrocolloids, 130*, 107694.
45. Loh, H. C., Seah, Y. K., & Looi, I. (2021). The COVID-19 pandemic and diet change. *Progress in Microbes and Molecular Biology, 4*.
46. Lonkila, A., & Kaljonen, M. (2022). Ontological struggle over new product category: Transition potential of meat alternatives. *Environmental Innovation and Societal Transitions, 42*, 1–11.
47. Marshall, D., Bano, F., & Banas, K. (2022). A meaty issue: The effect of meat-related label terminology on the willingness to eat vegetarian foods. *Food Quality and Preference, 96*.

48. McClements, D. J., & Grossmann, L. (2021). The science of plant-based foods: Constructing next-generation meat, fish, milk, and egg analogs. *Comprehensive Reviews in Food Science and Food Safety, 20*, 4049–4100.
49. Michel, F., Hartmann, C., & Siegrist, M. (2021). Consumers' associations, perceptions and acceptance of meat and plant-based meat alternatives. *Food Quality and Preference, 87*, 104063.
50. Moreira, M. N. B., da Veiga, C. P., da Veiga, C. R. P., Reis, G. G., & Pascuci, L. M. (2022). Reducing meat consumption: Insights from a bibliometric analysis and future scopes. *Future Foods, 5*, 100120.
51. Motoki, K., Park, J., Spence, C., & Velasco, C. (2022). Contextual acceptance of novel and unfamiliar foods: Insects, cultured meat, plant-based meat alternatives, and 3D printed foods. *Food Quality and Preference, 96*, 104368.
52. Murakami, K., Shinozaki, N., Yuan, X., Tajima, R., Matsumoto, M., Masayasu, S., & Sasaki, S. (2022). Food choice values and food literacy in a nationwide sample of Japanese adults: Associations with sex, age, and body mass index. *Nutrients, 14*, 1899.
53. Newton, P., & Blaustein-Rejto, D. (2021). Social and economic opportunities and challenges of plant-based and cultured meat for rural producers in the US. *Frontiers in Sustainable Food Systems, 5*, 624270.
54. Niva, M., & Vainio, A. (2021). Towards more environmentally sustainable diets? Changes in the consumption of beef and plant- and insect-based protein products in consumer groups in Finland. *Meat Science, 182*, 108635.
55. Ohlau, M., Spiller, A., & Risius, A. (2022). Plant-based diets are not enough? Understanding the consumption of plant-based meat alternatives along ultra-processed foods in different dietary patterns in Germany. *Frontiers in Nutrition, 9*, 35.
56. Onwezen, M. C. (2022). The application of systematic steps for interventions towards meat-reduced diets. *Trends in Food Science & Technology, 119*, 443–451.
57. Ortega, D. L., Sun, J., & Lin, W. (2022). Identity labels as an instrument to reduce meat demand and encourage consumption of plant-based and cultured meat alternatives in China. *Food Policy, 111*, 102307.
58. Pater, L., Kollen, C., Damen, F. W. M., Zandstra, E. H., Fogliano, V., & Steenbekkers, B. L. P. A. (2022). The perception of 8- to 10-year-old Dutch children towards plant-based meat analogues. *Appetite, 178*, 106264.
59. Ping, C., Xi, F., Blank, I., & Yuan, L. (2022). Strategies to improve

meat-like properties of meat analogs meeting consumers' expectations. *Biomaterials, 287*, 121648.

60. Ramos Diaz, J. M., Kantanen, K., Edelmann, J. M., Suhonen, H., Sontag-Strohm, T., Jouppila, K., & Piironen, V. (2022). Fibrous meat analogues containing oat fiber concentrate and pea protein isolate: Mechanical and physicochemical characterization. *Innovative Food Science & Emerging Technologies, 77*, 102954.
61. Rzymski, P., Kulus, M., Jankowski, M., Dompe, C., Bryl, R., Petitte, J. N., Kempisty, B., & Mozdziak, P. (2021). COVID-19 pandemic is a call to search for alternative protein sources as food and feed: A review of possibilities. *Nutrients, 13*, 150.
62. Saerens, W., Smetana, S., Van Campenhout, L., Lammers, V., & Heinz, V. (2021). Life cycle assessment of burger patties produced with extruded meat substitutes. *Journal of Cleaner Production, 306*, 127177.
63. Sakai, K., Sato, Y., Okada, M., & Yamaguchi, S. (2021). Improved functional properties of meat analogs by laccase catalyzed protein and pectin crosslinks. *Scientific Reports, 11*, 1–10.
64. Santo, R. E., Kim, B. F., Goldman, S. E., Dutkiewicz, J., Biehl, E. M. B., Bloem, M. W., Neff, R. A., & Nachman, K. E. (2020). Considering plant-based meat substitutes and cell-based meats: A public health and food systems perspective. *Frontiers in Sustainable Food Systems, 4*, 134.
65. Schreuders, F. K. G., Sagis, L. M. C., Bodnár, I., Boom, R. M., & van der Goot, A. J. (2022). Non-linear rheology reveals the importance of elasticity in meat and meat analogues. *Scientific Reports, 12*, 1–10.
66. Sha, L., & Xiong, Y. L. (2020). Plant protein-based alternatives of reconstructed meat: Science, technology, and challenges. *Trends in Food Science & Technology, 102*, 51–61.
67. Shahbazi, M., Jäger, H., Ettelaie, R., & Chen, J. (2021). Construction of 3D printed reduced-fat meat analogue by emulsion gels. Part I: Flow behavior, thixotropic feature, and network structure of soy protein-based inks. *Food Hydrocolloids, 120*, 106967.
68. Shen, Y. C., & Chen, H. S. (2020). Exploring consumers' purchase intention of an innovation of the agri-food industry: A case of artificial meat. *Foods, 9*, 745.
69. Siegrist, M., Bearth, A., & Hartmann, C. (2022). The impacts of diet-related health consciousness, food disgust, nutrition knowledge, and the Big Five personality traits on perceived risks in the food domain. *Food Quality and Preference, 96*, 104441.

70. Sucapane, D., Roux, C., & Sobol, K. (2021). Exploring how product descriptors and packaging colors impact consumers' perceptions of plant-based meat alternative products. *Appetite, 167*, 105590.

71. Szejda, K., & Parry, J. (2020). *Strategies to Accelerate Consumer Adoption of Plant-Based Meat: Recommendations from a Comprehensive Literature Review*. Research Report. Washington, DC: The Good Food Institute. Available at: go.gfi.org/plant-based-meat-consumer-adoption-recommendations (accessed on 22 October 2024).

72. Taufik, D., Bouwman, E. P., Reinders, M. J., & Dagevos, H. (2022). A reversal of defaults: Implementing a menu-based default nudge to promote out-of-home consumer adoption of plant-based meat alternatives. *Appetite, 175*, 106049.

73. Tonheim, L. E., Austad, E., Torheim, L. E., & Henjum, S. (2022). Plant-based meat and dairy substitutes on the Norwegian market: Comparing macronutrient content in substitutes with equivalent meat and dairy products. *Journal of Nutritional Science, 11*, 1–8.

74. Tonsor, G. T., Lusk, J. L., & Schroeder, T. C. (2022). Market potential of new plant-based protein alternatives: Insights from four US consumer experiments. *Applied Economic Perspectives and Policy, 45*(1040).

75. Toribio-Mateas, M. A., Bester, A., & Klimenko, N. (2021). Impact of plant-based meat alternatives on the gut microbiota of consumers: A real-world study. *Foods, 10*, 2040.

76. Tosun, P., Yanar, M., Sezgin, S., & Uray, N. (2021). Meat substitutes in sustainability context: A content analysis of consumer attitudes. *Journal of International Food & Agribusiness Marketing, 33*, 541–563.

77. Van Loo, E. J., Caputo, V., & Lusk, J. L. (2020). Consumer preferences for farm-raised meat, lab-grown meat, and plant-based meat alternatives: Does information or brand matter? *Food Policy, 95*, 428–437.

78. Vaskelainen, T., Siltaoja, M., & Hoskonen, H. (2022). Hypes and the birth of new sustainable market categories—A socio-cultural perspective on the emergence of the meat substitute category in Finland. *Technological Analysis & Strategic Management*, 1–13.

79. Vatanparast, H., Islam, N., Shafiee, M., & Dan Ramdath, D. (2020). Increasing plant-based meat alternatives and decreasing red and processed meat in the diet differentially affect the diet quality and nutrient intakes of Canadians. *Nutrients, 12*, 2034.

80. Wang, L., Xu, J., Zhang, M., Zheng, H., & Li, L. (2022). Preservation of soy protein-based meat analogues by using PLA/PBAT antimicrobial packaging film. *Food Chemistry, 380*, 132022.

81. Wang, Y., Tuccillo, F., Lampi, A. M., Knaapila, A., Pulkkinen, M., Kariluoto, S., Coda, R., Edelmann, M., Jouppila, K., Sandell, M., et al. (2022). Flavor challenges in extruded plant-based meat alternatives: A review. *Comprehensive Reviews in Food Science and Food Safety, 21*, 2898–2929.

82. Weinrich, R., & Gassler, B. (2021). Beyond classical van Westendorp: Assessing price sensitivity for variants of algae-based meat substitutes. *Journal of Retailing and Consumer Services, 63*, 102719.

83. Xia, S., Xue, Y., Xue, C., Jiang, X., & Li, J. (2022). Structural and rheological properties of meat analogues from Haematococcus pluvialis residue-pea protein by high moisture extrusion. *LWT, 154*, 112756.

84. Xie, Y., Cai, L., Zhao, D., Liu, H., Xu, X., Zhou, G., & Li, C. (2022). Real meat and plant-based meat analogues have different *in vitro* protein digestibility properties. *Food Chemistry, 387*, 132917.

85. Yuan, X., Zhu, X., Sun, R., Jiang, W., Zhang, D., Liu, H., & Sun, B. (2022). Sensory attributes and characterization of aroma profiles of fermented sausages based on fibrous-like meat substitute from soybean protein and *Coprinus comatus*. *Food Chemistry, 373*, 131537.

86. Yuliarti, O., Kiat Kovis, T. J., & Yi, N. J. (2021). Structuring the meat analogue by using plant-based derived composites. *Journal of Food Engineering, 288*, 110138.

87. Zhang, T., Dou, W., Zhang, X., Zhao, Y., Zhang, Y., Jiang, L., & Sui, X. (2021). The development history and recent updates on soy protein-based meat alternatives. *Trends in Food Science & Technology, 109*, 702–710.

88. Zhao, S., Wang, L., Hu, W., & Zheng, Y. (2022). Meet the meatless: Demand for new generation plant-based meat alternatives. *Applied Economic Perspectives and Policy*, 13232.

89. Zhou, H., Hu, Y., Tan, Y., Zhang, Z., & McClements, D. J. (2021). Digestibility and gastrointestinal fate of meat versus plant-based meat analogs: An *in vitro* comparison. *Food Chemistry, 364*,130439.

90. Zhou, H., Vu, G., Gong, X., & McClements, D. J. (2022). Comparison of the cooking behaviors of meat and plant-based meat analogues: Appearance, texture, and fluid holding properties. *ACS Food Science & Technology, 2*, 844–851.

Chapter 10

Practical Applications and Dietary Recommendations

Contents

Introduction 308

10.1. Incorporating Plant-Based Alternatives Into A Balanced Diet 308

10.2. Lower Animal to Plant Protein Intake Ratio Can Be Facilitated and Maintained with Plant-Based Meat Alternatives 313

10.3. Animal to Plant Protein Intake Ratios 317

10.4. Perspectives on PBMAs 324

10.5. Plant-Based Meat Alternatives Dietary Guidelines 330

10.6. Comparing Traditional Meat and Plant-Based Meat 333

References 337

INTRODUCTION

To promote balanced nutrition while taking cultural preferences, health concerns, and sustainable practices into consideration, practical applications and dietary recommendations for traditional meats are imperative. Traditional meats with high protein content and vital vitamins and minerals, such as beef, pork, chicken, lamb, and fish, are eaten by many people. However, careful consumption is necessary due to their diverse fat profiles, especially the saturated fat found in processed and red meats.

Depending on dietary needs and preferences, traditional meats can be included in a range of diets. For people looking for high-protein, low-fat options, such as in diets for weight management or muscle building, lean cuts of beef, pork, or chicken are appropriate. On the other hand, heart-healthy omega-3 fatty acids found in fatty fish, such as salmon, make them perfect for cardiovascular health. Fish and game meats are becoming more and more well-liked as lean, sustainable substitutes for commercial meats.

Meats can be braised, grilled, roasted, or slow-cooked in culinary applications, providing flexibility in terms of meal preparation. A balanced meal can be made by serving meats with vegetables and whole grains, or by marinating or grilling the meat to enhance flavor and lower fat content.

According to health guidelines, meat consumption should be moderated, with an emphasis on lean meats and fish and a reduction in processed and red meats because of their associations with chronic illnesses like cancer and heart disease. As part of a heart-healthy diet, the American Heart Association recommends 5–6 ounces (about 140–170 grams) of lean meat per day, and the Mediterranean Diet encourages eating fish twice a week. Combining traditional meats with plant-based proteins can also improve general health.

10.1. INCORPORATING PLANT-BASED ALTERNATIVES INTO A BALANCED DIET

Plant-based substitutes can be a nutrient-dense and satisfying addition to a balanced diet, with several health advantages including better digestion, lowered cholesterol, and a decreased chance of developing chronic illnesses. Here's how to apply them successfully.

10.1.1. Protein Sources

- **Legumes**: Beans, lentils, chickpeas, and peas are excellent sources of plant-based protein and fiber. They are versatile and can be used in soups, stews, salads, and grain bowls.

- **Tofu, Tempeh, and Seitan**: These soy-based and wheat gluten products are high in protein and can substitute meat in many dishes like stir-fries, curries, and sandwiches.
- **Nuts and Seeds**: Almonds, walnuts, chia seeds, hemp seeds, and flaxseeds provide not only protein but also healthy fats and essential vitamins.
- **Plant-based Meat Alternatives**: Products made from pea protein, soy, or other legumes can mimic the taste and texture of meat. Be mindful of their sodium and additive content, and opt for whole-food alternatives when possible.

10.1.2. Whole Grains

- **Quinoa, Brown Rice, and Barley**: These grains are nutrient-dense and rich in fiber, vitamins, and minerals. They help in maintaining energy levels and digestive health.
- **Whole-grain Bread and Pasta**: Swap refined grains for whole-grain options to ensure more nutrients and better blood sugar control.

10.1.3. Vegetables and Fruits

To obtain a range of vitamins, minerals, antioxidants, and fiber, include a variety of fruits and vegetables in your meals. Broccoli, cruciferous vegetables like bell peppers and sweet potatoes, and leafy greens like spinach and kale are good options. Berries, apples, and citrus fruits are good choices for snacks and desserts because they are naturally sweet and packed with nutrients.

10.1.4. Dairy Alternatives

- **Plant-Based Milks**: Almond, soy, oat, and coconut milk are good substitutes for dairy milk. They are often fortified with calcium, vitamin D, and vitamin B12.
- **Non-Dairy Yogurt and Cheese**: Look for alternatives made from almond, cashew, or coconut that are fortified with nutrients.

10.1.5. Healthy Fats

- **Avocados**: Rich in heart-healthy fats, avocados can be used in salads, spreads, or smoothies.
- **Olive Oil and Coconut Oil**: Use these oils in cooking and dressings to add essential fats to your meals.

10.1.6. Micronutrient Considerations

- **Vitamin B12**: Since B12 is primarily found in animal products, you may need to consume fortified foods like plant-based milks, and cereals, or take a supplement.
- **Iron**: Plant-based sources of iron (like spinach, lentils, and quinoa) are less easily absorbed, so pairing them with vitamin C-rich foods like bell peppers or oranges can enhance absorption.
- **Omega-3 Fatty Acids**: Include chia seeds, flaxseeds, walnuts, and algae-based supplements to ensure adequate omega-3 intake.

10.1.7. Balance and Variety

To ensure that a plant-based diet covers all necessary nutrients, a wide variety of foods should be included. You can make sure that you get a balance of carbohydrates, proteins, and fats by combining different food groups in each meal.

10.1.8. Plant-Based Meat Alternatives

Plant-based meat substitutes are food ingredients that are made with substances derived from plants with the specific goal of simulating the flavor, texture, taste, and most importantly, nutritional value of traditional meat. These products provide healthy, ethical, and sustainable alternatives for those who want to cut back on or avoid meat in their diet. Plant-based meat substitutes come in a variety of forms, including burgers, sausages, nuggets, curries, and stir-fries. They are made from soy, peas, lentils, and other plant proteins.

10.1.9. Five Plant-Based Meat Alternatives

10.1.9.1. Tofu and Tempeh

Soybeans, the ingredient in tofu and tempeh, are a fantastic plant-based source of protein that can be added to diets to increase the amount of protein in vegan and vegetarian dishes. In addition, soybeans are a good source of vitamins, minerals like calcium and iron, dietary fiber, and essential amino acids. Tempeh is easier to digest and absorb because of its fermentation process, and it has a higher protein and fiber content than tofu.

10.1.9.1.1. Ways to Add to Diet:

- *Tofu:* Add to soups, stir-fries, or scramble it as a substitute for eggs.

- *Tempeh:* Marinate and grill it, add it to tacos, or slice and add to sandwiches.
- *Side Effects:* A few individuals may be allergic to soy. Excessive consumption may cause bloating or discomfort in some people.

10.1.9.2. Seitan (Wheat Gluten)

Seitan is a fantastic plant-based meat substitute for heart patients because it is low in fat, cholesterol, and protein. Compared to most other options, it provides more protein per serving. In addition, it has vitamins, iron, and selenium, all of which improve physical health and wellbeing. It's a great substitute for animal meat products because it mimics the chewy texture of meat perfectly.

10.1.9.2.1. Ways to Add to Diet:

- *Stir-fries:* Sauté with vegetables and sauces.
- *Sandwiches:* Use in place of deli meat or as a filling in wraps.
- *Baked or Grilled:* Prepare as you would with chicken or beef.
- *Side Effects:* As it is obtained from wheat, it may not be suitable for those suffering from celiac disease or gluten sensitivity. It may also cause bloating or gas in sensitive individuals.

10.1.9.3. Pea Protein Products

The proteins found in peas that contain each of the nine essential amino acids that the body is unable to produce on its own is known as pea protein. It is most well-known that this protein source raises the nutritional value of the diet. In addition, it contains no common allergens like gluten or soy, making it gentle on the stomach.

10.1.9.3.1. Ways to Add to Diet

- *Burgers:* Use pea protein-based patties for a meatless burger option.
- *Smoothies:* Add pea protein powder to smoothies for an extra protein boost.
- *Meatballs:* Incorporate into recipes for vegetarian meatballs.
- *Side Effects:* It may cause mild digestive discomfort in some people.

10.1.9.4. Lentils

Rich in dietary fiber, iron, protein, and other essential nutrients, lentils are a plant-based source of protein. The fiber content improves heart health and lowers cholesterol. Proteins that are adaptable like lentils can be used in a wide variety of delicious dishes.

10.1.9.4.1. Ways to Add to Diet

- *Soups and Stews:* Add to soups for a hearty, protein-rich meal.
- *Salads:* Mix cooked lentils into salads for added texture and nutrition.
- *Lentil Loaf:* Use as a base for a vegetarian meatloaf.
- *Side Effects:* The high fiber content may cause gas or bloating. If not properly prepared, the antinutrient compounds may interfere with nutrient absorption.

10.1.9.5. Jackfruit

Jackfruit is a highly favored substitute for meat. It has abundant supplies of potassium, dietary fiber, vitamins C and A, and other nutrients. According to many studies, the flesh and seeds contain bioactive compounds that may reduce inflammation and prevent cardiovascular disease.

10.1.9.5.1. Ways to Add to Diet

- *Tacos:* Use shredded jackfruit as a filling for tacos.
- *Sandwiches:* Substitute for pulled pork in sandwiches.
- *Curries:* Add to curries for a meat-like texture.
- *Side Effects:*
  - A few individuals may be allergic to jackfruit.
  - Overconsumption can cause digestive discomfort due to high fiber content.

Plant-based meat substitutes have several advantages, ranging from enhanced general well-being to diminished ecological footprint. Including these substitutions in your diet can be a tasty and healthful way to still enjoy your favorite foods without using any animal products. However, be aware of possible adverse effects and select products based on your dietary requirements and preferences.

10.1.10. How Healthy and Safe Are Processed Plant-Based Meat Alternatives?

Processed meat substitutes such as plant-based sausages, burger patties, and nuggets are promoted as a healthier, more sustainable option to meat, but are they all wholesome or safe?

1. **Protein isolated and functionalized:** Proteins are extracted from plants and purified to produce flours, concentrates, or isolates. In some cases, they may be further processed to enhance their functionality.
2. **Product formulation:** Plant proteins are combined with other ingredients (like carbohydrates, lipids, salts, flavors, and colors) to form a plant-based product that has the required appearance, texture, chewiness, flavor, and cookability of a real meat product. In addition, nutrients may be added to match or exceed the nutrient profile of real meat.
3. **Processing:** The mixture of ingredients undergoes a series of processing operations that promote the formation of meat-like structures and properties. These operations may include mixing, extrusion, shearing, molding, and trimming.
4. **Storage:** The product, packaging materials, and environmental conditions must be designed so that the plant-based meat alternative remains safe and of high quality during storage, transportation, and distribution. This requires consideration and control of microbiological, chemical, and physical deterioration mechanisms.

10.2. LOWER ANIMAL TO PLANT PROTEIN INTAKE RATIO CAN BE FACILITATED AND MAINTAINED WITH PLANT-BASED MEAT ALTERNATIVES

Health organizations have been recommending consuming more whole-plant foods for the past few decades. These recommendations include eating more fruits and vegetables, whole grains, and foods high in fiber. Following these dietary recommendations has been linked, to varying degrees, to lower risks of diabetes, cancer, cardiovascular disease, and overall mortality. The advantages of switching from animal to plant protein for health reasons have been highlighted recently. The benefits of this exchange for the environment are also emphasized.

Animal sources account for a significant portion of the protein consumed in developed nations compared to plant sources. In the US, for instance, the animal-to-plant protein intake ratio is roughly 2:1 (54.88 g/d animal protein, 25.88 g/d plant protein), according to NHANES (2015–2018) data. Similar ratios have been reported in many developed nations.

The composition of plant protein i.e., the health advantages of eating a diet higher in plant protein are probably due, at least in part, to the nonprotein components of whole food sources of plant protein (such as amino acids). Thus, it is notable that, according to Perraud et al., one-third of dietary protein intake in France comes from plant sources, with refined grains accounting for 55% of this intake in the United States, where 46% of plant protein intake comes from this source. According to Mariotti and Huneau (2023) and other publications, protein food sources contribute other nutrients that tend to group together as a "protein package," e.g., protein intake profiles essentially define diets. Foods high in plant protein typically contribute fiber and polyunsaturated fatty acids, while red meat typically contributes bioavailable iron and saturated fatty acids.

The ideal ratio of animal to plant protein in the diet has not been determined and may vary depending on the availability of food. The Health Council of the Netherlands recommends a 1:1 animal-to-plant protein ratio, while the EAT-Lancet Commission advises getting the majority of your protein from plant sources. These two recommendations are in line with efforts to reduce, but not entirely eliminate, the dietary intake of animal products, especially meat. They are also more likely to be successful because relatively few people will ever completely eliminate animal protein from their diet, while many may be willing to significantly reduce their intake.

Although meat plays a significant role in most societies' cuisines and is widely regarded as an essential component of meals, current shifts in attitudes regarding the impact of meat consumption on human health, climate change, and animal welfare account for a large portion of the effectiveness of reductions in meat consumption (flexitarianism) and intentions to do so (proflexitarianism). As a result, even adhering to the Dutch recommendation of consuming equal amounts of protein from animal and plant sources will present challenges for populations in developed countries, despite the abundance of easily accessible plant sources of protein.

The hypothesis of research is that the recent wave of plant-based meat substitutes (PBMAs) may be crucial in helping people switch from consuming animal to plant-based sources of protein. Plant-based milk substitutes (PBMAs) have the potential to facilitate not only the shift towards a plant-based diet but also its maintenance on an altered diet. It's not a novel idea that foods high in plant-based macronutrients (PBMAs) can help shift a diet away from an animal-based diet. It is less well-known that PBMAs can help sustain a newly altered plant-based diet in the long run. According to a US study, 10% of adults (≥17 years old) were once vegetarians or vegans, while only 2% are currently in this category. This suggests that PBMAs may have a significant benefit in helping people stick to their plant-based diets. The functionality and orosensory qualities of the meat products that they are meant to replace are intended to be provided to the consumer by recently developed PBMAs. Accordingly, people in developed nations where meat has become a staple of the diet are more likely to accept these products.

It is crucial to weigh the nutritional and health benefits of plant-based meat alternatives (PBMAs) against whole plant foods (legumes, grains, nuts, and seeds) and the meat they replace, given their potential role in the shift to and maintenance of a diet higher in plant protein. The Canada Food Guide is an exception to the general lack of advice on the use of plant-based meat substitutes in dietary guidelines. It states that while many plant-based meat substitutes are highly processed and may contribute excessive amounts of sodium and saturated fat to the diet, it is still important to read food labels and make healthy decisions.

However, there is little evidence of the effect of reducing meat intake on nutrient adequacyein" in the context of the overall diet. As a result, the implications of reducing meat consumption on nutrient intake must be taken into consideration because meat contributes significantly to nutrient intake globally, including in Europe and the United States. That is, it has not been investigated how substituting other particular sources of calories for meat affects nutrient status. No sooner had Dussiot et al. (2022) conducted gradual reduction scenarios for meat intake in incremental steps of 10%. They discovered that certain food groups could both lead to generally healthier dietary patterns and replace meat's contribution to nutrient adequacyein" (i.e., fruits, vegetables, whole grains, and seafood). However, implementing these healthful eating habits necessitates a significant diet overhaul.

Vieux et al. (2022) estimate, that for French adults to meet nonprotein nutrient-based recommendations, roughly half of their total protein intake must come from animal sources. This would be in line with the 1:1 dietary animal:plant intake ratio that the Netherlands' Health Council is pushing. The results, however, seemed to be explained by a marked over-restriction of the possible solutions. Additionally, this estimate excluded the use of dietary supplements or fortified foods and did not take into consideration the types of more novel foods that may be consumed when meat intake is consciously reduced. The authors' optimization model and research strategy have been heavily criticized. Additionally, Fouillet et al. (2023) found that dietary supplementation and/or new food intake are necessary to meet nutrient requirements in adult French people only when plant protein makes up more than 80% of the total protein. Regardless, every diet that was modeled deviated significantly from the diets that are currently being observed, highlighting the need for significant changes to the majority of diets.

Lastly, it is acknowledged that in the near future, the dietary contribution of pasture-raised meat alternatives (PBMAs) will primarily be relevant to developed countries due to their high meat intake and ability to afford the relatively high cost of many PBMAs relative to more conventional meat alternatives like legumes. However, projections indicate that the global demand for animal products will rise by 60–70% by 2050, primarily due to the anticipated increased consumption in low- and middle-income countries. As expected, the price of PBMAs will decrease, and consumers will become more aware of the connection between diet and climate change as well as the relatively low environmental footprint of PBMAs.

This manuscript covers the following general topics: background information on plant-based milk alternatives (PBMAs), current ratios of animal to plant protein intake, the risk of chronic diseases associated with protein intake, the quality of protein and global protein needs, the use of legumes and PBMAs to reduce the animal to plant protein ratio and current knowledge about the health effects of PBMAs.

10.2.1. Historical Perspective on PBMAs

In certain Asian nations, meat alternatives have been around for much longer than in the West, where they have been for over a century. In this sense, notable creations include Dr. John Harvey Kellogg's Nuttose, a canned product primarily made of peanuts, which was introduced in France in 1896. Madison Foods in Tennessee created the first soy-based meat substitute in 1922. Nevertheless, the PBMA industry has been particularly inventive and active over the past 20 years, in part due to increased consumer demand and technological advancements.

People who followed plant-based diets used to get their protein from eating a combination of grains and legumes, which is a common practice in many cultures around the world. Because the indispensable amino acid (IAA) profiles of these food groups complement one another, the result is a protein with an IAA pattern that is more closely aligned with biological requirements. Due to the high-quality protein content of traditional soy foods like tempeh and tofu, which have been consumed for centuries in many parts of Asia, flexitarians, and vegetarians have come to embrace them over the past few decades. Another option is "veggie" burgers, like the black bean burger, which are made with beans and sometimes combined with grains and vegetables that may or may not be vegan. This kind of product appeals particularly to people who want a food that can be eaten similarly to a meat patty. Seitan, a product made from wheat (gluten), serves as a meat substitute as well. Though embraced by vegetarians, the new generation of plant-based meat alternative (PBMA) products are made of protein extracted from legumes like peas and soy and/or wheat gluten, and require new technology for production. However, none of the aforementioned products truly emulate the orosensory properties of meat, which has limited their entrance and acceptance by the mainstream consumer. In fact, US household data show that over a recent 2-year period, 86% of purchasers of PBMAs were consumers of beef.

Since numerous research teams have already published in-depth analyses on the nutritional makeup of a variety of PBMAs, this information won't be covered in-depth here. PBMAs have a wide range of nutrient contents that vary depending on the main protein sources, secondary ingredients (e.g., kind of fat added), and micronutrient addition. The focus of this manuscript is on legume-based PBMAs, or products like patties that are primarily made from extracts of legume protein. It is also acknowledged that meat alternatives made from algae, fungi, and insects are becoming more and more relevant, and that animal protein is being produced by *in vitro* methods and precision fermentation. It is noteworthy that a recent analysis of the French diet found that plant-based substitutes, including legumes, appear more nutritionally adequate to substitute for animal products than other alternatives. Surveys conducted thus far have shown that consumers are more accepting of PBMAs than other types of alternatives.

10.3. ANIMAL TO PLANT PROTEIN INTAKE RATIOS

Interestingly, crude estimates based on disappearance data indicate that in the United States in 1909, the amounts of protein derived from plant and animal sources were similar. However, over the past century, the ratio of animal to plant protein intake has steadily increased due to an increase in animal protein intake and a decrease in plant protein intake. This is in contrast to many developing countries, where the ratio is reversed—in many of them, much less than one-third of dietary protein is derived from animal sources. Take Bangladesh, for example, where 80% of the protein consumed is from plant sources.

10.3.1. Differences Between Animal and Plant Protein

10.3.1.1. Chronic Disease Risk

The majority of the evidence points to the nonprotein components of plant-based diets as the primary source of health benefits, rather than protein in and of itself. Additionally, the concurrent reduction in animal product intake is likely to play a role. Up until now, there has been debate over whether removing red and processed meat from the diet alone lowers the risk of CAD. In any case, the amount of risk reduction is probably determined by what is substituted for meat in the diet. For instance, replacing 5% of energy intake from saturated fats with equivalent energy intake from polyunsaturated fat, monounsaturated fat, or carbohydrates from whole grains was associated with a 25%, 15%, and 9% lower risk of CAD, respectively, according to data from the combined analysis of the Nurses' Health Study (n = 84,628 women) and the Health Professionals Follow-up Study (n = 42,908 men). However, replacing 5% of energy intake from refined starches/added sugars was not significantly associated with risk reduction. However, due to variations in their amino acid profiles, animal, and plant proteins may also differ in ways that impact the risk of chronic diseases. For instance, there is growing evidence that certain amino acids, like methionine and branched-chain amino acids, which are more abundant in animal protein than in plant protein, may have negative effects on metabolism, while certain amino acids, like glutamine, cysteine, and arginine, which are more abundant in plant protein than in animal protein, may have positive effects. Because of the variations in amino acid profiles and in the nutrients and nonnutrients contained in the various protein packages, diets with varying proportions of animal and plant protein may therefore have different metabolic effects. Phytate is an antioxidant that also inhibits the absorption of minerals. Other phytochemicals include isoflavones and glucosinolates, which have the potential to have both positive and negative effects. Animal protein is generally linked to higher risks of type 2 diabetes, cardiovascular mortality, and all-cause mortality, while plant protein is linked to lower risks of these outcomes, according to systematic reviews and meta-analyses of observational data conducted over the past few years.

Lastly, it would be negligent to overlook the fact that vegetarian and particularly vegan diets are deficient in certain nutrients. In this context, calcium, zinc, iodine, and vitamin B12 are noteworthy. Vegan diets necessitate the consumption of fortified foods and/or supplements. On the other hand, following the plant-based diet described here is far less likely to result in deficiencies than following a vegan diet. Furthermore, substituting a PBMA for a serving of legumes does not necessarily increase the likelihood of deficiencies. In fact, as will be mentioned later, PBMAs might be less common because they can be strengthened.

10.3.1.2. Protein Quality

Variations in protein quality are linked to variations in the amino acid profiles of plant and animal proteins. According to the digestible indispensable amino acid score and the protein digestibility corrected amino acid score, animal proteins typically have higher IAA contents and, consequently, higher protein quality scores. These variations are not likely to be clinically significant in populations that consume a combination of plant and animal protein. In fact, it's unclear if followers of plant-based diets need to consume more protein to meet their biological needs for nitrogen and amino acids due to the lower quality of some plant proteins. Currently, there is no US protein RDA for vegans, similar to the one for iron (1.8 versus 1.0 mg). Conversely, some experts—though not all—have suggested that vegans up their protein intake to compensate for the reduced digestibility of plant-based protein. To make up for the lower quality of plant protein, the Dutch Ministry of Health suggests that vegans consume 30% more protein than omnivores; nevertheless, this recommendation is far higher than that of the majority of experts.

10.3.1.3. Global Protein Needs

In three decades, it is expected that there will be nearly 10 billion people on the planet. It is imperative to ascertain the most effective means of fulfilling the protein and caloric requirements of the global populace by the year 2050. As stated in a recent report from the Stockholm Resilience Centre, "Meeting the demand for protein, within environmental limits, is one of the biggest challenges for the global food system in the 21st century," protein is by far the more significant factor.

It is unclear exactly how much protein will be required, but Irish researchers put forth five scenarios to try and find an answer. At the highest end of the spectrum, it was estimated that to meet the world's protein needs, protein production would need to increase by 78%, assuming that everyone would consume roughly 100 g of protein per day. At the low end, production could drop by 13% without affecting the availability of protein, assuming the population will only consume 50 g/d of protein. If availability permits, populations tend to consume approximately 16% of their calories in the form of protein, which is much greater than the approximately 10% of calories needed to meet the RDA. Although the types of protein consumed vary markedly among cultures and societies, the high estimate may be the more likely scenario.

Lastly, evidence suggests that over the next 30 years in developed countries, there is likely to be a shift toward a plant-predominant diet. This raises the question of how the world will produce the calories and plant protein needed by 2050 for a variety of reasons, including concern over the environment and animal welfare as well as personal health considerations. Global soybean production is sufficient to meet the need if some of the soy protein currently used for animal feed is instead used for direct human consumption. In contrast, if the plant protein is to come from non-soy legumes, production would need to significantly increase to keep up with demand.

10.3.1.4. Shifting the Animal to Plant Protein Intake Ratio

One obvious way to increase your intake of plant protein is to eat more legumes. In 2013, the United Nations proclaimed 2016 as the International Year of Pulses. The International Lipid Expert Panel rates soy, legumes, and nuts as the top (desirable) sources of protein for promoting cardiometabolic health. Legumes, which include beans and other pulses, are inexpensive, high in fiber, and low in environmental impact. Nevertheless, despite their many benefits, pulses are largely underutilized as food sources in most parts of the world. Pulses make up only 6% of the total protein consumed worldwide (Table 10.1).

Table 10.1. *World and Selected Region Pulse Intake*

Region	g/d	% Total Intake	
		Protein	kcal
World	21	6	3
Latin America and Caribbean	34	9	4
Sub-Saharan Africa	33	12	5
South Asia	33	11	5
North Africa	19	5	2
West Asia	19	6	3
Oceania	12	2	1
North America	11	2	1
Southeast Asia	9	3	1
Europe	7	2	1
East Asia	4	1	0
Caucasus and Central Asia	1	0	0

Source: Messina M, Duncan AM, Glenn AJ, Mariotti F. Perspective: Plant-Based Meat Alternatives Can Help Facilitate and Maintain a Lower Animal to Plant Protein Intake Ratio. Adv Nutr. 2023 May;14(3):392-405. doi: 10.1016/j.advnut.2023.03.003. Epub 2023 Mar 10. PMID: 36906147; PMCID: PMC10201680.

Only about 2% of total protein and 6% of plant protein consumed in the United States comes from eating dry beans. A recent study conducted in France revealed that even knowing about the health advantages of pulses did not encourage people to eat more of these foods (Reuzé et al., 2022). A recent summary of the factors linked to increased consumption of legumes among adult French people. Furthermore, the French public health authorities did not incorporate a specific guideline on legumes in the dietary guidelines until very recently (2017).

Pulses are the edible seeds of the legume plant, despite the fact that the terms legumes, pulses, and beans are frequently used synonymously. Beans, lentils, chickpeas, dry peas, and cow peas are examples of pulses. Pulses are not the only foods classified as legumes; soybeans, peanuts, snap beans, and snap peas are also included. Because of their high-fat content, soybeans, and peanuts are known as oilseed legumes. Countries and regions like Mexico (refried kidney beans), India (dhal and pappadums), the Mediterranean (navy bean soup and Greek fava), and the Middle East (falafel and hummus) have historically had higher percentages of legume in their cuisines.

Table 10.2 shows that, on a calorie basis, legumes contain about 29% protein, which is roughly twice as much as grains' 13% protein. Additionally, on average, legumes have more fiber than grains. If one were to replace the protein in one serving of cooked beans with an equivalent amount of animal protein each day, the animal-to-plant protein intake ratio would decrease from roughly 2.1:1 to 1.3:1 (for the US). As a result of this exchange, 8.6 g of plant protein would be consumed in addition. This would bring the total amount of plant protein consumed in the United States to 25.8 g/d. Deducting 8.6 g/d from the 54.8 g/d of animal protein consumed in the country would leave 46.2 g/d. As a result, the ratio of animal to plant protein in the diet would be 1:3 (46:2:34). However, fulfilling the current US Dietary Guidelines' recommendation for the amount of cooked beans, peas, and lentils—three servings per week, or 1.5 cups cooked—would obviously have a much smaller impact. If the approximate 3.7 g/d of protein from these plant foods were substituted for an equivalent amount of animal protein, the ratio would only drop from 2.1:1 to about 1.7:1 (51.1:29.5). Note that one serving of beans is defined by the US Department of Agriculture as a one-half cup or roughly 90 g; however, in a research paper, one serving is defined as 100 g (based on research by Marinangeli et al., 2017).

Table 10.2. *Protein and Fiber Content of Selected Legumes (per 100 g Cooked) From the USDA's Food Data Central*[1]

Legume	USDA FDC ID	Protein		Fiber
		(g)	% kcal	(g)
Soybeans	174271	18.21	42.35	6.0
Lupin	173804	15.57	53.69	2.8

Lentils	175254	9.02	31.65	7.9
Pinto beans	175200	9.01	25.20	9.0
Great Northern beans	173790	8.33	28.24	7.0
Kidney beans (red)	175242	8.67	27.31	7.4
Black beans	175237	8.86	26.85	8.7
Mung beans	175255	7.02	26.74	7.6
Peas (green)	170102	5.36	25.52	5.5
Navy beans	173794	8.23	23.51	10.5
Adzuki beans	173789	7.52	23.50	7.3
Lima beans	169316	6.81	22.15	5.3
Garbanzo beans	173799	8.86	21.61	7.6

[1]U.S. Department of Agriculture, Agricultural Research Service. Food Data Central, 2019. fdc.nal.usda.gov.

Source: Messina M, Duncan AM, Glenn AJ, Mariotti F. Perspective: Plant-Based Meat Alternatives Can Help Facilitate and Maintain a Lower Animal to Plant Protein Intake Ratio. Adv Nutr. 2023 May;14(3):392-405. doi: 10.1016/j.advnut.2023.03.003. Epub 2023 Mar 10. PMID: 36906147; PMCID: PMC10201680.

10.3.1.5. Is the Direct Consumption of Pulses Likely to Increase?

The likelihood that Americans and other Western populations will eat one serving of cooked pulses per day is high. Although there has been a call for increased public health initiatives to promote pulse consumption, there is little proof that these initiatives, if they are carried out, will be effective. Just 4% of respondents in NHANES (2013–2014; n=6048) reported consuming legumes on both survey days. This percentage was less than the 5.6% of respondents who reported consuming legumes on both survey days in the preceding two years. Even more recently, Tao, and co. (2022) found that the average daily intake of legumes among the 4058 participants in NHANES 2017–2018 was only 0.11 servings, which was not higher than the data from NHANES 2011–2012 (n = 4313, 0.12 servings/d). Data from the Continuing Survey for Individual Intakes (1994–1996) also show a downward trend in the consumption of legumes. Specifically, only 4.9% of participants in the 2017 Beans, Lentils, Peas Survey reported consuming beans on both survey days. This difference is even more striking when one considers that, based on grocery purchases, the US mean annual per capita expenditure on legumes was only $4.76 during 2017–2019.

Soybeans and peanuts are not classified as legumes in NHANES due to their higher fat content. Table 10.3 shows that while a serving of peanut butter (2 tbsp) has significantly more calories (~189 versus ~127), it contains nearly as much protein (7.27 g versus 8.62 g). Furthermore, male vegans were found to consume only about 65 g/d of legumes (excluding soy), or roughly three-quarters of a serving, according to the Adventist Health Study 2. However, they also consumed an average of 207 g/d of soy foods.

Table 10.3. *Energy, Macronutrient, and Fiber Content of Legumes and Foods Made from Legumes*[1]

	Pulses (*N* = 29)	**Tofu (*N* = 10)**	**Peanut butter (*N* = 3)**	**Peanuts (*N* = 3)**	**Soy-PBMAs (*N* = 5)**
Energy (kcal)	127 ± 14	100.10 ± 24.86	189.00 ± 1.73	161.33 ± 0.94	198 ± 83.28
Protein (g)	8.62 ± 1.54	10.95 ± 2.53	7.27 ± 0.38	7.23 ± 0.14	18.8 ± 5.08
% kcal protein	27.1	43.8	15.4	17.9	38.0
Fat (g)	0.66 ± 0.62	5.76 ± 1.91	16.07 ± 0.31	13.8 0.24	11.2 ± 6.97
Carbohydrate (g)	22.55 ± 3.49	2.20 ± 0.98	7.24 ± 0.40	5.03 ± 0.64	8.2 ± 2.23
Fiber (g)	7.4 ± 2.0	1.07 ± 0.60	1.99 ± 0.50	2.52 ± 0.12	4.2 ± 2.04

Nutrient database number for tofus: 16160, 16281, 16159, 16277, 16276, 16212, 16211, 16426, 16427, 16213; Peanut butter: 16097, 16398, 16167 and peanuts; 16091, 16093, 16095

PBMA, plant-based meat alternative.

[1]Values = mean ± standard deviation. Serving sizes: pulses, 100 g cooked tofu, 100 g raw; peanut butter, 2 tablespoons (32 g); peanuts, 1 oz (28 g); soy-PBMA (burgers, 1 patty).

Sources: Values for tofu, peanut butter, peanuts from U.S. Department of Agriculture, Agricultural Research Service. Food Data Central, 2019. fdc.nal.usda.gov. All tofus except one either hard or firm. Values for pulses and soy-PBMAs come from references and, respectively.

Source: Messina M, Duncan AM, Glenn AJ, Mariotti F. Perspective: Plant-Based Meat Alternatives Can Help Facilitate and Maintain a Lower Animal to Plant Protein Intake Ratio. Adv Nutr. 2023 May;14(3):392-405. doi: 10.1016/j.advnut.2023.03.003. Epub 2023 Mar 10. PMID: 36906147; PMCID: PMC10201680.

Intake of legumes is likewise low in the UK. Regular meat eaters (n = 2852) reported consuming only 26.7 g/d of pulses/legumes when standardized for an intake of 2000 kcal in the Oxford component of the European Prospective Investigation into Cancer and Nutrition. 68.4 g/d is still less than one serving, but vegans (n = 269) also consumed 59.6 g/d of vegetarian protein substitutes (e.g., Quorn, tofu, soymilk, and soy burgers), which is significantly more than the 4.9 g/d ingested by typical meat eaters.

The idea that beans are "poor man's food" or "poor man's meat," worries about antinutrients, the fact that beans are not a part of the traditional diet, worries about flatulence and abdominal discomfort, ignorance about preparation and cooking, disliked taste and texture, and high carbohydrate content are some of the obstacles to bean consumption in Western countries. A lot of work will be required to improve the general public's intake of whole legumes, such as more public education campaigns, alterations to the food environment, and the serving of more dishes based on legumes in hospitals, restaurants, and schools. However, consuming PBMAs made from legumes is a viable alternative if protein intake from the consumption of cooked legumes served traditionally is unlikely to increase significantly. Crucially, and as previously mentioned, PBMAs are a complement to legumes rather than an alternative; in other words, they are not either one or the other. You can use both of these foods to increase your intake of plant-based and legume proteins.

10.4. PERSPECTIVES ON PBMAS

Bianchi et al. (2022) recently provided evidence of the potential impact of meat substitutes in lowering dependence on and altering attitudes toward meat. According to these authors, a 4-week behavioral program that gave British study participants PBMAs, recipes, and information about the health benefits of eating less meat—as well as success stories of people who had cut back on their meat intake—spurred changes in psychosocial constructs that were consistent with a sustained reduction in meat intake and a decrease in meat intake.

There is ample evidence that PBMAs can significantly lower the ratio of animal to plant protein intake. As stated by Messina et al. In 2022, five soy-based patties had a protein content ranging from 14 to 27 g each, which is significantly higher than the average serving of pulses (~8.6 g). Even a conservative estimate suggests that replacing animal protein in the diet with just four servings (four patties) of soy-based patties per week, or roughly 75 g protein, would reduce the US animal-to-plant protein intake ratio to about 1:2:1. One could argue that because PBMAs have a higher protein content than legumes, they can help people switch to a diet that is more similar in terms of the amount of animal and plant protein they consume.

10.4.1. Skepticism About the Utility of PBMAs

Due to their growing popularity, PBMAs are being examined more closely for their nutritional content and their place in the diet. For instance, Toh et al. (2022) stated that, when comparing plant-based diets (PBBDs) with PBMAs, "there is a gap in our understanding of the long-term impacts of dietary patterns that characteristically feature PBMAs." In this instance, a balanced intake of grains, legumes, nuts, seeds, fruits, and vegetables is referred to as the PBDs. Fardet et al. (2018) assert that food degradation, such as that which happens when protein is extracted from beans, generally has a negative effect on the food's healthfulness, irrespective of the food's calorie and nutrient content. Additionally, according to correlations between the consumption of ultra-processed foods (UPF) and unfavorable health outcomes, Gehring et al. (2021) postulated that consuming more processed plant-based foods could mitigate or eliminate their potential health advantages.

10.4.2. Concerns About the Processing Involved in the Production of PBMAs

The processed nature of plant-based microwaveable meals (PBMAs) was brought to light by Macdiarmid, who concluded that the trend toward consuming more highly processed plant-based convenience foods raises questions for public health and the achievement of greenhouse gas emission reduction targets. In France, replacing meat with plant-based meat alternatives (PBMAs) was found to increase the percentage of energy from UPFs from 29% in observed diets to 34% on average and up to 40% depending on the type of substitution in modeled diets. This is about Macdiarmid's comment, which states that while it may be more convenient to eat a soy-based patty than to boil dry beans, it may not be more convenient than ground beef or canned beans. Furthermore, studies show that the environmental impact of plant-based burgers is minimal.

Since soy-based meat alternatives (or soy-based dairy alternatives) were all NOVA-classified as UPFs, the extensive discussion surrounding the NOVA food classification system and the risks associated with consuming foods classified as ultra-processed (UPFs) casts a negative light on the entire new generation of plant-based milk alternative (PBMA) products. However, a recent analysis found that none of the numerous common criticisms of UPFs applied to soy-based meat alternatives (or dairy products) as they were all NOVA-classified as UPFs, more so than they did on their animal-based counterparts, beef, and cow's milk, which are classified as unprocessed/minimally processed foods. The authors of this analysis addressed the following criticisms: (1) hyperpalatability; (2) high energy density; (3) increased energy intake rate; (4) low satiation; (5)

low cost/high snackability; and (6) high glycemic index (kinds of milk only). One issue that the authors did not address is the possibility that UPFs cause unfavorable changes in the microbiome; however, this does not seem to be the case for PBMAs.

In addition, several nutrient profiling models other than NOVA give PBMAs high ratings. For instance, Australian researchers recently discovered that the average Health Star rating for 50 plant-based burgers was 3.9 (5-point scale), while a meat burger received a rating of 2.9. According to a French analysis, plant-based meat and dairy substitutes had varying effects on nutrient security and adequacy, and they did not affect the overall quality of the French diet. Crucially, as was previously mentioned, it was discovered that plant-based alternatives, such as legumes, seem more nutritionally adequate to replace animal products than other alternatives, like cereal-based substitutes.

10.4.3. Clinical Data Supportive of PBMAs

The Study with Appetizing Plant Food-Meat Eating Alternative Trial (SWAP-MEAT) clinical trial provides the most direct information about the comparison between meat and PBMAs. For eight weeks, each participant (n = 36) received approximately 2.5 servings per day of PBMAs or similar meat products, accounting for 25% of total calories and 50% of total protein. The design was randomized and crossover. When compared to meat products, the results indicated that PBMA consumption significantly reduced body weight, circulating levels of trimethylamine oxide, and low-density lipoprotein cholesterol. Additionally, no differences in inflammatory markers were seen between the groups, and no negative effects were seen on any of the other endpoints that were examined.

The findings of the SWAP-MEAT Athlete study, which involved 24 athletes—12 recreational runners and 12 resistance trainers—who were randomized to three different diets—whole foods plant-based, plant-based with PBMAs, and an omnivorous diet—for four weeks each, are also noteworthy. As compared to a whole foods plant-predominant diet and an omnivorous diet, there were no differences in the running or lifting outcomes for the resistance trainers or runners at the end of the study, indicating that the inclusion of PBMAs in a plant-predominant diet may not affect athletes' fitness outcomes.

Furthermore, 164 individuals with type 2 diabetes participated in a recent 3-month parallel-arm clinical trial that examined the effects of a low-carb vegan diet versus a moderate-carb vegetarian diet; however, this trial did not directly compare the effects of meat and PBMAs. A lower carbohydrate content was attained by increasing the amount of fat (from canola oil) and protein in the vegan diet. Protein accounted for about 25% of total calories consumed; almost all of that protein was derived from plant sources, primarily from soy-based meat substitutes. Both diets resulted in significant drops in blood pressure, blood lipids, glycated hemoglobin, and body weight at the end of the study, with no side effects noted.

10.4.4. Legumes Compared with Legume-Based PBMAs

As mentioned earlier, PBMAs can be thought of as foods that help people make the shift to a plant-based diet by acting as gateway foods or transition foods. As Alae-Carews et al. (2022) put it, PBMAs are a crucial "stepping stone" for dietary modification. This viewpoint implies that while PBMAs are healthier than the meat they substitute, they may not be as healthy as less processed plant protein sources like whole grains and legumes. Many researchers have wondered if PBMAs made from legumes have the same nutritional advantages or ability to prevent chronic diseases as whole legumes. Van Vliet and associates., going one step further, declared in 2020 that "It is likely understated the true nutritional complexity of whole foods in their natural state, which contain hundreds to thousands of nutrients that impact human health when animal foods are mimicked using isolated plant proteins, fats, vitamins, and minerals."

Legumes indeed contain a variety of phenolic compounds in addition to nutrients and fiber, which may contribute to their health benefits. Studies show that eating legumes lowers the risk of colorectal cancer and all-cause mortality. Clinical evidence also suggests that pulses may help with weight management and glycemic control. It is unknown, though, how much bean phenolics contribute to these correlations.

There are few clinical data available for the new generation of PBMAs, and they haven't been commercially available long enough for observational studies to look into their intake. Conversely, the main proteins found in these products—soy and, to a lesser degree, pea—have undergone extensive testing in the same processed form (e.g., as that used in PBMAs (soy protein isolate and soy protein concentrate). For instance, soy protein is a premium protein that functions similarly to animal proteins, such as whey, in terms of promoting gains in muscle mass and strength and lowering blood cholesterol levels. Additionally, one feature of PBMAs mentioned by van Vliet et al. (2020) as a drawback—the use of isolated components—may, in fact, be a benefit when compared to eating legumes unprocessed (raw), since the manufacturing process makes it possible to fortify deficiencies in nutrients in diets that are primarily plant-based and animal-based, and can be customized to specific populations if necessary.

For instance, it is possible to imagine that PBMAs could be used to fortify a nutrient like iron in populations where the intake of that nutrient is minimal. When substituting plant protein for animal protein, industry could be encouraged to fortify their products with nutrients that could be of concern. Interestingly, a product must have the same nutritional value as the meat product it is meant to replace to be classified as a "simulated meat product" in Canada. A recent study that optimized the recipe for a legume-based patty to enhance nutritional content while accounting for technological limitations and applying nutritional constraints to limit the risk of overt deficiency in 12 essential nutrients is relevant to this discussion. Even in this instance, iron, zinc, and vitamin B_{12} fortification were still necessary.

Another benefit of plant-based meat alternatives (PBMAs) could be that when the protein is extracted from the beans, the amount of oligosaccharides—a component of legumes that can cause flatulence in some individuals—significantly decreases. The removal of substances that may hinder protein digestion can also improve the digestibility of the extracted protein. Additionally, compounds like phytate, which have historically been classified as antinutrients, can be removed or reduced through processing. Although PBMAs contain a range of nutrients, some varieties tend to be relatively high in sodium, even though they do not meet the U.S. Food and Drug Administration's (FDA) definition of a high-sodium food (≥400 mg per serving). Furthermore, to replicate the orosensory qualities of meat, some PBMAs have higher levels of saturated fat (although still lower than that found in beef).

Sodium can be added to plant-based ingredients to enhance their flavor and to cover up unwanted flavors like chalky and beany that are common in plant-based proteins. It is crucial to remember, though, that there might not be much of a sodium difference between PBMAs and beef as they are normally consumed because salt is frequently added to ground beef during cooking to improve texture and flavor. It is reasonable to assume that future PBMA iterations will be able to preserve or improve current orosensory properties while reducing sodium and saturated fat content given the recent improvements in taste and texture of PBMAs. As a matter of fact, the most recent iteration of a well-known soy-based burger has 8 grams of saturated fat instead of 6 grams per serving.

The possibility that PBMAs will be eaten in a manner akin to that of hamburgers—with a refined bun and maybe some fries or chips—is one reason for concern. While this is understandable, PBMAs come in a variety of forms besides patties these days. Products that mimic steak, chicken, and seafood, for instance, are either already on the market or will be shortly. As a result, there are now far more applications for PBMAs than there were in the past.

10.4.5. Hybrid Meat Products

Hybrid meat products—a term for products that combine animal and plant protein—might be a particularly alluring and successful strategy for reducing meat consumption because research indicates that dietary changes should not deviate too much from consumers' prior habits to be effective. These products lessen the amount of meat in the formulation while maintaining the flavor of the animal. For flexitarian consumers, these products offer convenience and a more distinctive flavor. In contrast to plant-based ingredients found in meat products, which are frequently added for their functionalities as binders, extenders, and fillers, the goal of hybrid meat products is to replace a portion of the meat with a more sustainable source. Since manufacturers have a lot of experience with these combinations, soy protein ingredients may help hybrid meat products succeed if they grow in popularity. This probably applies to other proteins found in legumes as well.

Research suggests that PBMAs are an easy, wholesome, and long-term solution for reducing the ratio of animal to plant protein intake in developed nations. It is unlikely that consuming fewer nutrients will be hampered by occasionally substituting PBMAs for meat, such as beef, especially if the PBMAs are fortified with micronutrients that are lacking. According to a conservative estimate, cutting the animal-to-plant protein intake ratio from roughly 2.1:1 to 1.2:1 only requires replacing 4 servings of meat per week with 4 servings of soy-based PBMAs. This is a reasonable goal considering the glacial pace at which dietary habit changes typically occur. More study on these goods will be required as new PBMAs are created to better comprehend their place in a balanced diet.

Subsequent versions of PBMAs ought to prioritize the enhancement of their nutritional composition while preserving their orosensory qualities, which render them advantageous choices for people transitioning from an animal-based diet to one that is more plant-based. In a way, it doesn't matter how healthy legumes are compared to PBMAs if there isn't a chance that bean intake will rise significantly; nonetheless, efforts to increase intake should still be made.

To effectively counsel patients and clients about the potential role that these foods can play in the diet, health professionals must possess a thorough understanding of PBMAs. It is important for consumers and health professionals to understand that PBMAs have significantly different nutrient contents. There is variation in plant-based kinds of milk as well; for instance, some have low protein content, and not all of them are fortified with essential nutrients like calcium and vitamin D. Recently, it was suggested that dietary guidelines should distinguish between plant-based substitutes that are suitable for regular consumption and those that need to be consumed in moderation or just for pleasure" (Klapp et al., 2022).

Legumes were recommended for vegans to eat at least three times a day in a recently released dietitian's guide on vegetarian diets. Although dry beans and pulses are both classified as legumes, soyfoods should be considered a distinct category due to the high-fat content of soybeans and the variety of ways they can be consumed. Peanuts (and peanut butter) can also be seen similarly to soy because of their high-fat content. Legumes-based PBMAs can include items from the categories of pulses/dry beans and soy/peanuts. Therefore, there are three different ways to include legumes in your diet: That is peanuts, foods made from soybeans, dry beans and pulses, and PBMAs.

Customers should think about selecting items from each of these three categories of legumes for convenience, variety in their diets, and improved adherence to plant-based diets by increasing the variety and facilitating adequate nutrient intake. When restricting calories, it is important to recognize that many PBMAs have a significantly higher calorie content per serving than traditional soyfoods like tofu and pulses. Consuming foods from the three categories of foods based on legumes can help you meet the recommended intake of legumes, but it will depend on a variety of personal and health-related factors, such as your overall lifestyle, economic status, and cultural beliefs.

PBMAs offer omnivores a practical way to switch to a diet with a lower proportion of animal protein to plant protein. PBMAs are probably going to become more and more commonplace as more of these products hit the market and their prices fall as predicted. Nutrient-dense, well-designed formulations of plant-based milk alternatives (PBMAs) should be able to fill in nutritional gaps for followers of plant-based diets. In addition, for many people, the increased availability of plant-based foods high in protein will probably improve adherence to the diet. Lastly, while a more comprehensive diet redesign that incorporates more conventional, minimally processed plant protein sources with demonstrated nutritional and health benefits should continue to be the main objective, PBMAs are helpful—indeed, practically indispensable—in helping a large portion of the population make the shift to a more plant-based diet and are likely to make sticking to it easier.

10.5. PLANT-BASED MEAT ALTERNATIVES DIETARY GUIDELINES

It's crucial to adhere to specific rules when adding plant-based meat substitutes to your diet to make sure they support a healthy, well-balanced diet. These plant-based substitutes, which are made from grains, legumes, or soy, are intended to resemble meat in both flavor and texture. But their nutritional value varies, so knowing how to pick and use them wisely is essential.

10.5.1. Nutritional Quality

- **Protein Content**: Look for plant-based meats that have a high protein content, similar to animal-based meats (about 15–20 grams per serving). Common protein sources in these alternatives include soy, peas, or wheat gluten.
- **Fiber**: Many whole-food plant-based alternatives, like lentil burgers, contain fiber, which is an advantage over traditional meats. Aim for options with at least 3–5 grams of fiber per serving.
- **Sodium Levels**: Some processed plant-based meat alternatives can be high in sodium due to flavor enhancers and preservatives. Choose products with lower sodium levels (under 400 mg per serving) or those labeled as "low-sodium."
- **Healthy Fats**: Check for healthy fat sources like canola, avocado, or olive oil, as some products may contain higher amounts of saturated fats from coconut oil or palm oil. Aim for less than 4 grams of saturated fat per serving.
- **Additives and Preservatives**: While many plant-based meats are minimally processed, others may contain additives like colorings, flavorings, and preservatives. Whenever possible, choose products with fewer artificial ingredients.

10.5.2. Frequency of Consumption

- **Limit Ultra-Processed Options**: While plant-based meat alternatives can be convenient, they are often processed, which may reduce their nutritional value compared to whole food sources. Aim to limit consumption of highly processed versions and prioritize whole-food alternatives, like bean burgers or homemade lentil-based patties.
- **Balance with Whole Foods**: Incorporate these alternatives as part of a diet rich in whole grains, vegetables, legumes, and nuts. Whole-food options are more nutrient-dense and can provide essential vitamins, minerals, and fiber that processed alternatives may lack.

10.5.3. Fortification and Supplementation

- **Vitamin B_{12}**: Some plant-based meats are fortified with vitamin B_{12}, an essential nutrient found naturally in animal products. If you're relying heavily on plant-based meats in your diet, ensure they are fortified with B_{12}, or consider a supplement to meet your needs.

- **Iron and Zinc**: Like B_{12}, plant-based sources of iron and zinc are less easily absorbed than their animal-based counterparts. Opt for fortified versions of plant-based meats or consume them alongside vitamin C-rich foods to enhance absorption.

10.5.4. Serving Sizes and Portion Control

- **Portion Awareness**: Plant-based meats can sometimes be calorie-dense, especially when made with added fats or oils. Stick to recommended serving sizes (typically around 4 oz or one patty) to avoid consuming excess calories or fat.
- **Balance with Other Nutrients**: Ensure your meals with plant-based meats are balanced with other nutrient-dense components like whole grains, vegetables, and healthy fats, rather than relying solely on the meat substitute for nutrition.

10.5.5. Sustainability Considerations

- **Environmental Impact**: One of the key reasons for choosing plant-based meat alternatives is their reduced environmental footprint compared to animal-based meats. However, not all products have the same level of sustainability, so opt for brands that prioritize organic, non-GMO ingredients, and minimal processing.

10.5.6. Types of Plant-Based Meat Alternatives

- **Soy-Based**: Soy is a popular source of protein in plant-based meats. Products like tofu, tempeh, and soy burgers are typically less processed and nutrient-dense. Be mindful of soy allergies and try to choose non-GMO, organic soy when possible.
- **Pea Protein-Based**: Many newer plant-based meats are made with pea protein, which is hypoallergenic and offers a high protein content. These can be a great alternative for those with soy or gluten sensitivities.
- **Grain- or Legume-Based**: Some alternatives are made from lentils, beans, or grains like quinoa. These options are often less processed and rich in fiber, making them a healthy alternative to highly processed soy or pea protein-based products.

10.5.7. Mindful Preparation

- **Cooking Methods**: Avoid deep-frying or heavily processing plant-based meats during cooking, as this can negate their health benefits.

Opt for grilling, baking, or sautéing them with minimal oil to maintain their nutritional value.

- **Pairing with Nutrient-Dense Foods**: Combine plant-based meats with nutrient-dense side dishes like roasted vegetables, whole grains, and salads to create a more balanced meal.
- Plant-based meat substitutes can be a part of a balanced, sustainable, and healthful diet by adhering to these dietary guidelines.

10.6. COMPARING TRADITIONAL MEAT AND PLANT-BASED MEAT

Plant-based meat substitutes have drawn a lot of attention lately. Some people assert—or misunderstand—that plant-based meat substitutes are just as nutritious as conventional meat substitutes, if not more so (Seehafer and Bartels, 2019). Both are well-known to be good sources of protein, but they differ significantly in terms of nutrition.

10.6.1. Plant-Based Protein

Plant-based meat substitutes are nothing new; veggie burgers and soy-extended burgers have been around for a while (Curtain and Grafenauer, 2019). However, due to issues with texture and flavor, consumers did not find these products to be very popular.

Figure 10.1. *The beyond burger® is a newer plant-based product.*

Source: By Pascal Shirley – Beyond Meat, CC BY-SA 4.0, https://commons.wikimedia.org/w/index.php?curid=110020263.

Though they never gained the same level of media attention as the current plant-based burgers, some products made with wheat protein have also fairly successfully replicated the texture of meat. In addition, some people experienced problems with the gluten present in the wheat protein. The more recent plant-based goods (like Beyond Meat®, and Impossible™ Burgers) are becoming more and more well-liked and are getting much closer to replicating the flavor and texture of conventional meat products (Figure 10.1).

10.6.2. Newer Plant-Based Meat Options

The producers of contemporary plant-based meat substitutes have done a fantastic job of balancing the macronutrient composition of their products with that of conventional meat substitutes. Moisture, fat, and protein are examples of macronutrients. However, they are using highly processed protein products (e.g., fermented, hydrolyzed, separated, extracted, etc.) from entire vegetables. How these processing techniques affect the nutritional effects (i.e., nutritionists are dubious because it is unknown how these vegetable proteins will digest (Curtain and Grafenauer, 2019). It is advised by nutritionists that people eat more vegetables, but not processed, plant-based meat substitutes (Table 10.4).

Table 10.4. *Level of Characterization of Public Health, Environmental, and Animal Welfare Implications in Research to Date on Meat Alternatives*

Implication category	Topic	Plant-based substitutes	Cell-based meat
Public health	Nutrients	Moderate	None
	Chronic disease risk	Limited	None
	Food safety	Limited	None
	Occupational health	None	None
	Community health	None	None
Environmental	GHG emissions	Moderate	Limited
	Water use	Limited	Limited
	Land use	Moderate	Limited
	Nutrient runoff	Limited	Limited
	Pesticide use	Limited	None
	Biodiversity	Limited	None
Animal welfare		Limited	Limited

This table classifies the extent to which key public health, environmental, and animal welfare implications explored in this paper have been characterized in existing academic literature on meat alternatives. Recognizing that many studies report or review secondary data, and that a significant proportion of the available environmental impact research on meat alternatives is from gray literature, designations were made based both on counting of available studies related to a specific product and implication as well as a qualitative judgment of the depth and quality to which these topics have been covered in the literature. Topics were classified based on the following criteria: none (no specific research on this product and implication), limited (covered by only a few studies), moderate (some research; more is needed); high (thoroughly researched).

Source: Santo RE, Kim BF, Goldman SE, Dutkiewicz J, Biehl EMB, Bloem MW, Neff RA and Nachman KE (2020) Considering Plant-Based Meat Substitutes and Cell-Based Meats: A Public Health and Food Systems Perspective. Front. Sustain. Food Syst. 4:134. doi: 10.3389/fsufs.2020.00134.

10.6.3. Nutrition of Traditional Meat Versus Plant-Based

Compared to plant proteins, animal proteins have a more optimal ratio of essential amino acids. Since essential amino acids are not produced by humans, adequate diet intake is necessary for human growth and maintenance. The balance of essential amino acids provided by meat proteins is particularly crucial for the healthy growth and development of young children. To make proteins for our growth and maintenance, we need essential amino acids. Certain proteins cannot be made if an amino acid is absent. This has an adverse effect on the development and upkeep of our bodies. Compared to plant proteins, meat proteins have a higher nutrient density due to their high quantity and quality. Iron is also abundant in meat proteins. Compared to iron in vegetables, the iron in meat is far more soluble. The only natural source of vitamin B_{12} is meat.

10.6.4. Key Ingredients in Both Products

Isolated soy protein (Impossible Burger) and isolated pea protein (Beyond Meat) are two examples of the vegetable proteins utilized in contemporary plant-based products. Methionine and lysine are among the amino acids that soy protein lacks, and it is included in the list of ingredients that some people should avoid due to allergies. Methionine is also absent from pea protein. Additionally, unprocessed pea protein is far less digestible than soy or meat proteins. Modern plant-based meat substitutes have a comparable fat content to conventional meat products aside from protein. Vegetable oils are added to the vegetable proteins to achieve that fat content. Saturated fatty acids make up 80% of coconut oil, which is frequently used in Beyond Meat and Impossible Burger products. A diet high in unsaturated fatty acids is considered to be more healthful than one high in saturated fatty acids, according to certain studies. It is known that 50% of saturated fatty acids are present in beef fat (Figure 10.2).

Figure 10.2. *Soybeans provide isolated soy protein for modern plant-based meat products, such Japanese soyabean meat.*

Source: By DryPot – Own work, CC BY 2.5, https://commons.wikimedia.org/w/index.php?curid=16989887.

Commercial plant-based burgers are made with 18–20 ingredients that are meant to mimic the flavor and texture of traditional meat products. There is only beef in traditional ground beef. Plant-based burgers have a fairly high sodium content. But unlike traditional meat products, plant-based foods also contain fiber. The one ingredient that some would argue is detrimental to traditional meat products is their cholesterol content, especially for those who already have high cholesterol (Figure 10.3).

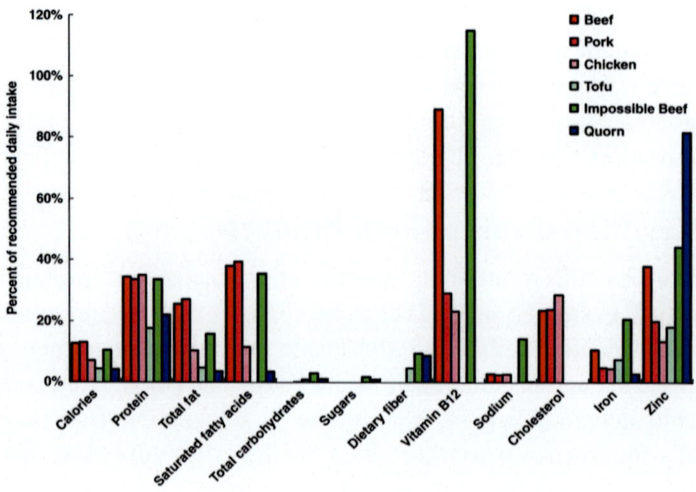

Figure 10.3. *A nutritional comparison between the Impossible burger and a traditional beef burger.*

Source: Rubio, N.R., Xiang, N. and Kaplan, D.L. Plant-based and cell-based approaches to meat production. Nat Commun ***11****, 6276 (2020). https://doi.org/10.1038/s41467-020-20061-y.*

REFERENCES

1. Agarwal, S., & Fulgoni, V. L., 3rd. (2022). Contribution of beef to key nutrient intakes in American adults: An updated analysis with NHANES 2011-2018. *Nutrients Research, 105*, 105–112. https://doi.org/10.1016/j.nutres.2022.06.009.
2. Alae-Carew, C., Green, R., Stewart, C., Cook, B., Dangour, A. D., & Scheelbeek, P. F. D. (2022). The role of plant-based alternative foods in sustainable and healthy food systems: Consumption trends in the UK. *Science of The Total Environment, 807*(3). https://doi.org/10.1016/j.scitotenv.2021.151041.
3. Bianchi, F., Stewart, C., Astbury, N. M., Cook, B., Aveyard, P., & Jebb, S. A. (2022). Replacing meat with alternative plant-based products (RE-MAP): A randomized controlled trial of a multicomponent behavioral intervention to reduce meat consumption. *American Journal of Clinical Nutrition, 115*(5), 1357–1366. https://doi.org/10.1093/ajcn/nqab414.
4. Curtain, F., & Grafenauer, S. (2019). Plant-based meat substitutes in the flexitarian age: An audit of products on supermarket shelves. *Nutrients, 11*(11), 2603. https://doi.org/10.3390/nu11112603.
5. Dussiot, A., Fouillet, H., Perraud, E., Salomé, M., Huneau, J. F., Kesse-Guyot, E., et al. (2022). Nutritional issues and dietary levers during gradual meat reduction: A sequential diet optimization study to achieve progressively healthier diets. *Clinical Nutrition, 41*(12), 2597–2606. https://doi.org/10.1016/j.clnu.2022.09.017.
6. Fardet, A., & Rock, E. (2018). Perspective: Reductionist nutrition research has meaning only within the framework of holistic and ethical thinking. *Advances in Nutrition, 9*(6), 655–670. https://doi.org/10.1093/advances/nmy044.
7. Fardet, A., & Rock, E. (2022). Chronic diseases are first associated with the degradation and artificialization of food matrices rather than with food composition: Calorie quality matters more than calorie quantity. *European Journal of Nutrition, 61*(5), 2239–2253. https://doi.org/10.1007/s00394-021-02786-8.
8. Fouillet, H., Dussiot, A., Perraud, E., Wang, J., Kesse-Guyot, E., & Mariotti, F. (2023). Plant to animal protein ratio in the diet: Nutrient adequacy, long-term health, and environmental pressure. *medRxiv*. https://doi.org/10.1101/2022.05.20.22275349.
9. Gehring, J., Touvier, M., Baudry, J., Julia, C., Buscail, C., Srour, B., et al. (2021). Consumption of ultra-processed foods by pesco-vegetarians, vegetarians, and vegans: Associations with duration and age at diet

initiation. *Journal of Nutrition, 151*(1), 120–131. https://doi.org/10.1093/jn/nxaa196.

10. Hu, F. B., Otis, B. O., & McCarthy, G. (2019). Can plant-based meat alternatives be part of a healthy and sustainable diet? *JAMA, 322*(16), 1547–1548. https://doi.org/10.1001/jama.2019.13187.
11. Klapp, A. L., Feil, N., & Risius, A. (2022). A global analysis of national dietary guidelines on plant-based diets and substitutions for animal-based foods. *Current Developments in Nutrition, 6*(11). https://doi.org/10.1093/cdn/nzac144.
12. Marinangeli, C. P. F., Curran, J., Barr, S. I., Slavin, J., Puri, S., Swaminathan, S., et al. (2017). Enhancing nutrition with pulses: Defining a recommended serving size for adults. *Nutrition Reviews, 75*(12), 990–1006. https://doi.org/10.1093/nutrit/nux058.
13. Messina, M., Sievenpiper, J. L., Williamson, P., Kiel, J., & Erdman, J. W. (2022). Perspective: Soy-based meat and dairy alternatives, despite classification as ultra-processed foods, deliver high-quality nutrition on par with unprocessed or minimally processed animal-based counterparts. *Advances in Nutrition, 13*(3), 726–738. https://doi.org/10.1093/advances/nmac026.
14. Perraud, E., Wang, J., Salomé, M., Mariotti, F., & Kesse-Guyot, E. (2023). Dietary protein consumption profiles show contrasting impacts on environmental and health indicators. *Science of The Total Environment, 856*(1). https://doi.org/10.1016/j.scitotenv.2022.159052.
15. Reiley, L. (2019). Impossible Burger: Here's what's really in it. *The Washington Post.*
16. Reuzé, A., Méjean, C., Carrère, M., Sirieix, L., Druesne-Pecollo, N., Péneau, S., et al. (2022). Rebalancing meat and legume consumption: Change-inducing food choice motives and associated individual characteristics in non-vegetarian adults. *International Journal of Behavioral Nutrition and Physical Activity, 19*(1), 112. https://doi.org/10.1186/s12966-022-01317-w.
17. Seehafer, A., & Bartels, M. (2019). Meat 2.0—The regulatory environment of plant-based and cultured meat. *European Food and Feed Law Review, 14*(4), 323–331.
18. Tao, M. H., Liu, J. L., & Nguyen, U. D. T. (2022). Trends in diet quality by race/ethnicity among adults in the United States for 2011–2018. *Nutrients, 14*(19), 4178. https://doi.org/10.3390/nu14194178.
19. Toh, D. W. K., SRV, A., & Henry, C. J. (2022). Unknown impacts of plant-based meat alternatives on long-term health. *Nature Food, 3*(2), 90–91. https://doi.org/10.1038/s43016-022-00463-5.

20. van Vliet, S., Kronberg, S. L., & Provenza, F. D. (2020). Plant-based meats, human health, and climate change. *Frontiers in Sustainable Food Systems, 4*, 1–17. https://doi.org/10.3389/fsufs.2020.00128.

21. Vieux, F., Rémond, D., Peyraud, J. L., & Darmon, N. (2022). Approximately half of total protein intake by adults must be animal-based to meet non-protein, nutrient-based recommendations, with variations due to age and sex. *Journal of Nutrition, 152*(11), 2514–2525. https://doi.org/10.1093/jn/nxac150.

Chapter 11

Current Research and Innovations

Contents

Introduction..342
11.1. Recent Studies on Nutritional Comparisons....................................342
11.2. Innovations in Plant-Based Meat Technology....................................344
11.3. Future Trends in Meat Alternatives....................................347
References..353

INTRODUCTION

A growing number of people are looking for plant-based meat substitutes due to worries about animal welfare, environmental sustainability, and health. Enhancing these products' flavor, texture, nutritional makeup, and general consumer acceptability are the main goals of this field of research. Innovations are being made in a number of areas, such as the source of ingredients, processing methods, and flavor augmentation. Innovations aimed at sustainability are also being introduced into traditional meat production, such as regenerative agriculture techniques that improve biodiversity and soil health while lowering greenhouse gas emissions related to livestock farming. Furthermore, improvements in animal husbandry techniques are made to improve animal welfare without sacrificing productivity. To put it briefly, there are major research-driven changes taking place in both the plant-based substitutes and traditional meat sectors to satisfy consumer demands for healthier options and address environmental issues.

11.1. RECENT STUDIES ON NUTRITIONAL COMPARISONS

Recent studies have focused on the nutritional profiles of traditional meats and plant-based meat substitutes, especially as consumer interest in health, sustainability, and ethical eating keeps growing. When making dietary decisions, consumers must be aware of the variations in nutrient content, bioavailability, and overall health impacts.

11.1.1. Protein Content and Quality

Traditional meats and meat substitutes made from plants are both excellent sources of protein. However, there are differences in the quality of protein. All the essential amino acids are present in sufficient amounts in traditional meats, which offer complete proteins. Conversely, while many plant-based substitutes might not contain all nine essential amino acids, they can still be enhanced or combined with additional protein sources (e.g., legumes along with grains) to obtain a comprehensive profile of amino acids. Certain plant-based products, especially those made from soy or pea protein, have been shown in recent studies to have protein contents that are either equal to or higher than those of traditional meats (Gibbs and Cappuccio, 2022).

11.1.2. Fat Composition

Traditional meats vary greatly in fat content based on the kind of meat (e.g., meat versus. chicken). Conventional meats frequently contain saturated fats, which, if consumed in excess, may have an adverse effect on cardiovascular health. However, a lot of plant-based substitutes use oils like canola or coconut oil, which might include unsaturated fats that are better for you. Studies reveal that although plant-based alternatives have less total fat, they might also have additional oils added to enhance flavor and texture, which could raise their calorie density.

11.1.3. Vitamins and Minerals

Zinc, selenium, heme iron, vitamin B_{12}, and other vitamins and minerals are all abundant in traditional meats. Given that vitamin B_{12} is mostly found in animal products, plant-based substitutes frequently lack it unless they are fortified. Non-heme iron, or iron from plants, is not as bioavailable as heme iron from meat, so people who only eat plant-based diets must take in more iron or include foods high in vitamin C to improve absorption. According to recent studies, there is still variation in the fortifications of plant-based products compared to traditional meats, even though some of them are fortified with nutrients like iron and B_{12}.

11.1.4. Fiber Content

The high fiber content of many plant-based meat substitutes is one of their main advantages. While most plant-based meat substitutes include components like whole grains or legumes, which add healthy fiber, traditional meats don't. A higher fiber diet has been linked to several health advantages, such as better digestive function and a lower risk of developing chronic illnesses.

11.1.5. Sodium Levels

Although there can be significant differences in sodium content between the two groups, processed plant-based meats typically have higher sodium content because of the addition of flavor enhancers and preservatives. Studies show that because high sodium intake is associated with hypertension and cardiovascular disease, consumers should exercise caution when consuming both types of products.

11.1.6. Health Implications

The long-term health effects of ingesting either kind of product as part of a balanced diet are now being studied. According to certain research, diets heavy in processed and red meats may put people at higher risk of developing certain illnesses, like colorectal cancer. On the other hand, diets high in plant-based foods, such as fruits, vegetables, nuts, seeds, and legumes, are typically associated with lower risk profiles for disease. Recent research indicates that adding moderate amounts of carefully prepared plant-based meat substitutes may have comparable advantages while adding extra nutrients like fiber.

Finally, it should be noted that although traditional meats and plant-based meat substitutes both offer essential nutrients for human health, they have different nutrient profiles that make them more or less advantageous. Research on these variations and their role in general dietary patterns intended to promote health is still ongoing, with the goal of improving our understanding.

11.2. INNOVATIONS IN PLANT-BASED MEAT TECHNOLOGY

For many years, extrusion technology has been a mainstay in the food processing industry. The method, which is perfect for mass-producing food, entails forcing ingredients through a die to form a continuous shape. This technology has been further developed by high-moisture extrusion, which allows producers of plant-based foods to mimic the soft, chewy texture of meat and seafood. Extrusion has undoubtedly expanded the possibilities for alternative proteins, but there are still a ton of cutting-edge technologies on the horizon that are revolutionizing plant-based food processing. These include electrospinning, 3D printing, shear-cell technology, and ohmic heating, and they have allowed businesses to produce real goods at a speed and precision never seen before. This section will discuss some of the alternative protein companies that are setting the standard for their industry's adoption as well as the most recent advancements in plant-based food processing.

11.2.1. Ohmic Heating

Ohmic heating, sometimes called power heating, is the process of rapidly and steadily raising the temperature of a food product by passing an alternating current through it. It is frequently used in the food manufacturing process to pasteurize and sterilize food items. It hasn't been thoroughly investigated for the synthesis of substitute proteins until recently. Ohmic heating has the advantage of providing accurate temperature control, which makes it a useful tool for processing proteins derived from plants that may denature and coagulate when subjected to heat processing. Ohmic heating is a useful tool for enhancing the emulsification, gelation, and foaming capabilities of plant-based proteins by changing their structural and techno-functional properties. To create a more genuine culinary experience, uniform heating can also lower antinutrient levels, maintain nutritional quality and sensory attributes, and enhance the mouthfeel, flavor, and texture of alternative protein products. Compared to chemical and biological protein treatments, ohmic heating and other physical treatments like pressure application and radiation require fewer inputs and leave fewer unwanted residues; as a result, they may be less expensive options. However, since the use of ohmic heating is relatively new in the alternative protein industry, more research is required.

11.2.2. Shear Cell

By applying high pressure to raw materials to break down cell walls, shear cell technology—also known as high-pressure processing, or HPP—improves the texture, digestibility, flavor, nutritional profile, and shelf life of alternative protein products. Shear cell technology offers more controlled and consistent deformation of plant-based proteins compared to extrusion processes, making it possible to produce fibrous meat analogs that maintain their structural integrity and stability even after cooling. The goal of further technological advancement is to increase accuracy and productivity. One example of this is the design of the Couette shear cell. Shear cell technology is used by companies like The Vegetarian Butcher and Riverside Foods to produce plant-based meat, dairy, and seafood substitutes, among many other applications in the alternative protein market.

11.2.2.1. Rival Foods

- **Headquarters:** Wageningen, Gelderland, The Netherlands.
- **Founded:** 2019.
- **Founders:** Birgit Dekkers and Ernst Breel.
- **Funding:** $6M.

Shear cell technology is being used by Rival Foods to create whole-cut specialties that have the true flavor, texture, and juiciness. The company is working on a commercially feasible shear cell machine that can better understand the potential of shear cell technology in plant-based meat production by structuring proteins derived from plants.

11.2.3. Electrospinning

One kind of plant-based scaffolding technology being investigated for the creation of substitute proteins is electrospinning. It involves melting or combining a solution of a polymer or protein with an electric field to create nanofibers. Real plant-based food products can then be made with the nanofibers that are produced. With electrospinning, structures that closely resemble the texture of meat and seafood derived from animals could be produced in a highly precise and customizable manner. To maximize the application of this technology in the alternative protein market, more study is necessary. Meatable is one company creating plant-based scaffolding technology. Meatable will investigate the viability of non-animal-derived electrospun scaffolds for use in the alternative protein industry in partnership with contract manufacturer and product developer VIVOLTA.

11.2.4. 3D Printing

Using a layer-by-layer printing technique, 3D printing technology is an additive manufacturing method for creating substitute proteins. To produce the desired plant-based food product, a 3D model must be created and printed using a specialized printer. Its uses include the creation of bioprinted meat, but its exact control also makes it possible to produce plant-based food substitutes with distinctive flavors, textures, and forms that resemble those of animal products. Take 3D-printed steaks, for instance. Additionally, it makes it possible to produce goods that are specifically suited to the preferences of each customer, like dietary supplements or plant-based nutrition bars. Even though there are still many obstacles to overcome, most notably those related to scalability, 3D printing technology has the potential to completely transform the food industry by offering an effective and sustainable method of processing plant-based protein products. Revo Foods, Redefine Meat, and SeaSpire are a few businesses that are producing plant-based alternative proteins using 3D printing technology.

11.2.4.1. Revo Foods

- **Headquarters:** Vienna, Wien, Austria.
- **Founded:** 2020.
- **Founders:** Manuel Lachmayr, Robin Simsa, and Theresa Rothenbücher.
- **Funding:** $4M.

A company called Revo Foods uses 3D printing technology to create whole-cut salmon with authentic flavor, texture, and juiciness as well as tuna spreads. To promote research into the application of 3D-printed mycoprotein to enhance taste and texture and encourage widespread consumer adoption, the company is collaborating with Mycorena.

11.2.5. Environmental Impact of These Technologies

- Decreased greenhouse gas emissions: One of the main causes of greenhouse gas emissions is traditional livestock farming. Plant-based meat substitutes have a lower carbon footprint than meat.

- Conservation of land and water: Compared to conventional livestock farming, meat substitutes require less land and water, providing a more sustainable method of producing food.
- Biodiversity: Meat substitutes can help to preserve biodiversity by lowering the need for deforestation to make pastureland.

But it's imperative to take the life cycle analysis into account. For example, although meat produced in laboratories may require less land, their energy needs may outweigh any environmental advantages.

A few of the many innovative technological advancements that are transforming the manufacturing of real alternative proteins include electrospinning, 3D printing, shear cell technology, and ohmic heating. These methods of processing food made from plants have many advantages, such as better nutritional profiles, improved organoleptic qualities, and increased functionality. For instance, shear cell technology can give plant-based meat substitutes a fibrous texture, and ohmic heating can make plant-based proteins more soluble. Technologies like electrospinning and 3D printing enable the creation of precisely structured, highly customizable products. Furthermore, these technologies add to the ever-expanding collection of scalable and reasonably priced plant-based food processing methods (Sridhar et al., 2022).

11.3. FUTURE TRENDS IN MEAT ALTERNATIVES

Because they contain the nine essential amino acids required for optimal body growth, repair, and energy needs, proteins are fundamental to our health. Maintaining good health and optimal bodily functions requires making sure that one consumes an adequate amount of high-quality proteins. The hunt for substitute sources of protein is gaining traction, even though we have historically depended on foods like meat, fish, dairy, nuts, grains, and eggs, which reflect a variety of cultural preferences and food availability. Alternative protein sources provide exciting new ways to meet our nutritional needs while protecting the environment. These sources range from plant-based options like legumes and quinoa to inventive lab-grown meats and even insect protein. These substitutes not only suit a wide range of dietary requirements, but they also hold the key to solving some of the world's most pressing issues related to food security and sustainability. Nearly 70% of the protein consumed in the US comes from animal sources. In the meantime, more than 2 billion people, or 25% of the global population, depend on insects to meet their protein needs in other regions. This demonstrates our extraordinary flexibility in obtaining necessary nutrients. Our natural ability to adapt is fueling a new movement in the direction of alternative protein production.

Global macrotrends towards the production of alternative proteins are being driven by multiple factors. These include the strains posed by population growth, the effects of climate change, the environmental impact of animal agriculture, and the rising importance of ethical consumption. We'll delve deeper into these trends. Producing alternative proteins is becoming more and more feasible thanks to developments in AgTech and Food Tech. Many startups are springing up to meet the public's growing demand for a variety of protein sources due to the expanding market for plant-based proteins. Major food producers are also willing to invest in new goods and food technologies, and they are very open to novel plant-based protein solutions.

Let's delve into the top alternative protein trends that are not only redefining our plates but also paving the way for a sustainable future.

11.3.1. Embracing Sustainability and Ethical Consumption

Over 8 billion people are on the planet now. By 2050, it is expected to reach 9.8 billion, and by 2100, 11.2 billion, according to UN estimates. It is impossible to provide food security for all of these people using traditional agricultural (as well as food processing and distribution) methods without causing irreversible environmental harm. Extreme weather conditions like drought and flooding, as well as geopolitical events like wars, can disrupt supply chains and reduce the availability of essential fertilizers, restrict access to agricultural land, or block shipping lanes.

The quick development of AgTech products and solutions has already made it possible for us to witness the urgent transition to sustainable farming. Among these is an increasing interest in producing alternative proteins locally. Affordably priced, premium protein products could be produced near the consumer base with industrialized production processes that can be scaled from large plants to cottage industries and operate on hybrid clean energy grids. These can include anything that produces food without requiring acres of land, such as fermentation plants, tanks for algae or seaweed, lab-cultured meat, and so on.

11.3.2. Highlighting Innovative Protein Sources on the Rise

Raised in affluent Western European and North American societies, baby boomers grew up on meat, cheese, and eggs as well as milk. The agricultural revolution of the post-war era produced an unparalleled quantity of proteins derived from animals. Traditional mainstays like oats and peas became largely unfashionable, while soy products and nuts like almonds were viewed as novel and unusual "health foods." The average consumer in the 1950s would have thought that 21st-century innovations like lab-grown (cell-based) meats, proteins derived from air, fungi, seaweeds, and other plant varieties were something out of a science fiction film. Contemporary consumers are far more accepting of new technologies and are used to seeing food items from other countries on store shelves. However, the success of the alternative protein market still depends on how consumers view and accept these new trends.

11.3.3. Prioritizing Health, Wellness, and Nutritional Value

The public's desire for foods that promote overall wellness, including mental and physical health, is benefiting food producers. Science-based consumer preferences have gradually replaced superstitious and folkloric ideas about what makes for a nutritious diet in these requirements. Global dietary trends toward low-fat foods, high-fiber diets, and low-carb diets emerged in the latter half of the 20th century, and knowledge of the significance of protein and its role in achieving physical health has since grown.

It's indisputable that most people, given the option, will choose healthful foods the majority of the time. The importance of alternative proteins as a key element of a diet rich in nutrients is becoming more widely recognized. For many years, food manufacturers have added vitamins and minerals to foods like bread and cereals. To increase the nutritional value of alternative proteins and satisfy consumer demands, they are now supplementing them with vital nutrients.

11.3.4. Advancing Technological Innovations in Food Production

From the time that our ancestors discovered methods like crop rotation, selective breeding, smoking, and salting, and fertilization, technology has been essential to the production of food. Cellular agriculture, fermentation, and genetic engineering are examples of contemporary innovations.

11.3.4.1. Next-Generation Plant-Based Proteins: Whole-Muscle Solutions

The majority of people who consume meat do so because they find it to be delicious. The average consumer will not compromise on taste, texture, or odor because they are crucial components of the experience. Producers of plant-based substitutes for whole-muscle meats, like steaks and chicken breasts, have had difficulty imitating the flavor. The general belief that plant-based substitutes for whole-muscle meats are inferior is opening up a lot of doors for creative thinkers who can provide workable answers.

11.3.4.2. Hybrid Meats

Consumer interest in hybrid meats is one of the most intriguing trends in alternative protein sources. They usually blend plant-based proteins with meat material produced in a lab. People who choose a flexitarian diet (mostly fruits, vegetables, legumes, and whole grains) and are concerned about their health and the environment find hybrid meats appealing because they are lower in meat, fat, and cholesterol than conventional products. Rather than completely removing meat from their diets, vegans prefer to consume less of it overall. The same concerns about taste, texture, and odor that suppliers of only plant-based meat substitutes face must be addressed by food producers if they hope to gain the loyalty of consumers.

11.3.4.3. Precision Fermentation

Simply put, precision fermentation is an advancement over antiquated technology. It uses microorganisms (grown in carefully regulated environments) to make certain proteins that are identical to those in animal products. Precision fermentation is effectively producing sustainable, high-quality protein substitutes, including alternatives to dairy and egg whites. Industrialized protein fermentation facilities might eventually proliferate in our cities on par with breweries. There is a great deal of potential benefits for nutrition, food security, and sustainability, especially when plants use clean energy and recycle waste.

11.3.5. Enhancing Transparency and Traceability

Additionally, consumers who actively search for food products with alternative proteins are also far more likely to have high standards for traceability and transparency. Ethical and health concerns are two factors that contribute to the desire to find out where food originates. Transparency and traceability are becoming standard features of packaging (and marketing) thanks to blockchain and related technologies. Growing awareness of the need for greater transparency at every stage of the supply chain—from the source of seeds to the finished product—is awakening producers. Other factors to take into account are a product's overall green rating (carbon footprint), whether or not workers are paid a living wage, whether or not the product is made locally, and even how well the parent company does on environmental, social, and governance (ESG) scales.

11.3.6. Making Alternative Proteins More Affordable

The food industry is a fiercely competitive one. Shoppers will likely fill their baskets and make decisions based on health and ethical considerations when the economy is strong. Cell-cultured meat and other alternative protein products are currently expensive, which limits their appeal to a wider audience. Producers must reach a larger market and make audacious investments in cutting-edge technology. One of the main objectives of innovators is to lower the cost of alternative proteins so that a wider range of consumers can afford them. All areas of production technologies, sourcing raw materials, and distribution have their share of difficulties and opportunities. These products may be able to hit the market sooner if quicker regulatory developments and cheaper costs are implemented. Although there is a potential demand, scalability is still a problem.

11.3.7. Rising Interest in Flexitarian Diets

In the context of hybrid meat development, we have already discussed the phenomenon of flexitarian diets. Any business hoping to capitalize on the popularity of alternative protein foods needs to be cognizant of the flexitarian movement. Hybrid meat, which blends lab-created (zero cruelty) meat with plant-based proteins, is a compromise that many will welcome. Even vegans and vegetarians occasionally want the flexibility to consume animal-based proteins. Even those who identify as essentially omnivores may nevertheless have ethical and environmental concerns regarding the production of staple foods for animal feed, as well as concerns regarding animal welfare. If hybrid meats are as cheap and of the same quality as farmed meat, flexitarians may choose them on a regular basis. Flexibility and the freedom to select and adhere to one's own preferences—which are subject to frequent change—are the fundamental components of a flexitarian diet. Producing food requires a foothold in this expanding market.

11.3.8. Innovating Packaging and Reducing Waste

Consumers who choose diets and foods in an ethical and ecologically responsible manner also frequently anticipate sustainable packaging options. With eco-friendly packaging and recycling programs, food producers and retailers have a significant chance to shape consumer preferences and foster brand loyalty. As part of a circular economy, there are many new packaging technologies to embrace and opportunities to create new ones. Gaining support from the public and cultivating brand loyalty also depend on reducing waste in the manufacturing and retail stages. Intriguing options exist for using advanced phosphates to increase the shelf life of substitute protein products. Businesses that can turn production byproducts into something edible or useful will have a lot of marketing and publicity opportunities.

11.3.9. Leveraging Generative AI for Alternative Protein Development

The alternative protein market is changing as a result of generative AI, which improves everything from product development to market readiness. AI outperforms conventional techniques by evaluating vast amounts of data, and improving the nutrition, texture, and flavor of proteins. This technology effectively addresses particular customer needs by speeding up research and introducing precision in production and customization.

The Internet of Things (IoT) in conjunction with smart production techniques is leading the way in more accurate and waste-free processes. Anticipating consumer trends, optimizing inventory, and cutting surplus are all made possible by predictive analytics. To preserve customer trust, however, the adoption of AI necessitates close consideration of data privacy and transparency. AI has enormous potential for alternative proteins, providing a route to more consumer-focused and sustainable products. AI has the potential to completely transform food innovation as we continue to refine its applications, which will help us achieve global sustainability and health goals.

People's desire for more affordable, environmentally friendly food that is also healthier is driving the rapid growth of the alternative protein market. The need to find ways to feed everyone without harming the environment or abusing animals is growing as the world's population approaches 10 billion. A significant portion of the answer may lie in substituting proteins. Although the market for alternative proteins is still in its infancy, innovation, and new technology are helping to grow it. To expedite the integration of these high-tech foods into mainstream diets, industry collaboration, regulatory support, dynamic innovation, and effective marketing to improve consumer education and awareness are all necessary. A strategic chance exists to sculpt a future food system that is safe, sustainable, and compliant with international health and wellness objectives as more people begin to request these goods (Alcorta et al., 2021).

REFERENCES

1. Adesogan, A. T., Havelaar, A. H., McKune, S. L., et al. (2020). Animal source foods: Sustainability problem or malnutrition and sustainability solution? Perspective matters. *Global Food Security, 25*, 100325. https://doi.org/10.1016/j.gfs.2019.100325.
2. Alae-Carew, C., Green, R., Stewart, C., et al. (2022). The role of plant-based alternative foods in sustainable and healthy food systems: Consumption trends in the UK. *Science of The Total Environment, 807*, 151041.
3. Alcorta, A., Porta, A., & Tárrega, A. (2021). Foods for plant-based diets: Challenges and innovations. *Foods, 10*(2), 293. https://doi.org/10.3390/foods10020293.
4. Anderson, C. A. M., Thorndike, A. N., Lichtenstein, A. H., et al. (2019). Innovation to create a healthy and sustainable food system: A science advisory from the American Heart Association. *Circulation, 139*, 1025–1032.
5. Aschemann-Witzel, J., Gantriis, R. F., Fraga, P., et al. (2021). Plant-based food and protein trend from a business perspective: Markets, consumers, and the challenges and opportunities in the future. *Critical Reviews in Food Science and Nutrition, 61*, 3119–3128. https://doi.org/10.1080/10408398.2020.1793730.
6. Attwood, S., & Hajat, C. (2020). How will the COVID-19 pandemic shape the future of meat consumption? *Public Health Nutrition, 23*, 116–120.
7. Bakhsh, A., Lee, S. J., Lee, E. Y., et al. (2021). Traditional plant-based meat alternatives, current, and future perspective: A review. *Journal of Agricultural and Life Sciences, 55*, 1–11.
8. Banta, J. E., Lee, J. W., Hodgkin, G., et al. (2018). The global influence of the Seventh-Day Adventist Church on diet. *Religions, 9*(9), 251. https://doi.org/10.3390/rel9090251.
9. Barnard, N. D., Levin, S. M., & Yokoyama, Y. (2015). A systematic review and meta-analysis of changes in body weight in clinical trials of vegetarian diets. *Journal of the Academy of Nutrition and Dietetics, 115*, 954–969. https://doi.org/10.1016/j.jand.2014.11.016.
10. Boukid, F. (2020). Plant-based meat analogues: From niche to mainstream. *European Food Research and Technology, 247*, 297–308.
11. Bouvard, V., Loomis, D., Guyton, K. Z., et al. (2015). Carcinogenicity of consumption of red and processed meat. *The Lancet Oncology, 16*, 1599–1600.

12. Bryant, C. J. (2022). Plant-based animal product alternatives are healthier and more environmentally sustainable than animal products. *Future Foods, 6*, 100174. https://doi.org/10.1016/j.fufo.2022.100174.

13. Bryant, C., Szejda, K., Parekh, N., et al. (2019). A survey of consumer perceptions of plant-based and clean meat in the USA, India, and China. *Frontiers in Sustainable Food Systems, 3*, 11. https://doi.org/10.3389/fsufs.2019.00011.

14. Choudhury, D., Singh, S., Seah, J. S. H., et al. (2020). Commercialization of plant-based meat alternatives. *Trends in Plant Science, 25*, 1055–1058. https://doi.org/10.1016/j.tplants.2020.08.006.

15. Culliford, A., & Bradbury, J. (2020). A cross-sectional survey of the readiness of consumers to adopt an environmentally sustainable diet. *Nutrition Journal, 19*, 1–13. https://doi.org/10.1186/s12937-020-00644-7.

16. Dinu, M., Abbate, R., & Gensini, G. F. (2017). Vegetarian, vegan diets and multiple health outcomes: A systematic review with meta-analysis of observational studies. *Critical Reviews in Food Science and Nutrition, 57*, 3640–3649. https://doi.org/10.1080/10408398.2016.1138447.

17. European Commission. (2020). *Farm to Fork Strategy for a Fair, Healthy and Environmentally-Friendly Food System*. https://food.ec.europa.eu/system/files/2020-05/f2f_action-plan_2020_strategy-info_en.pdf (accessed on 22 October 2024).

18. Farsi, D. N., Uthumange, D., Munoz, J., et al. (2021). The nutritional impact of replacing dietary meat with meat alternatives in the UK: A modelling analysis using nationally representative data. *British Journal of Nutrition, 127*, 1–11.

19. Fiorentini, M., Kinchla, A. J., & Nolden, A. A. (2020). Role of sensory evaluation in consumer acceptance of plant-based meat analogs and meat extenders: A scoping review. *Foods, 9*(9), 1334. https://doi.org/10.3390/foods9091334.

20. Fresán, U., Mejia, M. A., Craig, W. J., et al. (2019). Meat analogs from different protein sources: A comparison of their sustainability and nutritional content. *Sustainability, 11*(11), 3231.

21. García-Oliveira, P., Fraga-Corral, M., Pereira, A. G., et al. (2022). Solutions for the sustainability of the food production and consumption system. *Critical Reviews in Food Science and Nutrition, 62*, 1765–1781.

22. Gastaldello, A., Giampieri, F., De Giuseppe, R., et al. (2022). The rise of processed meat alternatives: A narrative review of the manufacturing, composition, nutritional profile, and health effects of newer sources of protein, and their place in healthier diets. *Trends in Food Science &*

Technology, 127, 263–271. https://doi.org/10.1016/j.tifs.2022.07.005.

23. Gibbs, J., & Cappuccio, F. P. (2022). Plant-based dietary patterns for human and planetary health. *Nutrients, 14*(1), 1–11.
24. González, N., Marquès, M., Nadal, M., et al. (2020). Meat consumption: Which are the current global risks? A review of recent (2010–2020) evidence. *Food Research International, 137*, 109341.
25. Graça, J., Oliveira, A., & Calheiros, M. M. (2015). Meat, beyond the plate: Data-driven hypotheses for understanding consumer willingness to adopt a more plant-based diet. *Appetite, 90*, 80–90.
26. Harguess, J. M., Crespo, N. C., & Hong, M. Y. (2020). Strategies to reduce meat consumption: A systematic literature review of experimental studies. *Appetite, 144*, 104478. https://doi.org/10.1016/j.appet.2019.104478.
27. Harland, J., & Garton, L. (2016). An update of the evidence relating to plant-based diets and cardiovascular disease, type 2 diabetes, and overweight. *Nutrition Bulletin, 41*, 323–338. https://doi.org/10.1111/nbu.12235.
28. Hartmann, C., & Siegrist, M. (2017). Consumer perception and behavior regarding sustainable protein consumption: A systematic review. *Trends in Food Science & Technology, 61*, 11–25. https://doi.org/10.1016/j.tifs.2016.12.006.
29. Huang, R. Y., Huang, C. C., Hu, F. B., et al. (2016). Vegetarian diets and weight reduction: A meta-analysis of randomized controlled trials. *Journal of General Internal Medicine, 31*, 109–116. https://doi.org/10.1007/s11606-015-3390-7.
30. Ishaq, A., Irfan, S., Sameen, A., et al. (2022). Plant-based meat analogs: A review with reference to formulation and gastrointestinal fate. *Current Research in Food Science, 5*, 973–983.
31. Jafari, S., Hezaveh, E., Jalilpiran, Y., et al. (2022). Plant-based diets and risk of disease mortality: A systematic review and meta-analysis of cohort studies. *Critical Reviews in Food Science and Nutrition, 62*, 7760–7772. https://doi.org/10.1080/10408398.2021.1918628.
32. Jahn, S., Furchheim, P., & Strässner, A. M. (2021). Plant-based meat alternatives: Motivational adoption barriers and solutions. *Sustainability, 13*, 1–17.
33. Jairath, G., Mal, G., Gopinath, D., et al. (2021). A holistic approach to assess the viability of cultured meat: A review. *Trends in Food Science & Technology, 110*, 700–710. https://doi.org/10.1016/j.tifs.2021.02.024.
34. Joyce, A., Dixon, S., Comfort, J., et al. (2012). Reducing the environmental

impact of dietary choice: Perspectives from a behavioral and social change approach. *Journal of Environmental and Public Health*. https://doi.org/10.1155/2012/978672.

35. Kahleova, H., Levin, S., & Barnard, N. (2017). Cardio-metabolic benefits of plant-based diets. *Nutrients, 9*(1), 1–13.
36. Kim, H., Caulfield, L. E., & Rebholz, C. M. (2018). Healthy plant-based diets are associated with lower risk of all-cause mortality in US adults. *The Journal of Nutrition, 148*(4), 624–631. https://doi.org/10.1093/jn/nxy019.
37. Kwasny, T., Dobernig, K., & Riefler, P. (2022). Towards reduced meat consumption: A systematic literature review of intervention effectiveness, 2001–2019. *Appetite, 168*, 105739. https://doi.org/10.1016/j.appet.2021.105739.
38. Kwok, C. S., Umar, S., Myint, P. K., et al. (2014). Vegetarian diet, Seventh Day Adventists, and risk of cardiovascular mortality: A systematic review and meta-analysis. *International Journal of Cardiology, 176*, 680–686. https://doi.org/10.1016/j.ijcard.2014.07.080.
39. Lanou, A. J., & Svenson, B. (2011). Reduced cancer risk in vegetarians: An analysis of recent reports. *Cancer Management and Research, 3*(1), 1–8.
40. Lonnie, M., & Johnstone, A. M. (2020). The public health rationale for promoting plant protein as an important part of a sustainable and healthy diet. *Nutrition Bulletin, 45*, 281–293.
41. MacDiarmid, J. I. (2021). The food system and climate change: Are plant-based diets becoming unhealthy and less environmentally sustainable? *Proceedings of the Nutrition Society, 80*(3), 1–6.
42. Malek, L., & Umberger, W. J. (2021). How flexible are flexitarians? Examining diversity in dietary patterns, motivations, and future intentions. *Clean Responsible Consumption, 3*, 100038. https://doi.org/10.1016/j.clrc.2021.100038.
43. McClements, D. J., & Grossmann, L. (2021). A brief review of the science behind the design of healthy and sustainable plant-based foods. *npj Science of Food, 5*, 17. https://doi.org/10.1038/s41538-021-00099-y.
44. Naghshi, S., Sadeghi, O., Willett, W. C., et al. (2020). Dietary intake of total, animal, and plant proteins and risk of all-cause, cardiovascular, and cancer mortality: Systematic review and dose–response meta-analysis of prospective cohort studies. *BMJ, 370*, m2412. https://doi.org/10.1136/bmj.m2412.
45. Orlich, M. J., Singh, P. N., Sabaté, J., et al. (2013). Vegetarian dietary

patterns and mortality in Adventist Health Study 2. *JAMA Internal Medicine, 173*(12), 1230–1238. https://doi.org/10.1001/jamainternmed.2013.6473.

46. Reynolds, C. J., Horgan, G. W., Whybrow, S., et al. (2019). Healthy and sustainable diets that meet greenhouse gas emission reduction targets and are affordable for different income groups in the UK. *Public Health Nutrition, 22*(8), 1503–1517. https://doi.org/10.1017/S1368980018003774.
47. Richi, E. B., Baumer, B., Conrad, B., et al. (2015). Health risks associated with meat consumption: A review of epidemiological studies. *International Journal for Vitamin and Nutrition Research, 85*(2), 70–78. https://doi.org/10.1024/0300-9831/a000224.
48. Rosell, M., Appleby, P., Spencer, E., et al. (2006). Weight gain over 5 years in 21,966 meat-eating, fish-eating, vegetarian, and vegan men and women in EPIC-Oxford. *International Journal of Obesity, 30*, 1389–1396. https://doi.org/10.1038/sj.ijo.0803305.
49. Rust, N. A., Ridding, L., Ward, C., et al. (2020). How to transition to reduced-meat diets that benefit people and the planet. *Science of the Total Environment, 718*, 137208. https://doi.org/10.1016/j.scitotenv.2020.137208.
50. Sabaté, J., Sranacharoenpong, K., & Harwatt, H. (2015). The environmental cost of protein food choices. *Public Health Nutrition, 18*(12), 2067–2073. https://doi.org/10.1017/S1368980015001349.
51. Scarborough, P., Kaur, A., Cobiac, L., et al. (2016). Eatwell guide: Modelling the dietary and cost implications of incorporating new sugar and fibre guidelines. *BMJ Open, 6*(9), e011632.
52. Segovia-Siapco, G., & Sabaté, J. (2019). Health and sustainability outcomes of vegetarian dietary patterns: A revisit of the EPIC-Oxford and the Adventist Health Study-2 cohorts. *European Journal of Clinical Nutrition, 72*, 60–70.
53. Singh, M., Trivedi, N., Enamala, M. K., et al. (2021). Plant-based meat analogue (PBMA) as a sustainable food: A concise review. *European Food Research and Technology, 247*, 2499–2526. https://doi.org/10.1007/s00217-021-03810-1.
54. Song, M., Fung, T. T., Hu, F. B., et al. (2016). Animal and plant protein intake and all-cause and cause-specific mortality: Results from two prospective US cohort studies. *JAMA Internal Medicine, 176*(11), 1453–1463.
55. Spencer, E. A., Appleby, P. N., Davey, G. K., et al. (2003). Diet and body mass index in 38,000 EPIC-Oxford meat-eaters, fish-eaters, vegetarians, and vegans. *International Journal of Obesity, 27*, 728–734. https://doi.org/10.1038/sj.ijo.0802300.

56. Sridhar, K., Bouhallab, S., Croguennec, T., et al. (2022). Recent trends in the design of healthier plant-based alternatives: Nutritional profile, gastrointestinal digestion, and consumer perception. *Critical Reviews in Food Science and Nutrition, 1*, 1–16. https://doi.org/10.1080/10408398.2022.2081666.
57. Statista. (2022). Meat consumption worldwide from 1990 to 2021, by meat type. Retrieved from: https://www.statista.com/statistics/274522/global-per-capita-consumption-of-meat/#statisticContainer (accessed on 22 October 2024).
58. Steenson, S., & Buttriss, J. L. (2021). Healthier and more sustainable diets: What changes are needed in high-income countries? *Nutrition Bulletin, 46*(4), 279–309. https://doi.org/10.1111/nbu.12518.
59. Szenderák, J., Fróna, D., & Rákos, M. (2022). Consumer acceptance of plant-based meat substitutes: A narrative review. *Foods, 11*(9), 1274.
60. Tantamango-Bartley, Y., Jaceldo-Siegl, K., Fan, J., et al. (2013). Vegetarian diets and the incidence of cancer in a low-risk population. *Cancer Epidemiology, Biomarkers & Prevention, 22*(2), 286–294. https://doi.org/10.1158/1055-9965.EPI-12-1060.
61. Tonstad, S., Butler, T., Yan, R., et al. (2009). Type of vegetarian diet, body weight, and prevalence of type 2 diabetes. *Diabetes Care, 32*(5), 791–796. https://doi.org/10.2337/dc08-1886.
62. Tran, E., Dale, H. F., Jensen, C., et al. (2020). Effects of plant-based diets on weight status: A systematic review. *Diabetes, Metabolic Syndrome and Obesity: Targets and Therapy, 13*, 3433–3448.
63. Tso, R., & Forde, C. G. (2021). Unintended consequences: Nutritional impact and potential pitfalls of switching from animal- to plant-based foods. *Nutrients, 13*(1), 1–16.
64. Turner-McGrievy, G. M., Davidson, C. R., Wingard, E. E., et al. (2015). Comparative effectiveness of plant-based diets for weight loss: A randomized controlled trial of five different diets. *Nutrition, 31*(3), 350–358.
65. Weinrich, R. (2019). Opportunities for the adoption of health-based sustainable dietary patterns: A review on consumer research of meat substitutes. *Sustainability, 11*(14), 4028.
66. Wellesley, L., Happer, C., & Froggatt, A. (2015). Changing climate, changing diets: Pathways to lower meat consumption. Retrieved from: https://www.chathamhouse.org/2015/11/changing-climate-changing-diets-pathways-lower-meat-consumption (accessed on 22 October 2024).

67. Willett, W., Rockström, J., Loken, B., et al. (2019). Food in the Anthropocene: The EAT–Lancet commission on healthy diets from sustainable food systems. *The Lancet, 393*(10170), 447–492.

68. Zhang, R., Tang, X., Liu, J., et al. (2022). From concept to action: A united, holistic, and one health approach to respond to the climate change crisis. *Infectious Diseases of Poverty, 11*(1), 4–9. https://doi.org/10.1186/s40249-022-00941-9.

INDEX

A

Abdominal discomfort 324
Acid-base balance 49
Active ingredients 220
Advanced physical extraction methods 220
Agricultural crops 226
Agricultural inputs 258
Alcohol glycerol 46
Allergenicity 216
Alpha-linolenic acid 176
Alpha-linolenic fatty acid 48
Alternative protein sources 347
Alzheimer's disease 199
Amino acid composition 45
Amino acids 310
Animal agriculture 265
Animal-based diets 91, 97
Animal-derived products 220
Animal husbandry techniques 342
Animal products 350
Animal sources 313
Animal-to-plant protein 313, 314, 321, 324, 329
Animal welfare 2, 33, 35, 282
Anisotropic fiber structures 133
Antinutrient compounds 312
Antinutrient levels 344
Antinutrients 228
Antinutritional factors 88, 103
Antioxidant effects 202
Apricot seeds 226
Aroma formation 8
Aspartic acid 46
Athletic performance 246

B

Bacillus species 225
Balanced electrolytes 202
Balanced meal 333
Balanced nutrition 308
Bean flours 4
Beef minced meat (BMM) 139
Beyond Meat Burger 198
Binding agents 3, 7, 12
Bioactive compounds 312
Bioavailability 50, 78
Biochemicals 246
Biodiversity loss 217
Bioengineered ingredients 92
Black Bean Burger (BBB) 198
Blood coagulation 167
Blood lipids 327
Blood pressure 327
Bologna-style sausage (BS) 142
Bottom-up method 259

Brain development 167
Brand loyalty 351
Breast cancer 223

C

Calorie density 343
Canola oil 326
Carbohydrates 41, 42, 43, 46, 52, 54
Cardioprotective impact 201
Cardiovascular diseases 41, 48, 62
Carnitine 198, 201, 204
Carnosine 198, 201, 204
Cell-cultured meat 351
Cell division 49
Cell technology 345
Cellular agriculture 349
Central nervous system function 202
Chaotropic ions 5
Chemical meat composition 57
Chicken analogue 137
Chickpea flour (CF) 141
Choice architecture 289
Cholesterol content 336
Chronic fatigue 199
Chronic illnesses 343
Circular economy 351
Clean energy 350
Climate change 261
Clinical trial 326
Clostridium species 225
Coarse-particle meatless products 132
Cognitive impairment 50
Colorectal cancer 343
Color stability 133
Comfort food 285
Communication channel 294
Conjugated fatty acids (CLAs) 170
Conjugated linoleic acid (CLA) 49, 73
Consumer acceptability 342
Consumer acceptance 284
Consumer demands 216, 342
Consumer segments 294
Consumption habits 282
Contemporary food processing technologies 2
Conventional dry extrusion 229
Conventional livestock farming 265
Conventional meat 102, 283
Conventional production methods 183
Cooking methods 130
Coronary heart disease (CHD) 169, 173
Creatine 198, 200, 204
Creatine supplements 200
Crop rotation 349
Crystalline particles 191
Culinary applications 308
Cultured meat 131
Customary meats 40
Customers right 276
Cutting-edge technologies 344
Cysteinesulfinic acid decarboxylase (CSD) 99

D

Defense mechanism 108
Deforestation 258
Deoxyribonucleic acid (DNA) 52
Derogatory sense 264
Dietary change 278
Dietary decisions 40, 342
Dietary deficiencies 234
Dietary fats 166
Dietary iron 95
Dietary optionality 233
Dietary patterns 344
Dietary protein 92

Dietary requirements 347
Dietary tryptophan 203
Diet-related disease 88
Diffusion of innovations theory (DOI) 280
Digestible indispensable amino acid score (DIAAS) 92, 148
Digestive discomfort 312
Digestive function 343
Docosahexaenoic acid (DHA) 91, 198, 204
Docosahexaenoic acid (DHA)— 244
Dry fractionated pea protein (PPDF) 140
Dynamic descriptive 293
Dynamic innovation 352

E

Eco-friendly packaging 351
Ecosystem degradation 259
Ecosystems 262
Edible fungi 4
Edible seeds 321
Egg albumen 7
Egg white 3
Eicosapentaenoic acid 47
Eicosapentaenoic Acid (EPA) 198, 202
Emerging technologies 288
Emulsification 3, 36
Emulsified meat 166
Endogenous estrogens 223
Energy metabolism 40, 78
Enterobacteriaceae 229, 231
Environmental aspects 258
Environmental degradation 247
Environmental footprint 233
Environmental risks 225
Environmental, social, and governance (ESG) scales 350
Environmental sustainability 2, 35, 342
Enzymatic conversion 91
Enzyme treatments 222
Essential fatty acids (EFAs) 243
Ethical considerations 350
European Food Safety Authority (EFSA) 170
Evidence-based strategy 276
Extracted protein 328
Extreme weather conditions 348
Extrusion technology 219, 229

F

Faba bean protein 135, 139, 141, 162
Fat binders 7
Fat composition 47, 63, 65
Fat mimetics 186, 187
Fat-soluble vitamin 54
Fatty acid composition 166, 175, 177, 179, 195, 196
Fatty meat 200
Fatty tissue 47
Fermentation-based mycoproteins 131
Fermentation-derived sources 206
Fermentation plants 348
Fermentation process 310
Fibrous structural developments 4
Fibrous structure 166, 195, 259
Fine-particle meatless products 132
Fitness instructors 294
Flavoring agents 221
Foam stabilization capabilities 5
Food and agriculture organization (FAO) 258
Foodborne pathogens 216
Food choice 285
Food engineering techniques 228
Food environment 324
Food industry 216

Food ingredients 310
Food items 289
Food materials 42, 49
Food microbiology 225
Food neophobia 279, 287
Food processing industry 344
Food processing procedures 225
Food production 2, 35, 88
Food proteins 3, 17
Food-related allergies 224
Food science 2
Food security 88
Food sources 314
Food systems 263
Food trends 2
Formaldehyde 48
Framing theory 294
Fungal proteins 131

G

Gains-focused framing 291
Game meat 57, 83
Gastrointestinal tract (GIT) 146
Gastronomic cultures 287
Gelatinization activity 5
Generally Recognized as Safe (GRAS) 187
Genetically modified (GM) sources 224
Genetic variability 234
Genetic variations 91
Global New Product Database (GNPD) 27
Global warming potentials (GWP) 249
Glucosinolates 318
Glutamic acid 46, 104
Gluten sensitivities 332
Glycated hemoglobin 327
Glycogen reservoir 58
Good Food Institute (GFI) 101
Greenhouse gas emissions 346
Greenhouse gas production 260
Grocery stores 289
Ground Pork (GP) 198
Growth-promoting agents 177
Gut health 88

H

HDL (High-density lipoproteins) 62
Health coaches 294
Health guidelines 308
Health organizations 313
Health profile 279
Healthy diet 130
Heart disease 168, 169, 170, 171, 173, 183, 221
Heart failure 175
Heart-healthy diet 308
Heat-labile colorants 8
Heat treatment 42, 67
Heme-based molecules 2
Heme iron 95, 96, 97, 103, 198, 201, 204
Heme iron intake 221
Hemoglobin 50, 52
Hemp seed protein isolate (HPI) 152
Hemp seed protein meal (HPM) 151
Heterocyclic compounds 8
High blood pressure 221
High-fidelity products 283
Highly processed meat 262
High-pressure process (HPP) 154
High-quality proteins 41
High sodium content 221
High-value animal protein 166
Human consumption 319
Humane treatment 258
Human growth 41, 49
Human health 2
Human omnivory 91

Hybrid food products 136
Hybrid meat development 351
Hybrid meat products 328

I

Immune regulation 202
Immunization 49
Impossible Foods Burger 198
Indispensable amino acid (IAA) 316
Induce celiac disease 135
Industrial meat production 2
Inflammation management 167
Inflammatory markers 326
Integrative model 295
Intergenerational neurological disorders 199
International Agency for Research on Cancer (IARC) 73, 78
Intracellular energy 203
Intracellular enzyme systems 52
Ionization radiations 46
Isolated proteins 266
Isolated soy protein 335
Isolation process 219

K

Kentucky Fried Chicken (KFC) 132
Knowledge gaps 262

L

Lactic acid 42, 43, 58
Lactic acid bacteria 117
Lamb loin muscle 185
Lamb meat 185
LCA analyses 249
LDL (Low-density lipoproteins) 62
Leghemoglobin 221
Legume-based PBMAs 317
Legume proteins 5, 12, 13
Legumes 228
Linolenic acid 47, 64
Livestock farming 2, 11, 258, 262, 342
Low-density lipoprotein (LDL) 183
Low moisture meat analogs (LMMAs) 143
Low-moisture texturization technique 133
Low-molecular-weight organogelators (LMOG) 190
Low-quality protein 88

M

Macro-minerals 49
Macronutrient composition 218, 334
Maillard reactions 8
Market demand 288
Marketing campaign 296
Market saturation 281
Market strategy 264
Meal preparation 308
Meat alternative (MA) 145
Meat analogs 5, 12, 13, 17, 22, 23, 24, 25, 26, 32
Meat-based ingredients 3
Meat composition 60, 61
Meat consumption 308
Meat extenders 3
Meat production 342
Meat reduction 293
Meat substitutes 2, 3, 4, 6, 7, 8, 9, 10, 11, 12, 14, 15, 16, 17, 18, 23, 25, 27, 28, 29, 31, 34, 35, 342
Medical professionals 294
Medium-chain triglycerides (MCT) 186
Menu design 290
Methionine 335
Micro-algae 4
Microbial contamination 230

Microbiological contamination 133
Microbiome 326
Micronutrient composition 88, 89
Micronutrients 41
Microparticulated protein products (MPP) 187
Minced meat 166
Minced meat substitutes (MMS) 139
Mixed alimentary (MA) 140
Mock meat 264
Molecular characteristics 217
Monounsaturated fats 167
Moral dilemmas 287
MUFA (Monounsaturated fatty acids) 62
Muscle-based foods 131
Muscle tissues 42, 48, 50
Mushroom fungus 135
Mutton sheep 185
Mycoprotein 264
Mycotoxins 226
Myoglobin 50

N

National Association of Game Meat Producers 56
Natural toxins regulations 227
Nerve function 206
Net protein utilization (NPU) 20
Neurological function 40
Neurological problems 199
Night blindness 203
Non-GMO ingredients 332
Non-starch polysaccharides 108
Non-textured proteins 134
Non-vegetarian groups 285
Normative beliefs 296
Nutraceuticals 219
Nutrient absorption 208, 312
Nutrient adequacy 315
Nutrient contents 317, 329
Nutrient-dense foods 198
Nutrient profiling models 326
Nutritional composition 40, 74, 84, 85, 216
Nutritional makeup 262
Nutritional needs 240
Nutritional profiles 342
Nutritional quality 344
Nutritional value 89, 103, 267
Nutrition considerations 277

O

Oat fiber 5
Oat fiber concentrate (OFC) 139
Oat protein (OP) 140
Ohmic heating (OH) 154, 158
Oilseed legumes 321
Oil seeds 4
Omnivory's optionality 233
Organoleptic properties 220
Osteoporosis 221

P

Pale, soft, and exudative condition (PSE) 43
Paracrine lipid mediators 48
Partially hydrogenated oils (PHOs) 189
Pasture-raised meat alternatives (PB-MAs) 316
Pea protein concentrate (PPC) 137
Pea protein isolate (PPI) 137
Per capita expenditure 322
Phosphatase enzyme 89
Physical activity 167
Physical health 349
Physical strains 265
Phytic acid 108, 109, 110, 111, 114, 119, 124
Plant-based burgers 2, 10, 34
Plant-based diets 89, 90, 91, 96, 97,

122, 314, 343
Plant-based extracts 221
Plant-based food processing 344
Plant-based food replacements 90
Plant-based ingredients 2, 88, 102, 228
Plant-based macronutrients (PBMAs) 314
Plant-based meat 2, 5, 6, 9, 10, 14, 15, 16, 17, 22, 27, 28, 29, 30, 31, 32, 35, 37
Plant-based meat alternatives (PBMAs) 198
Plant-based meat substitutes 2, 15
Plant-based milk alternatives 217
Plant-based products 342
Plant-based proteins 3, 9, 36
Plant-based sausages 312
Plant-based scaffolding technology 345
Plant-based substitutes 342
Plant compounds 91, 95
Plant farming 253
Plant lipids 186
Plant-predominant diet 326
Plant proteins 2, 17, 21, 24, 89, 92, 93, 94, 98, 99, 104, 113, 127
Polymer colloidal systems 166
Polyunsaturated fats 167
Polyunsaturated fatty acids (PUFAs) 91, 170, 175
Post-consumption allergic 225
Post-mortem inspection 57
Precision fermentation 350
Product accessibility 292
Product choice 292
Production technologies 351
Prostaglandins 48
Protein binding 7
Protein components 290
Protein-containing substance 9
Protein desulfation bonding 4
Protein digestibility 92, 108, 113, 117, 120
Protein digestibility corrected amino acid score (PDCAAS) 148
Protein digestibility-corrected amino acid score (PDCAAS) 92, 113
Protein digestibility-corrected amino acid scores (PDCAAS) 43
Protein fermentation 350
Protein Foods, Environment, Technology, and Society program (PROFETAS) 132
Protein functionalization 218
Protein-hydrogen interaction 5
Protein malnutrition 44
Protein market 352
Protein quality 318
Protein sources 342
Proximate composition 59, 60, 61, 62
Psychiatric problems 199
Public awareness 289
Public education campaigns 324
Pulsed Electric Field (PEF) 159
Purchasing behavior 293

R

Rapeseed protein concentrate (RPC) 142
Recommended daily allowance (RDA) 92
Red blood cell formation 237
Registered dietitians 294
Resource-intensive nature 2
Riboflavin 52, 203
Rice flour (RF) 144
Rice protein isolate (RPI) 144
Risk assessments 227
Roasted vegetables 333
Rumen biohydrogenation 174, 176,

182
Ruminant methane emissions 258

S

Salad dressings 187
Saturated fat 293, 328
Saturated fatty acids (SFA) 170, 171, 172
Savory scents 7
Selective breeding 349
Self-efficacy 297
Sensory qualities 277, 283, 285, 286, 287, 292
Sexual hormones 67
Shear cell technology 133, 135, 142, 347
Social influence theories 295
Social norms 287
Social norms messaging techniques 293
Social pressure 293
Social settings 282
Socioeconomic circumstances 253
Soy allergies 332
Soy-based meat alternatives 325
Soybean production 319
Soy flour 4
Soy isoflavone compound 222
Soy leghemoglobin 89, 96, 231
Soy protein 3, 4, 7, 8, 15, 17, 18, 34, 35, 37
Stomach cancer 221
Sucrose fatty acid ester (SFE) 188
Sucrose fatty acid polyester (SPE) 188
Supermarkets 281
Sustainable food systems 2
Synthetic toxins 227
Synthetic vitamins 92

T

Target audience 294
Taurine 198, 202, 204
Technological advancements 347
Textured vegetable protein (TVP) 15
Texturized Vegetable Protein (TVP) 133
Thermal insulation 166
Thermal processing 222
Thermostable coloring agents 8
Thiamine 51, 52, 70, 71
Thin-looking fibers 5
Toxic substances 226
Traditional meat consumption 2
Traditional meats 40, 342
Trans-fatty acids (TFA) 170
Triacylglycerols 171, 174
Type 2 diabetes 326

U

Ultra-processed foods (UPF) 325
Unsaturated fatty acids 47, 49, 64, 86

V

Vegan diet 318
Vegan dishes 3
Vegetable-based protein 8
Vegetable oil 6
Vegetable textured protein 3
Vegetarian meals 291
Video communication 294
Viscoelasticity 133
Vitamin C-rich foods 266

W

Waste management 262
Water binding capacity (WBC) 141
Water scarcity 258
Weight management 308

Wheat gluten 3, 4, 7, 15, 17, 18, 21, 34
Wheat gluten products 309
Wound healing 49

X

Xanthan gum 7
Xanthosoma sagitifolium powder (XSP) 145

Y

Yellow pea isolates (YPI) 142

Z

Zelpaterol hydrochloride 177
Zinc deficiency 50